Applied Maple for Engineers and Scientists

The Artech House Computer Science Library

ISBN: 0-89006-853-4 *Applied Maple for Engineers and Scientists,* Chris Tocci and Steve Adams

ISBN: 0-89006-682-5 *ATM Switching Systems,* Thomas M. Chen and Stephen S. Liu

ISBN: 0-89006-510-1 *Authentication Systems for Secure Networks,* Rolf Oppliger

ISBN: 0-89006-691-4 *Client/Server Computing: Architecture, Applications, and Distributed Sytems Management,* Bruce Elbert and Bobby Martyna

ISBN: 0-89006-757-0 *Computer-Mediated Communications: Multimedia Applications,* Rob Walters

ISBN: 0-89006-660-4 *Computer Telephone Integration,* Rob Walters

ISBN: 0-89006-614-0 *Distributed and Multi-Database Systems,* Angelo R. Bobak

ISBN: 0-89006-812-7 *A Guide to Programming Languages: Overview and Comparison,* Ruknet Cezzar

ISBN: 0-89006-552-7 *Heterogeneous Computing,* Mary M. Eshagian, editor

ISBN: 0-89006-492-X *Introduction to Document Image Processing Techniques,* Ronald G. Matteson

ISBN: 089006-799-6 *Managing Computer Networks: A Case-Based Reasoning Approach,* Lundy Lewis

ISBN: 0-89006-654-X *Networks and Imaging Systems in a Windowed Environment,* Marc R. D'Alleyrand

ISBN:0-89006-865-8 *Practical Guide to Software Quality Management,* John W. Horch

ISBN: 0-89006-831-3 *Survival in the Software Jungle,* Mark Norris

ISBN: 0-89006-778-3 *UNIX Internetworking,* Second Edition, Uday O. Pabrai

ISBN: 0-89006-609-4 *Wireless LAN Systems,* A. Santamaría and F. J. López-Hernández

ISBN: 0-89006-717-1 *Wireless: The Revolution in Personal Telecommunications,* Ira Brodsky

ISBN: 0-89006-740-6 *X Window System User's Guide,* Uday O. Pabrai

For further information on these and other Artech House titles, contact:

Artech House
685 Canton Street
Norwood, MA 02062
617-769-9750
Fax: 617-769-6334
Telex: 951-659
email: artech@artech-house.com

Artech House
Portland House, Stag Place
London SW1E 5XA England
+44 (0) 171-973-8077
Fax: +44 (0)171-630-0166
Telex: 951-659
email: artech-uk@artech-house.com

Applied Maple for Engineers and Scientists

Chris Tocci • Steve Adams

Artech House
Boston • London

Library of Congress Cataloging-in-Publication Data
Tocci, Christopher.
Applied Maple for engineers and scientists / Chris Tocci and Steve Adams.
p. cm.
Includes bibliographical references and index.
ISBN 0-89006-853-4 (alk. paper)
1. Engineering mathematics—Data processing. 2. Science—Mathematics—Data processing. 3. Maple (Computer file) I. Adams, Stephen, 1959- . II. Title.
TA345.T63 1996
620'.00285'53—dc20 96-19492
CIP

British Library Cataloguing in Publication Data
Tocci, Chris
Appled Maple for engineers and scientists
1. Maple (Computer program) 2. Algebra—Computer programs
3. Engineering mathematics—Computer programs
I. Title II. Adams, Steve, 1959-
620'.00285'5369

ISBN 0-89006-853-4

Cover and text design by Darrell Judd

685 Canton Street
Norwood, MA 02062

International Standard Book Number: 0-89006-853-4
Library of Congress Catalog Card Number: 96-19492

10 9 8 7 6 5 4 3 2 1

To my wife, Mercedes, my dad, Anthony and the 'leggy' Woolfie

C.T.

To my long suffering wife, Fiona and my daughter Sophie who will have to leer at this stuff

S.A.

Contents

Chapter 3

Chapter 4

Chapter 5

Chapter 6

Foreword

ABOUT 7 YEARS AGO, after some 35 years as a computational physicist, I opened up my first version of the computer algebra system (CAS), Maple IV. Maple seemed to be an answer to my need for an error-free and fast system to derive long involved algebra and calculus solutions. My delight turned to a somber realization that Maple did not replace all the mathematical and applied mathematical skills that I had developed. I still had to think like a computational physicist and to learn how to use Maple as a tool to extend my capabilities to achieve solutions more quickly and with greater accuracy. Maple is not a mathematician in a box!

My usual approach to solving, numerically, physics (and electrical engineering) problems using computers involved developing a sequence of models:

Physics → Mathematical → Numerical Analytical → Software → Computer

Each of these models represents an approximation to the actual physical process. In fact, the above chain represented the *setting up* of the problem for a solution. The actual running of the solution on a computer was the trivial part. When I started to use Maple IV to develop first the mathematical and then the software model, I realized I would have to learn a new language to effectively use this tool. I had to rethink the way I solved the problem of a mathematical representation of the physics or engineering process I was trying to solve. However, I had no guide, no handbook, no Morse and Feshbach (*Methods of Theoretical Physics*), the *bible* for mathematical physicists of my generation! So, I had to learn by trying, by experimenting. It was a long and difficult process, especially since Maple was continuously being improved and its capabilities extended. After several years, I am still not finished with this process.

However, Steve Adams and Chris Tocci have made the road much smoother and more level with this book, *Applied Maple for Engineers and Scientists*. Set around Maple V, Release 4—the latest version of this software—they show how to solve a variety of problems using Maple as the principal tool. Ranging from linear active filters through curve fitting to ODEs, they show how to set up the problem using Maple. Most importantly, even if none of the applications covered in the book is germane to the reader's specific problems, Adams and Tocci demonstrate how an engineer or scientist should to think about a problem when using Maple as a tool.

The authors do not leave the reader hanging if they are not already proficient in Maple—they include a tutorial on Maple V, Release 4, which contains the principal features of this system. Of equal importance is the discussion of the physics or engineering processes and of the important mathematical functions used in each example. These discussions, plus the plotting of the solutions using the Maple graphic engine, are critical elements in making this book an almost-self-contained reference and teaching text.

Thomas N. Casselman
Casselman Computational Consultants (C3)
Dublin, CA
July 1996

Preface

Motive for using this book

MAPLE IS ONE OF THE most powerful mathematics computer algebra systems or computer algebra system (CAS) packages on the market today. Considering today's economic realities, it makes perfect sense that an initial indepth computer analysis of a quantitative problem could save many person-hours and material resource costs, hence making you and your organization more competitive. *Applied Maple for Engineers and Scientists* will get readers thinking about their specific problems by using what the authors call "template" application case studies.

Who needs to use this book

The more timely and accurate a professional needs to be about his or her decisions, the more proficient that professional needs to be with a Maple-type software package. Maple affords modern professionals the ability to visualize the dynamics via Maple's graphics. For students, Maple provides valuable insight into the underlying dynamics, which, in turn, imparts an important understanding of the process under study not evident before the advent of CAS tools. In previous years, both students and professionals had to wait hours for batch loading and, later, dumb terminal-type centralized computers to perform what Maple can do very quickly today on the omnipresent PC, PowerPC, or UNIX workstation platforms.

Purpose of the book

Applied Maple for Engineers and Scientists was written with the purpose of creating template applications for student and practicing technical/ business professionals. Templating serves the reader and authors by showing different examples on how the Maple symbolic and numerical mathematics system can be generally used in solving a very wide range of everyday quantitative problems. Even though the reader may never need a single one of the specific examples discussed, the concepts, syntax, and approaches shown to problem solving with Maple can be extremely helpful in getting the user up and running quickly in his or her particular problem area.

Philosophy of the book

The text is geared toward technical professionals and students who have an understanding of the technical principles of their respective fields, but are not cognizant of how a CAS software package such as Maple can be used to facilitate timely and understandable solutions. In no way do the authors pretend that this is a text on any of the engineering, scientific, or quantitative business disciplines described during any template session. There are references given, as needed, in the individual chapters if the reader is interested in more advanced aspects of the particular application.

Chapter structure and order

The chapters have no particular order and this was done by choice. Each chapter is fairly "stand alone" in its content and offers the reader a reinforcement of software approaches that should become apparent as one wanders among the different applications. This reinforcement is both within each application and throughout all of them in the form of reiteration of command lines and trying to stay to the more common syntax approaches (though at times *not* the most elegant) used by the Maple engine. Efforts to minimize "exotic" or "highly efficient" hard-to-comprehend coding will keep the reader on-track with the fundamental usage of the Maple language. This thinking, we believe, will give the new Maple user a more robust ability to develop work sessions more quickly and accurately. Obviously, in time, the user will develop his or her own personal syntax forms and approaches to problem solving with the Maple engine. For this reason also, the authors wanted to minimize their syntax coding "fingerprint" among the applications by keeping the syntax methods very general.

The text contains the following chapters:

Chapter 1: Introduction (Adams)

An explanation is given of what a computer algebra system is and how Maple fulfills the requirements of a CAS. A brief tutorial is given to get the novice Maple user up and running along with a description of Maple's online help file system.

Chapter 2: Active filter design and analysis (Tocci)

A detailed analysis and design approach is discussed that deals with common active filter circuits. In particular, two separate types of filters are dealt with in this chapter. The first part describes the continuous Butterworth low-pass filter, and the second deals with a switching or sampled data approach to bandpass filtering using charge-coupled device (CCD) technology.

Chapter 3: Curve fitting (Tocci)

One of the most common functions performed by statistical packages is curve fitting of raw experimental data. Maple has a very strong curve-fitting capability and the first example used is derived from a real-world situation of peak detection associated with a spectrophotometer. The chapter also gives data sets that can cause severe problems for conventional curve-fitting programs and shows how a Maple nonlinear regression program can obtain a reasonable result. Finally, the chapter gives a vivid example of how badly "blind" curve fitting can "lie" if the user is not aware of what he or she is doing when arbitrarily assigning a high-order polynomial to the variable set.

Chapter 4: Mathematical models: working with differential equations (Adams)

Chapter 4 gets into one of the fundamental aspects of the Maple program, namely, ordinary differential equations. The chapter gives several basic template applications on Maple's capability to analyze the dynamic behavior of real rotational and translational mechanical systems.

Chapter 5: Continuous control application and theory (Tocci)

Chapter 5 describes the basic applied principles of how Maple is applied in continuous control systems. Two approaches are examined, namely, frequency (Laplace) and time (state space) domains. For purposes of comparison, a real-world third-order template controller problem is solved using both approaches.

Chapter 6: Discrete control applications (Adams)

Chapter 6 delves into template applications associated with both the pulse and Z-transform methods and comparatively with the discrete time state space techniques for discrete control design and analysis.

Chapter 7: Discrete data processing (Adams)

Chapter 7 describes some basic digital signal processing associated with 1-D and 2-D information. The concept of image conversion is described and exemplified as are several approaches using classical linear digital FIR and IIR filters.

Chapter 8: Switching topologies (Tocci)

Chapter 8 shows the reader how Maple can solve rather complex boundary problems associated with periodic signals. One of the most common applications of this analysis is used on a buck-type switching power supply. An associated template application depicts how Maple is used in solving and describing the dynamics involved with a pulse-width modulator (PWM) used for signal acquisition.

As the reader can see, the applications areas are very diverse (filter, control, data manipulation, and signal and systems applications) and basic (both continuous and discrete systems) in their importance to most engineering and applied science disciplines. Consequently, the reader can jump into any chapter to abstract whatever information is needed to solve their particular problem with Maple. Chapter 1 is for those who are unfamiliar with Maple and should be looked over and studied. However, if you are an experienced Maple user, you should find the application templates useful for any unfamiliar engineering-type problem you encounter.

The authors would like to acknowledge the following individuals for assistance during the development of this manuscript: The technical staff at Waterloo Maple Software, specifically, Drs. Stan Devitt, Jerome Lang,

Tom Lee, and David Pintur for their help during the developmental phases of release 4 and the preparation of the manuscript. The editorial staff at Artech House and particularly Theron Shreve and Kimberly Collignon for their "polite" approach to helping the authors stay fairly timely in the different stages of manuscript production and delivery.

Chapter 1

Introduction

IN 1969 THE LABORATORY of Computer Science at the Massachusetts Institute of Technology released what is regarded as the first commercial computer algebra system (CAS), Macsyma. Since then the number of computer algebra systems available has grown to include Derive, Reduce, Theorist, Mathematica, Maple, and others. In this book we will be looking at Maple and how it can be used to help engineers and scientists investigate a diverse set of problems numerically, symbolically, and graphically. The major emphasis of this book is applications and how to effectively use Maple as an analysis tool.

The first chapter is both a demonstration and a tutorial that shows how Maple functions as a numerical and symbolic calculator, a powerful visualization tool, and a programming language. It is the only portion of the book in which the Maple syntax is discussed for its own sake. Of course, syntax is discussed elsewhere but only as part of the general discussion surrounding the application being considered. If you are already comfortable with the Maple system, feel free to skip this chapter. We would, however, suggest that even the most advanced users should give this first chapter a cursory glance as a means of introducing the authors' style. Chapters 2 through 8 concentrate on how to apply Maple to problems in the areas of analog and discrete control theory, analog and digital filtering, ordinary differential equations, power supply design, and curve fitting to data.

What is a CAS ?

A computer algebra system is just a calculator that does mathematics, but unlike a conventional calculator it does more than just manipulate floating-point numbers. A CAS can perform numerical computations using either exact or floating-point arithmetic, and it can manipulate symbolic quantities and display both functions and data graphically. Some, but not all, CASs are also sophisticated programming environments ideally suited for the representation and manipulation of mathematical quantities and objects.

Numbers

The majority of modern CASs are not limited to the number of significant digits set by the floating-point hardware so they are capable of providing solutions that are exact or to a user-specified level of precision. Maple supports arbitrary precision arithmetic, which means that it is capable of storing and using numbers having in excess of 500,000 digits. This is in stark contrast to a normal calculator or spreadsheet that can only support 10 digits. Here we calculate the product of the cubes of the first 40 odd numbers,

$$\prod_{I=0}^{39} (2i+1)^3:$$

```
product((2*x+1)^3, x=0..39);
```

$$\begin{aligned}&507748306245828599556221571230208944802988775653262\\&\quad 350807652219192332706057011255827529763014747456\\&\quad 487009593823229001624355876374611035376614633415 4\\&\quad 133527362719178199768066406 25\end{aligned}$$

This returns the floating-point approximation.

```
evalf(product((2*x+1)^3, x=0..39));
```

$$.5077483062\ 10^{177}$$

In the preceding example we have used Maple to calculate a large integer both exactly and approximately, Maple can also represent other mathematical quantities exactly: $\sqrt{2}$, ⅓, γ, π, e, etc. The advantage of being able to manipulate exact quantities is demonstrated below:

```
1/3 * 3;
```

$$1$$

```
ANS:=evalf(1/3);
```

$$ANS := .3333333333$$

```
ANS * 3;
```

$$.9999999999$$

It is obvious that these results are not equal. This is because the second expression is merely an approximation of the first exact expression. Regardless of how many significant digits are used, the second expression will never be equal to the first. Although this is a trivial example it is not difficult to imagine a case where such a simple error would be unacceptable.

Symbols

A symbol can just as easily represent a known quantity such as π, e, or γ as it can an unknown one such as x. The ability to define and manipulate symbolic quantities is where the true power of a CAS such as Maple lies. We can define mathematical formulas symbolically, operate on them, and then substitute for known values when appropriate. For example, take the expression for compound interest,

$$\text{interest} = \frac{\text{principal} \cdot \text{time} \cdot \text{rate}}{100}$$

Once we have entered this in our current CAS session, we can manipulate it to make the subject of the expression time and hence calculate the time necessary to accrue a target amount of interest given a certain rate and principal. The following is how we would tackle this using Maple. If the syntax does not immediately make sense, do not worry because we will cover it shortly. First we define the equation relating the amount of interest to the amount of principal, the time, and the prevailing interest rate.

```
EQN:= i = p*t*r/100;
```

$$\text{EQN} := i = \frac{1}{100}\, ptr$$

In this particular instance we will make time the object of the equation. We do this with the `isolate` function. This function is one of many Maple functions that is `readlib` defined, which means that although it is part of the Maple library the function is not part of the main package and must be loaded explicitly using `readlib` as shown.

```
EQN1:=readlib(isolate)(EQN, t);
```

$$\text{EQN1} := t = 100\,\frac{i}{pr}$$

```
subs(i=5004/10, p=1000, r=51/5, EQN1);
```

$$t = \frac{417}{85}$$

Here are two more examples of symbolic computation: defining the volume of a sphere in terms of the radius r and obtaining the Mellin transform of the expression $ln(y)e^{(-3y\verb|^|2)}$.

```
volume_of_a_sphere := (4*Pi*r^3)/3;
```

$$\text{volume\_of\_a\_sphere} := \frac{4}{3}\pi\, r^3$$

```
with(inttrans,[mellin]):
mellin(ln(y)*exp(-3*y^2), y, s);
```

$$-\frac{1}{4}\left(\frac{1}{3}\right)^{(1/2\,s)}\ln(3)\,\Gamma\left(\frac{1}{2}\,s\right)+\frac{1}{4}\left(\frac{1}{3}\right)^{(1/2\,s)}\Psi\left(\frac{1}{2}\,s\right)\Gamma\left(\frac{1}{2}\,s\right)$$

More about Maple

Maple Vr4 is a comprehensive CAS that can perform numerical, symbolic, and graphical computations on PCs (DOS, Windows, Mac, Amiga, etc.) and workstations (Apollo, HP, Sun, DEC, etc.). It is an interactive, easy-to-use system comprising more than 2500 functions that can be used either on its own or in conjunction with many commonly used numerical analysis and office applications. It is also its own programming language: 95% of Maple Vr4 is written in Maple, which means that its scope is essentially limitless. If an application area is not initially supported, Maple is quickly extended through the addition of new functions and procedures. It is therefore hardly surprising that Maple is the first choice for many mathematicians, engineers, scientists, educators, and others who need an easy-to-use environment in which to perform mathematics.

Maple was first conceived late in 1980 at the University of Waterloo as a research project. Since then it has steadily grown to become the largest, most robustly tested general mathematical software system currently available. It consists of three distinct parts: the kernel, the library, and the worksheet. The worksheet, with its typeset mathematics, text, and graphics support is a natural medium for entering and manipulating mathematical expressions, models, prototypes, and test data. The kernel, which is written in C, contains the system core functions and accounts for approximately 5% of the Maple Vr4 system. The library, on the other hand, is where the majority of the system resides. Functions held in the library are written in Maple and are user accessible.

Maple: a tutorial

The goal of this tutorial is twofold: First, it introduces you to the Maple syntax and, second, it gives you a quick tour of Maple itself. We should, at this point, reiterate that this tutorial gives only the briefest of introductions to the Maple syntax and may be skipped by the experienced user.

Help

Maple's Help database consists of more than 8 MB of data organized into manual style pages containing examples and hypertext links. Every page has the same basic format: Function, Calling Sequence, Parameters, Description, Arguments, Examples, and See Also sections. Sections containing pointers to additional information, for example, the See Also section, contain hyperlinks to other Help pages. This wealth of information is searchable in a variety of ways via the **Help** menu or the Worksheet—contents, topic/keyword, or full-text searches from the menu—and the manual pages and hypertext links are directly accessible from the Worksheet. If we access Maple's Help system from the **Help** menu we do so via dialog boxes, whereas when using the Worksheet we can use the either the question mark notation (`?`, `??`, and `???`) or the `help` function. Either method of accessing the help system will result in a Help Worksheet being displayed for the topic or keyword used in the search if a match is found. Because the entries in the **Help** menu are self-explanatory, we will concentrate on how to use the Help system from within the Worksheet. The Help system is accessed from the Worksheet by using the question mark as follows (note that the semicolon terminator is not required):

```
> ?help
```

The brackets on the left-hand side of the Worksheet shown in Figure 1.1 are collapsible regions and the underlined text indicates hypertext. The requested help page is displayed as an inert Worksheet (the Maple commands cannot be executed) in its own window, which means that all or part of it can be copied into the currently active Worksheet using the normal copy-and-paste procedure. The Examples section of a Help page can be copied separately, without highlighting them beforehand, using the Edit:Copy Examples menu item. The selected items are then copied into the active Worksheet using either Edit:Paste or Edit:Paste Maple Text. The first method will result in inclusion of both the Maple input and output regions, whereas the second only copies the Maple input regions into the active Worksheet, as shown in Figure 1.2, in which the Examples section for the Help topic `simplify[power]` are used as an example.

The Help system will also look for partial matches, so, for example, entering `?A` will prompt us by providing pointers to the Maple functions that begin with A.

```
> ?A
```

Function: help - descriptions of syntax, datatypes, and functions

Calling Sequence:

?topic or ?topic,subtopic or ?topic[subtopic] or

help(topic) or help(topic,subtopic) or help(topic[subtopic])

Description:

```
intro              introduction to Maple
index              list of all help categories
index[category]    list of help files on specific topics
topic              explanation of a specific topic
topic[subtopic]    explanation of a subtopic under a topic
distribution       for information on how to obtain Maple
copyright          for information about copyrights
```

- Note 1: The recommended way to invoke help is to use the question mark.
- Note 2: When invoking help using the function call syntax, help(topic), Maple keywords (reserved words) must be enclosed in backquotes. For example, help(quit) causes a syntax error. Use help(`quit`) instead. Note that the string delimiter is the backquote (`), not the apostrophe ('), nor the double quote ("). When using the question mark syntax for help, no quotes are required.
- Note 3: A command must end with a semicolon, followed by RETURN or ENTER, before Maple will execute it and display the result. The semicolon can appear on the next line if you forget to end the command with it, but it must appear. There can be multiple commands on one line, separated by semicolons or colons. An exception to this is when a line starts with a question mark in which case help is invoked and no semicolon is required.
- To contact Waterloo Maple Software, see distribution. To contact the authors of Maple, see scg.

See Also: keyword, quotes, colon, quit, example, scg, distribution, TEXT, makehelp

Figure 1.1

Function: sin, cos, ... - The Trigonometric functions

Function: sinh, cosh, ... - The Hyperbolic functions

Calling Sequence:

sin(x cos(x) tan(x)

sec(x) csc(x) cot(x)

sinh(x) cosh(x) tanh(x)

sech(x) csch(x)coth(x)

Parameters:

x - an expression

Description:

Examples:

See Also: invtrig, invfunc, inifcns

Figure 1.2

There are no matching topics. Try one of the following:

AFactor AFactors Airy AngerJ

The inclusion of an Examples section in the manual is a very useful feature. By using `???` or its functional equivalent `example`, this section can be accessed for any Maple topic that has a Help page. In the example of Figure 1.3 we get the Examples section for the rand function directly using the `???` but `example(rand)` would produce the same result.

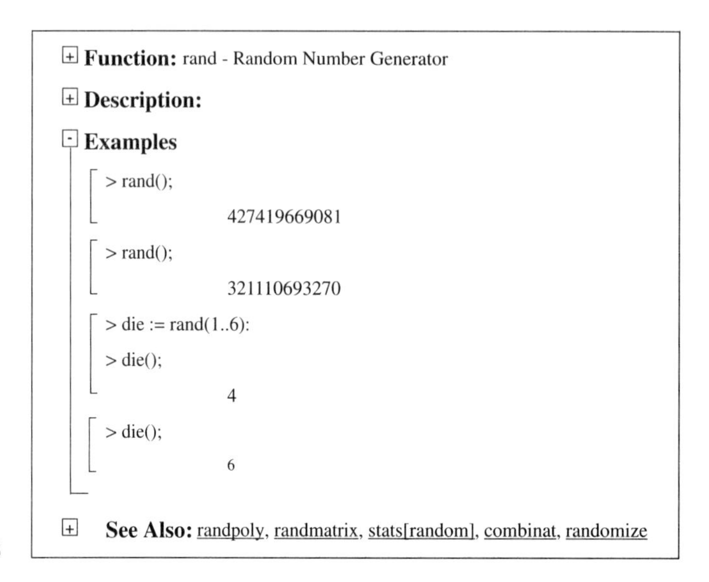

Figure 1.3

Similarly, the function description and the calling sequence section can be accessed directly by using `??`. In Figure 1.4, we get the function description and the calling sequence for the `sin` and other trigonometric functions.

```
> simplify((a^b)^c,power);
> simplify((a^b)^c,power, symbolic);
> simplify(x^a*x^b, power);
> simplify(exp(5*ln(x)+1), power);
> simplify(ln(x*y),power);
> simplify(ln(x*y),power, symbolic);
```

Figure 1.4

Finally, hyperlinks to functions and information related to the current topic of interest can also be easily obtained using the function `related` (see Figure 1.5).

```
related(dsolve[numeric]);
```

⊞ **Function:** dsolve/numeric - numerical solution of ordinary differential equations

⊞ **Description:**

⊞ **Examples:**

⊞ **See Also:** dsolve[rkf45], dsolve[dverk78], dsolve[classical], dsolve[gear], dsolve[mgear], dsolve[lsode], dsolve[taylorseries], plots[odeplot], DEtools[DEplot], DEtools[DEplot3d], DEtools[PDEplot]

Figure 1.5

Maple as a calculator

Numeric

Here we use Maple to perform exact and approximate arithmetic:

```
1234+5678;
```

$$6912$$

```
1/2 - 3/4 + 5/6 - 7/8 + 9/10;
```

$$\frac{73}{120}$$

```
1357^24;
```

$$1520254661801097535145921172472595185185444\\2067283222407924852791532889176 01$$

```
2468*1357^(-1/2);
```

$$\frac{2468}{1357}\sqrt{1357}$$

The following examples use the ditto (") pointer to calculate the floating-point approximations to the previous two expressions:

```
evalf(");
```

66.99698076

```
evalf("",25);
```

66.99698076106124289230084

The previous result stack in only three deep so more than three double quotes have no meaning. We can see that in the first case the default number of significant digits is used, whereas in the second 25 digits are used. This type of calculation is made possible because Maple supports arbitrary precision arithmetic as standard. The default number of digits (10) used in a floating-point calculation is held in the global variable `Digits`.

If we access some of Maple's 2,500 standard functions we can easily perform more complex calculations:

```
erf(infinity);
```

1

```
120!;
```

66895029134491270575881180540903725867527463331380298102956713523016355724496298936687416527198498130815763789321409055253440858940812185989848111438965000596496052125696000000000000000000000000000000

```
evalf(log(Pi/exp(1)),200);
```

.14472988584940017414342735135305871164729481291531157151362307147213776988482607978362327027548970770200981222869798915904820552792345658727908107881028682527639391426634590290248477335886993778920313

```
> exp(1);
```

$$e$$

```
> evalf(",40);
```

$$2.718281828459045235360287471352662497757$$

```
> evalf(sqrt(3.56*log(Pi))*12345.67,5);
```

$$24922.$$

Complex arithmetic Maple operates over both the real and complex domains, *I* being the complex constant: $\sqrt{-1}$.

```
> (1+7*I)/((2+I)*conjugate(-3+5*I));
```

$$-\frac{46}{85}+\frac{3}{85}I$$

Name aliasing is also supported. This means that j, commonly used in engineering to mean $\sqrt{-1}$, can be aliased to *I*.

```
> alias(I=I,j=sqrt(-1));
```

$$j$$

```
> j*j;
```

$$-1$$

Trigonometric functions Maple supports all of the standard trigonometric and hyperbolic functions:

```
> sin(Pi/4);
```

$$\frac{1}{2}\sqrt{2}$$

```
evalf(");
```

$$.7071067810$$

```
tanh(88.5);
```

$$1.$$

```
sec(-1.);
```

$$1.850815718$$

```
evalf(csch(exp(3.*Pi)));
```

$$.4750727074\ 10^{-5381}$$

Symbolic

Maple's great strength lies in its ability to perform symbolic calculations:

```
expand((x - 3*y^3)^5);
```

$$x^5 - 15x^4y^3 + 90x^3y^6 - 270x^2y^9 + 405xy^{12} - 243y^{15}$$

```
normal(1/a-1/b*1/c);
```

$$-\frac{-b\ c + a}{a\ b\ c}$$

```
factor(diff((x - 3*y^3)^5,y));
```

$$-45\left(x - 3\ y^3\right)^4\ y^2$$

Series, sums, and products Maple can generate series expansions (such as Taylor and power), summations, and products of arbitrary expressions quickly and without error:

```
series(sin(x),x=h,5);
```

$$\sin(h) + \cos(h)\,(x-h) - \frac{1}{2}\sin(h)\,(x-h)^2 - \frac{1}{6}\cos(h)(x-h)^3 + \frac{1}{24}\sin(h)\,(x-h)^4 + O\left((x-h)^5\right)$$

```
series(GAMMA(x),x=0,3);
```

$$x^{-1} - \gamma + \left(\frac{1}{12}\pi^2 + \frac{1}{2}\gamma^2\right)x + \left(-\frac{1}{3}\zeta(3) - \frac{1}{12}\pi^2\gamma - \frac{1}{6}\gamma^3\right)x^2 + O(x^3)$$

```
normal(sum(1/(1-N)^x,x=1..4));
```

$$-\frac{-4 + 6N - 4N^2 + N^3}{(-1 + N)^4}$$

If no range is specified, the indefinite form is returned, for example, here we compute the indefinite product $\pi \dfrac{(1-N)^x}{(x+1)}$ in x:

```
product((1-N)^x/(x+1),x);
```

$$\frac{(1 - N)^{(\frac{1}{2}x\,(x-1))}}{\Gamma(x+1)}$$

Constants Maple uses the globally accessible sequence `constants` to maintain the list of currently recognized constants:

```
constants;
```

false γ, ∞, true, Catalan, E, FAIL, π

```
type(123, constant);
```

true

```
type(pi, constant);
```

false

```
type(A_VARIABLE, constant);
```

false

The list can be added by the user very easily, as shown here:

```
constants := constants, A_VARIABLE;
```

constants := false, γ, ∞, true, Catalan, E, FAIL, π, A_VARIABLE

A constant added to the constants list is treated in the same way as the initially known constants:

```
type(2*A_VARIABLE/3.5, constant);
```

true

Variables A variable can be any name other than a reserved word. Maple treats variables in a way that mimics the normally understood mathematical variable. Maple variables can either have values assigned to them or they can remain unassigned. An assigned variable behaves in the same way as a variable in any other programming language in that it can be viewed, set, reset, and used in calculations. We can set variables and look at them:

```
x:=10!:
x;
```

3628800

Or x can be used in a calculation and be reassigned:

```
x:=ifactor(x);
```

$$x := (2)^8\ (3)^4\ (5)^2\ (7)$$

An unassigned variable is treated as a symbol that can be used in calculations and have a value assigned to it at some later time:

```
NOT_ASSIGNED;
```

NOT_ASSIGNED

Equation solver

Maple can be used to investigate the solutions to systems of algebraic and differential equations.

Numerical solutions

```
fsolve(cos(alpha)=alpha^2,{alpha});
```

$$\{\alpha = .8241323123\}$$

```
solve(X^7+X+1=0, X);
```

$$\mathrm{RootOf}\left(\_Z^7 + \_Z + 1\right)$$

Maple has returned the `RootOf` placeholder as a solution. This and other placeholders (`DESol, Limit, Product, ...`) are used by Maple to store, manipulate, and display data in a concise way. By using `allvalues` we force Maple to return the roots of the polynomial.

```
allvalues(");
```

$-.7965443541 \quad .7052980879 \quad +.6376237698\ I \quad .7052980879$
$+.6376237698\ I \quad .9798083845 + .5166768838\ I$

Here, by means of a Fehlberg four-five-order Runga-Kutta solver, we solve numerically a first-order ode. We then plot the result (see Figure 1.6).

```
f:=dsolve({diff(y(t), t)=sin(t)^2, y(0)=3}, y(t),
 type=numeric);
f := proc(rkf45_x) ... end
```

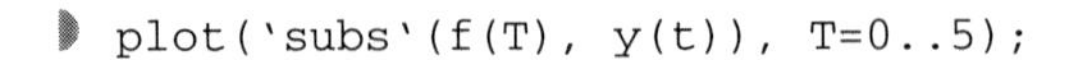

```
plot('subs'(f(T), y(t)), T=0..5);
```

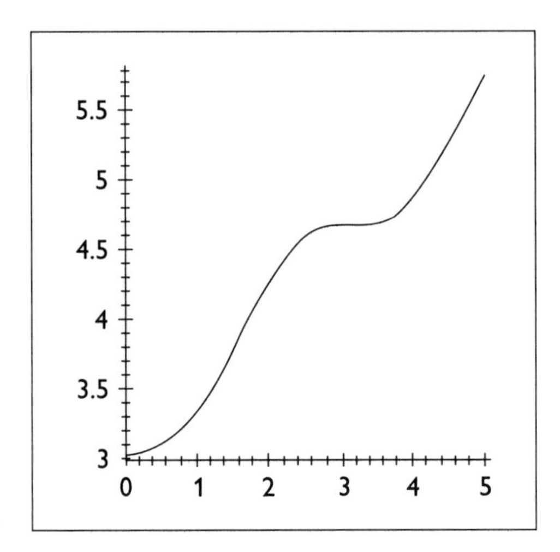

Figure 1.6

Symbolic solutions Find the points of intersection of a circle of radius r centered at the origin and an ellipse with major and minor axes a and b, respectively:

```
circle := x^2 + y^2 = r^2;
ellipse := x^2/a + y^2/b = 1;
```

$$\text{circle} := x^2 + y^2 = r^2$$
$$\text{ellipse} := \frac{x^2}{a} + \frac{y^2}{b} = 1$$

```
pts := [solve({circle, ellipse}, {x, y})];
```

$$\text{pts} := \left[\left\{y = \text{RootOf}\left((-b+a)\_Z^2 + br^2 - ab\right),\ x = \text{RootOf}\left((-b+a)\_Z^2 + ab - r^2 a\right)\right\}\right]$$

We can also investigate systems of inequalities:

```
solve( x^3+x^2-5>5, x );
```

$$\mathrm{RealRange}\left(\mathrm{Open}\left(\frac{1}{3}(134+3\sqrt{1995})^{1/3}+\frac{1}{3}\frac{1}{(134+3\sqrt{1995})^{1/3}}-\frac{1}{3}\right),\infty\right)$$

Here we solve a second-order ode. The solution contains two constants of integration because no initial conditions are specified.

```
dsolve(diff(y(t), t$2) + y(t) = 12*cos(t)^2, y(t));
```

$$y(t) = -4\cos(t)^2 + 4\cos(t) + 8 + \_C1\cos(t) + \_C2\sin(t)$$

Calculus

Maple supports single and multivariate calculus. First let us look at the process of differentiation:

```
diff(sin(x)*x^2, x);
```

$$\cos(x)\,x^2 + 2\sin(x)\,x$$

```
diff(expand((sin(x)+tan(y))^2), x, y);
```

$$2\cos(x)\,(1+\tan(y)^2)$$

Both definite and indefinite integration can be performed. Integration is performed through a combination of knowledge-based programming, simplification, and the application of the Riche algorithm. Numerical integration can be easily performed if a symbolic closed-form solution cannot be found:

```
int(1/(1-x^3), x);
```

$$-\frac{1}{3}\ln(x-1)+\frac{1}{6}\ln\left(x^2+x+1\right)+\frac{1}{3}\sqrt{3}\arctan\left(\frac{1}{3}(2x+1)\sqrt{3}\right)$$

```
Int(2*x*sin(x), x=0..Pi/a) = int(2*x*sin(x), x=0..Pi/a);
```

$$\int_{0}^{\frac{\pi}{a}} 2\,\sin(x)\,x\,dx = 2\,\frac{\sin\left(\frac{\pi}{a}\right)a - \cos\left(\frac{\pi}{a}\right)\pi}{a}$$

In this particular example we use the inert form of the integrate function (`Int`) with `evalf` to calculate the numerical solution. By using the inert form Maple does not waste time searching for a symbolic solution, which in this case does not exist.

```
evalf(Int(exp(exp(exp(-x))), x=0..3));
```

$$13.77323428$$

Graphics

An increasingly important part of any research and development tool is its ability to display data. Maple possesses an expansive set of both two- and three-dimensional plotting functions, tools, and animation routines. The style of a plot can be filled, point, or wireframe, axes can be included, and labels and a title can be added. All of these options can be set either at the command line or via the graphics menu items. Version Vr4 also illuminates three-dimensional surfaces with user-selectable lighting models. Color functions can be added as a means of adding an extra dimension to any plot. Figures 1.7 and 1.8 give the plots of the following two operations:

```
plot((x^3-5*x^2+2*x+8)/(x^3-3*x^2-25*x+75), x=-10..10,
  discont=true, view=[DEFAULT, -10..10],
  labels=['x','f(x)']);
```

```
t:=(x,y) - sqrt(x^2+y^2):
plot3d(cos(t((x+3), (y+3))*Pi)*exp(-t((x+3),
  (y+3)))+3*cos(t((x-1),
  (y-1))*Pi)*exp(-t((x-1), (y-1))), x=-3..3, y=-3..3,
  style=HIDDEN, color=BLACK, numpoints=40^2,
  orientation=[18, 50]);
```

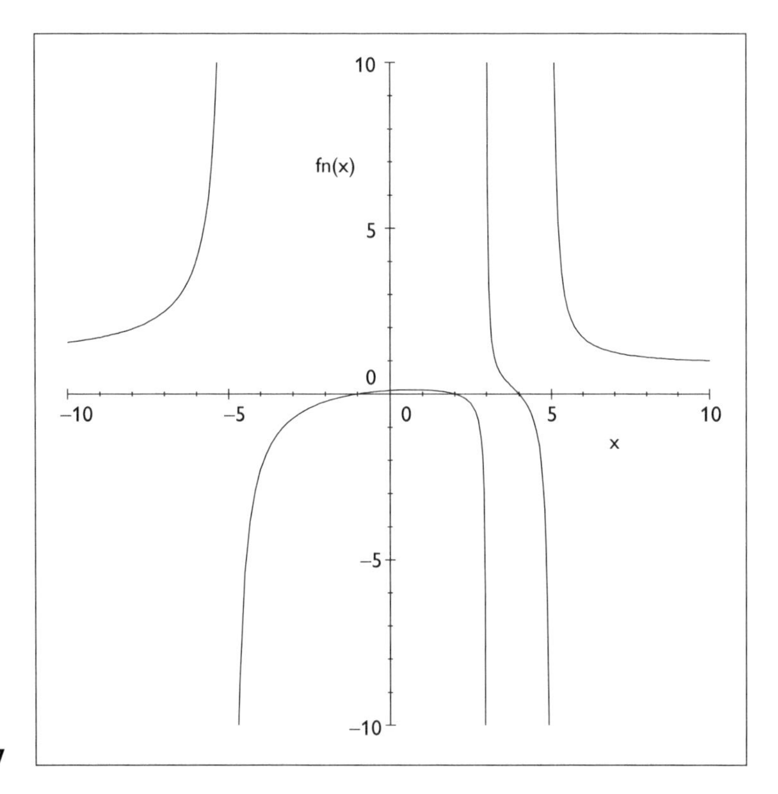

Figure 1.7

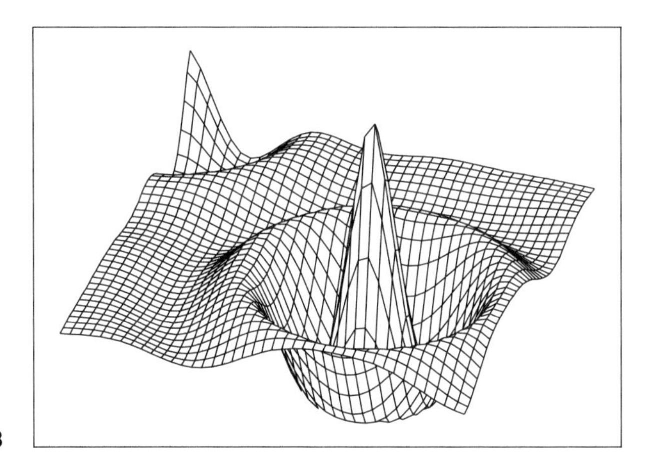

Figure 1.8

Before leaving plots we should mention that version Vr4 of Maple supports animation. `Animate` and `animate3d` functions are available as a standard sequence of plot structures, hence giving the perception of animation. In Figure 1.9, we animate a bouncing ball using this second method:

```
path := t -> 5*exp(-0.5*Pi*t)*abs(cos(t)):
ball:= (x, y) -> plottools[disc](x, y+1, color=BLACK):
seq(plots[display]([path(x), ball(x, path(x))], x=0..5):
plots[display](", insequence=true);
```

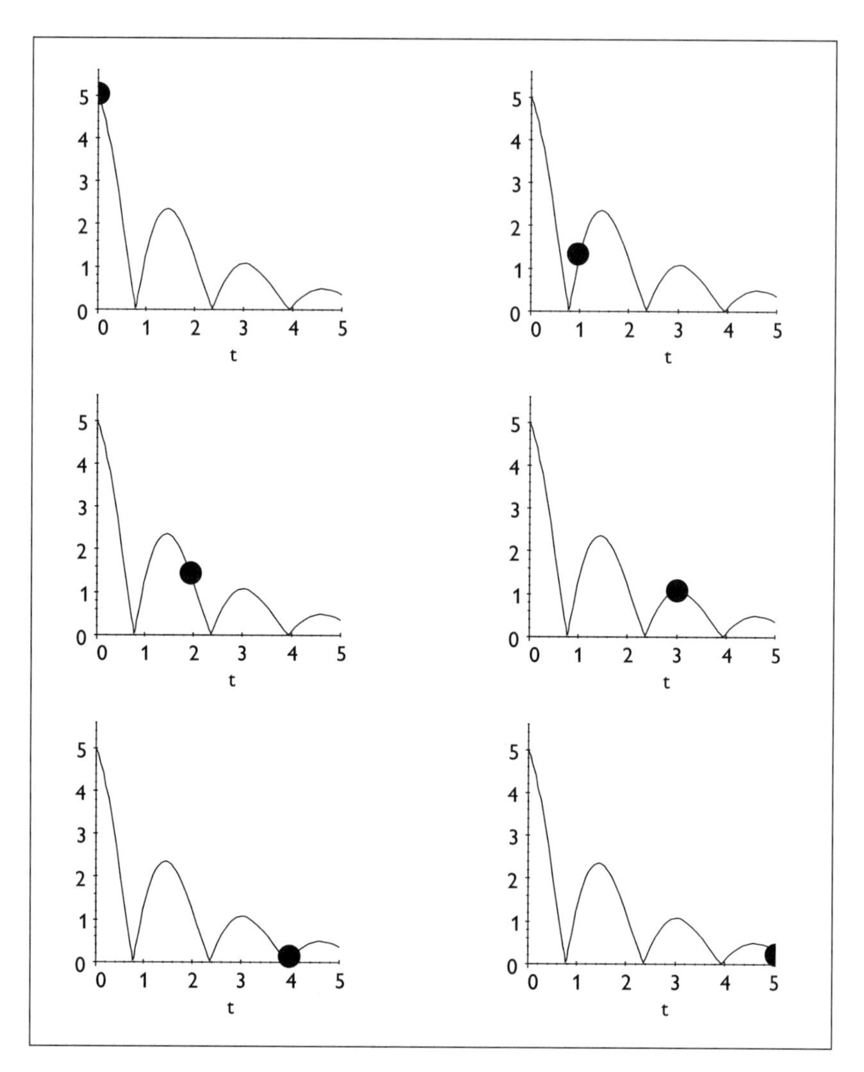

Figure 1.9

Data structures

Maple can store numeric, symbolic, graphic, or programming data in sets, lists, tables, vectors, arrays, and matrices. This extensive range of data structures allows us to both store and manipulate our data in a way that is appropriate to the specific problem.

Lists and sets Lists and sets are common structures used to store and manipulate linear data. The major differences between the two are (1) a list is ordered whereas a set is not and (2) sets will automatically remove repeated entries. Lists are defined using square brackets [...] while sets use curly braces {...}. The elements of either can be any valid Maple expression.

```
A_LIST := [1, 2, 4/5, GAMMA, 'the last element' ];
```

$$A\_LIST := \left[1,\ 2,\ \frac{4}{5},\ \Gamma,\ \text{the last element}\right]$$

```
A_SET:= {1, 2, 4/5, GAMMA, 'the last element' };
```

$$A\_SET := \left\{1,\ 2,\ \Gamma,\ \text{``the last element''}\right\}$$

Note that the element "the last element" is a string. Strings are denoted using string quotes '...'. These data structures can be easily manipulated using `op`, `nops`, `subsops`, `intersect`, `union`, `member`, and `minus`.

```
THE_LENGTH:=nops(A_LIST);
```

$$THE\_LENGTH := 5$$

```
THE_2ND_ELEMENT:=op(2, A_SET);
```

$$THE\_2ND\_ELEMENT := 2$$

```
ELEMENTS_2_TO_4 := op(2..4, A_SET);
```

$$ELEMENTS\_2\_TO\_4 := \{2,\ \Gamma,\ \text{the last element}\}$$

```
A_SET intersect {4/5, Zeta, 'the last element'};
```

$$\left\{\text{the last element},\ \frac{4}{5}\right\}$$

Now remove elements:

```
A_SET minus {1, GAMMA};
```

$$\left\{2, \text{ the last element}, \frac{4}{5}\right\}$$

```
subsop(3=NULL, A_LIST);
```

$$[1, 2, \Gamma, \text{ the last element }]$$

or

```
A_LIST[3]:=NULL;
```

$$[1, 2, \Gamma, \text{ the last element }]$$

Here we test for list and set membership:

```
member(Pi, A_LIST);
```

$$\text{false}$$

```
member( `the last element`, A_SET);
```

$$\text{true}$$

Lists and sets can be joined.

```
[op(A_LIST), op([`a`, [`new`], `list`])];
```

$$\left[1, 2, \frac{4}{5}, \Gamma, \text{ the last element}, a, [new], list\right]$$

```
A_SET union { {`a`}, `new`, `set`};
```

$$\left\{1, 2, set, \Gamma, \text{ the last element}, \{a\}, new, \frac{4}{5}\right\}$$

Arrays, matrices, and vectors Arrays, matrices, and vectors in Maple mimic their mathematical namesakes. Arrays and matrices are ideally suited to the storage and manipulation of two-dimensional data, whereas one-dimensional data can be manipulated using vectors. The major difference between an array and a matrix is that matrix indices must start at zero, whereas an array can be indexed from any integer starting point.

Here we define and manipulate some arrays. First we define an empty array and then set the element A2,2 equal to four:

```
A := array(1..2, 1..3);
```

$$A := \text{array}\,(1\,..\,2,\ 1\,..\,3,\ [\ ])$$

```
A[2,2]:=4;
```

$$A_{2,2} := 4$$

To look at the contents of an array we use `eval`, `evalm`, or `print`. This step is necessary because arrays conform to the principle of last name evaluation. This means that for an array printed in the normal way (by entering the array name followed by a carriage return), only the array's name will be printed because it is the last name evaluated. To see the contents of the array, full evaluation needs to be forced:

```
A;
```

$$A$$

```
evalm(A);
```

$$\begin{bmatrix} A_{1,1} & A_{1,2} & A_{1,2} \\ A_{2,1} & 4 & A_{2,3} \end{bmatrix}$$

Here we define and operate on some matrices. Maple's linear algebra tools are grouped together in the Maple package `linalg` (see `?linalg` for more details). All or some of the functions in the `linalg` package, this goes for Maple's other packages as well, can be loaded using `with`. An alternative method of accessing a particular function is to use its long name:

`package name[function name]`. In the following examples, we use the long name method to access a selection of `linalg` functions.

```
B:=linalg[matrix](2, 2, [a, b, c, d]);
```

$$B := \begin{bmatrix} a & b \\ b & d \end{bmatrix}$$

```
C:=linalg[matrix](2, 2, [[1, 2],[3, 4]]);
```

$$C := \begin{bmatrix} 1 & 2 \\ 3 & 4 \end{bmatrix}$$

```
evalm(B &* C);
```

$$\begin{bmatrix} a + 3b & 2a + 4b \\ c + 3d & 2c + 4d \end{bmatrix}$$

In a similar fashion we define and operate on some vectors:

```
linalg[vector](3);
```

$$\begin{bmatrix} ?_1 & ?_2 & ?_3 \end{bmatrix}$$

```
linalg[vector]([1, 2, 3]):
d:=linalg[transpose](");
```

$$d := \text{transpose}\left(\begin{bmatrix} 1 & 2 & 3 \end{bmatrix}\right)$$

The previous result, although correct, is not displayed in the column form we expected. Maple can manipulate row and column vectors correctly as we will see later.

```
e:=linalg[vector]([sin(a), Pi, 3]);
```

$$e := \begin{bmatrix} \sin(a) & \pi & 3 \end{bmatrix}$$

```
linalg[crossprod](e, linalg[transpose](d));
```

$$[3\,\pi - 6 \quad 3 - 3\sin(a) \quad 2\sin(a) - \pi]$$

Tables Maple is capable of storing and manipulating multidimensional data in tables. A table is a generalized form of the more common structures of arrays and matrices. Unlike these specialized structures, a table has some significant advantages: A table can be n-dimensional (note that an array with more than two dimensions is automatically cast into a table by Maple) and its index can be any Maple expression.

Here we create a table implicitly by making assignments to an indexed variable.

```
A_DIFF_TABLE[sin(x)]:= cos(x); A_DIFF_TABLE[cos(x)]:=
-sin(x);
```

$$\text{A\_DIFF\_TABLE}_{\sin(x)} := \cos(x)$$
$$\text{A\_DIFF\_TABLE}_{\cos(x)} := -\sin(x)$$

Here we define a two-dimensional table explicitly:

```
A_2D_TABLE := table([(-1,-1)='first row', (-1,0)=zero,
(-1,1)=one, (1,-1)='second row', (1,0)=zero,
(1,1)='last entry']);
```

A_2D_TABLE := table([
(−1, 1) = one
(1, −1) = second row
(1, 0) = zero
(−1, −1) = first row
(−1, 0) = zero
(1, 1) last entry
])

This adds an entry to our first table:

```
> A_DIFF_TABLE[tan(x)]:=1/cos(x);
```

$$A\_DIFF\_TABLE_{\tan(x)} := \frac{1}{\cos(x)}$$

The previous entry is in error so it needs to be altered:

```
> A_DIFF_TABLE[tan(x)]:=1+tan(x)^2;
```

$$A\_DIFF\_TABLE_{\tan(x)} := 1 + \tan(x)^2$$

We will make one final entry in our test table and then we will take a look at it:

```
> A_DIFF_TABLE[x^n]:=n*x^(n-1);
```

$$A\_DIFF\_TABLE_{x^n} := n x^{(n-1)}$$

```
> A_DIFF_TABLE;
```

$$A\_DIFF\_TABLE$$

This is not what we expected but it is correct. Maple evaluates table names in the same way as it does an array name using last name evaluation. To view the contents of a table, `eval` or `print` must be used:

```
> print(A_DIFF_TABLE);
```

table ([
cos(x) = −sin(x)
sin(x) = cos(x)
tan(x) = 1 + tan(x)2
$x^n = nx^{(n-1)}$
)]

Maple as a programmable calculator

Maple's extensive list of functions and procedures makes it a powerful mathematical tool that can be applied easily to problem areas as diverse as virology and cosmology. However, ideas and techniques change and advance—as is true of all fields of human endeavor—so Maple would soon become obsolete if it were not able to adapt. Maple can adapt because in addition to being a CAS it is also a powerful programming language.

Functions

A function is a Maple object that performs work, taking arguments as inputs and returning an answer. The general form of a function definition is:

$$\mathrm{tag} := (\ \mathrm{arg}_1,\ \mathrm{arg}_2,\ \ldots,\ \mathrm{arg}_n\ ) \rightarrow \mathrm{body}$$

where `tag` is associated with the function definition, sometimes called the head, `arg1`, `arg2`, ..., `argn` is the list of formal parameters passed to body. The formal parameter names appearing in the argument list on the left-hand side of the arrow match with the corresponding variables in body of the definition. The body is an expression that yields the value of the function when the formal parameters are replaced with the values of the actual arguments.

The function f(x, y, z) is a function of three variables x, y, and z and returns the sum of the first two arguments raised to the third argument: $f(x, y, z) = x^z + y^z$. This simple function is represented in Maple as:

```
f := (x, y, z) -> x^z + y^z;
```

$$f := (x, y, z) \rightarrow x^z + y^z$$

The function is invoked as follows:

```
f(A, 4, Pi);
```

$$A^\pi + 4^\pi$$

Pure functions The Maple name space contains every function and variable name that has had an assignment made to it by either the user or the system. A pure function does not have a tag associated with it and hence reduces the loading on the Maple name space. A pure function is defined as follows:

$$(\ \mathrm{arg}_1, \mathrm{arg}_2, ..., \mathrm{arg}_n\ ) \rightarrow \mathrm{body}$$

In the following examples, we see how pure functions, based on `map` and `map2`, can be easily used when manipulating data. The functions `map` and `map2` are extremely powerful Maple functions that enable complex operations to be performed easily on the elements of data structures like tables, lists, arrays, and matrices. The required operation is mapped onto every element in the target data structure.

```
map(x -> x^2 , array([[x, x^2], [1+x, x^d]]));
```

$$\begin{bmatrix} x^2 & x^4 \\ (1 + x)^2 & (x^d)^2 \end{bmatrix}$$

```
map2((x,y)->x^y, n, {1,2,3});
```

$$\{n, n2, n3\}$$

Unapply Maple expressions can be transformed into functions easily using `unapply`. In the following example we find the closed-form solution for the eigenvalues of a two-by-two matrix and then transform the solution into a Maple function. First we define an empty array.

```
A:=array(1..2, 1..2);
```

$$A := \mathrm{array}\,(1\ ..\ 2,\quad 1\ ..\ 2,\quad [\ \ ])$$

Next we calculate its characteristic equation in λ. We do this by creating a weighted two-by-two identity matrix using `linalg[band]`, subtracting it from the matrix A1, and computing the determinant of the resulting matrix:

```
A1 := evalm(A - linalg[band]([lambda], 2));
```

$$A1 := \begin{bmatrix} A_{1,1}-\lambda & A_{1,2} \\ A_{2,1} & A_{2,2}-\lambda \end{bmatrix}$$

```
A2 := linalg[det](A1);
```

$$A2 := A_{1,1}A_{2,2} - A_{1,1}\lambda - \lambda A_{2,2} + \lambda^2 - A_{1,2}A_{2,1}$$

This solves the characteristic equation for l.

```
SOLS := [solve(A2, lambda)];
```

$$\left[\frac{1}{2}A_{1,1} + \frac{1}{2}A_{2,2} + \frac{1}{2}\sqrt{A_{1,1}{}^2 - 2A_{1,1}A_{2,2} + A_{2,2}{}^2 + 4A_{1,2}A_{2,1}}\,,\right.$$
$$\left.\frac{1}{2}A_{1,1} + \frac{1}{2}A_{2,2} - \frac{1}{2}\sqrt{A_{1,1}{}^2 - 2A_{1,1}A_{2,2} + 4A_{1,2}A_{2,1}}\right]$$

Here we use `unapply` to form a function that takes an array as input and returns a list of eigenvalues:

```
E_VALUES := unapply(SOLS,A);
```

$$E\_VALUES := A \to$$
$$\left[\frac{1}{2}A_{1,1} + \frac{1}{2}A_{2,2} + \frac{1}{2}\sqrt{A_{1,1}{}^2 - 2A_{1,1}A_{2,2} + A_{2,2}{}^2 + 4A_{1,2}A_{2,1}}\,,\right.$$
$$\left.\frac{1}{2}A_{1,1} + \frac{1}{2}A_{2,2} - \frac{1}{2}\sqrt{A_{1,1}{}^2 - 2A_{1,1}A_{2,2} + A_{2,2}{}^2 + 4A_{1,2}A_{2,1}}\right]$$

The eigenvalues can now be calculated for any two-by-two matrix:

```
E_VALUES(array([[1,2],[3,4]]));
```

$$\left[\frac{5}{2} + \frac{1}{2}\sqrt{33},\ \frac{5}{2} - \frac{1}{2}\sqrt{33}\right]$$

Control statements

Maple supports basic control structures to govern the flow of evaluation (`if` and `elif`), as well as repetition (`for` and `while`). Although it is more usual to use these control structures inside Maple procedures and functions, we will use them interactively in the following examples for clarity.

If Here is a simple conditional statement to test an input to see whether it is odd or even.

```
x:=4:
if x mod 2 = 0 then
print(cat(x,' is even'));
else
print(cat(x,' is odd'));
fi;
```

4 is even

In this and the following example, the infix form of the mod function is used. A function is said to be invoked in infix form when the following syntax is used: *arg1 function name arg2*. Note that not every Maple function has a standard infix form. The infix form of a function is used for this situation: If nested `if` statements are needed, then the `elif` statement is used in conjunction with `if` as shown. Here we extend the above example to test for zero:

```
x:=-3:
if x=0 then
 print(`zero entered')
elif x mod 2 = 0 then
 print(cat(x,' is even'));
else
 print(cat(x,' is odd'));
fi;
```

–3 is odd

Looping

There are two basic repetition constructs, `for` and `while`. However, looping can also be accomplished with `seq` and `$`.

For The general syntax of the for construct is `for |count| from |start| by |increment| to |end| do body od`. Any expression of the form `|keyword|` is optional. If omitted they default to unity with the exception of end, which defaults to infinity. The following examples demonstrate how the `for-loop` operates:

```
ANS:=NULL:
for x to 5 by 2 do ANS:=ANS, sin(x) od:
ANS;
```

$$\sin(1), \sin(3), \sin(5)$$

```
ANS:=NULL:
for x from -1 to 1 by 0.2 do ANS:=ANS, x^2 od:
ANS;
```

$$1, .64, .36, .16, .04, 0, .04, .16, .36, .64, 1.00$$

```
ANS:=NULL:
for x from -5 to 5 do ANS:=ANS, x^(1/2) od:
ANS;
```

$$\sqrt{-5}\,,\ \sqrt{-4}\,,\ \sqrt{-3}\,,\ \sqrt{-2}\,,\ I,\ 0,\ 1,\ \sqrt{2}\,,\ \sqrt{3}\,,\ \sqrt{4}\,,\ \sqrt{5}$$

The `for` construct does have one other form that is useful when operating on objects, such as lists or sets of unknown length, the `for-in` construct.

```
ANS:=NULL:
for x in [a, b, c, d, e, f] do ANS:=ANS, (x+1)/x od:
ANS;
```

$$\frac{1+a}{a}, \frac{1+b}{b}, \frac{1+c}{c}, \frac{1+d}{d}, \frac{1+e}{e}, \frac{1+f}{f}$$

While The `while` construct is an alternate way of performing repetitive operations and has this basic form: `while` condition is true `do` body `od`.

```
ANS:=NULL: x:=1:
while ithprime(x) 12 do
 ANS:=ANS, cat(ithprime(x)<' is prime');
 x:=x+1;
od:
ANS;
```

2 is prime, 3 is prime, 5 is prime, 7 is prime, 11 is prime

Seq and $ Maple provides two alternative methods of looping: `seq` and `$`. Both of these functions are basically equivalent to: `for count from start to end by 1 do .. od`. A couple of points, however, should be noted: The variable used as the count will maintain its final value on exit and only `seq` will operate over sets and lists.

```
seq(n^x, x=0..5);
```

$$1, n, n^2, n^3, n^4, n^5$$

The count variable x now has a value:

```
x;
```

$$6$$

The function `seq` also works over the elements of a list:

```
seq(f^2, f=[1,2,a,b]);
```

$$1, 4, a^2, b^2$$

Both constructs will operate on floats, but not on symbolic constants such as π. In such cases `evalf` is simply used to convert to a float. The `$` symbol is the infix form of `seq` and is used as follows:

```
n^y$(y=1.1..5.7);
```

$$n^{1.1}, n^{2.1}, n^{3.1}, n^{4.1}, n^{5.1}$$

The alternative form of `$` can also be used to create variables of the form a1, a2, ..., or copies easily:

```
`a.n`$(n=1..3);
```

$$a1, a2, a3$$

```
A_COPY$4;
```

A_COPY, A_COPY, A_COPY, A_COPY

Procedural

Procedural programming is deeply entrenched in Maple and is supported by the most popular programming languages and tools such as BASIC, FORTRAN, C, and Pascal.

The general form of a procedure is:

```
procedure_name :=proc(arg1, arg2, ..., argn )
 local var1, var2, ..., varn;
 global var1, var2, ..., varn;
 options opt1, opt2, ..., optn;
 body
end;
```

The formal parameters `arg1, arg2, ..., argn` correspond to variables found within the body of the procedure with the same names. If no operands are specified then, like C, the argument list can be any length. In this case the global variables `args` and `nargs` are used to retrieve the argument list and its length. In the next example we construct a custom plotting procedure that takes a function and a range and plots the function and its integral. The integral is computed and stored using the local variable `the_int`. It should also be noted that the function `op` is used to extract the independent variable from the range `r`.

```
MY_PLOT := proc(x, r)
 local the_int;
 the_int:= int(x, op(1, r));
 plot({x, the_int}, r, color=BLACK);
end:
```

Next we call the procedure defined above:

```
MY_PLOT( sin(X)/X, X=-10..10);
```

A plot of this procedure is shown in Figure 1.10. For further information on procedure options see `?procedure[options]`.

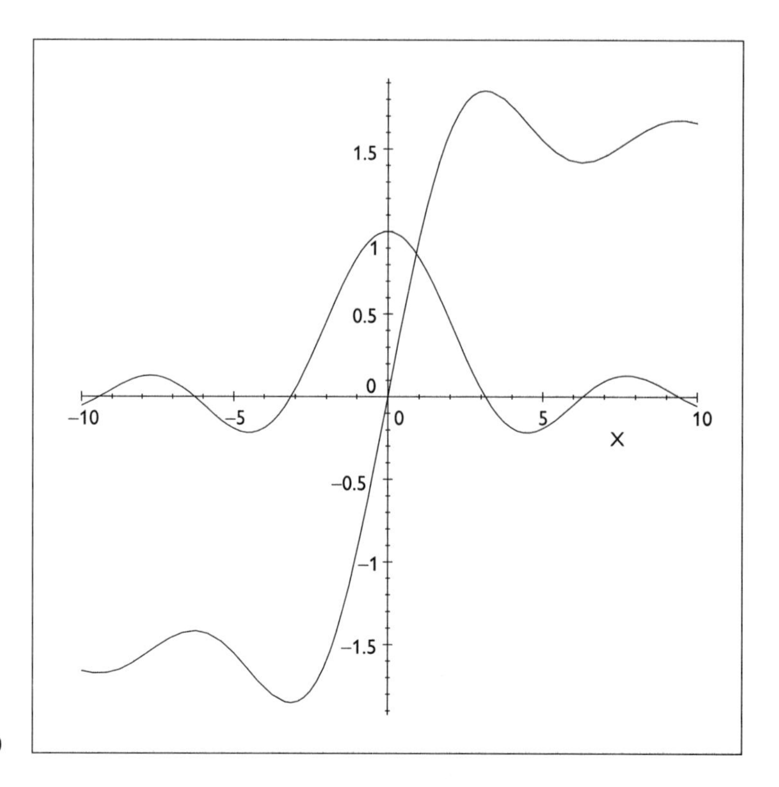

Figure 1.10

Pure procedures In the same way that Maple supports pure functions it also supports pure procedures. A pure procedure is exactly the same internally as a normal procedure but does not have a tag assigned to it. The first example uses `zip`. This function is very useful in many applications where the corresponding elements of two data structures have to be combined to produce the result. The lengths of the component data structures do not have to be the same as the example shows. Although in many cases pure procedures and interchangeable with pure functions, these examples show how pure procedures are used.

```
zip( proc(x, y) x^y end,[tan(1), cot(2), coth(3)],
  [a,b], filler);
```

$$\left[\tan(1)^a, \cot(2)^b, \coth(3)^{\text{filler}}\right]$$

```
map( proc( x ) x^2 end, [a, b, c] );
```

$$\left[a^2, b^2, c^2\right]$$

Types

Maple strongly types language. The following parameter types are some of the more commonly used ones. For more information see `?type`.

^	*	**	+	<	<=
<>	=	PLOT	PLOT3D	RootOf	algebraic
anything	array	complex	constant	equation	float
fraction	function	indexed	linear	list	listlist
matrix	negative	numeric	positive	procedure	ange
rational	series	set	string	string	vector

Table 1.1

```
whattype(2/37);
```

fraction

```
whattype(series(exp(beta), beta));
```

series

```
type(10!, integer);
```

true

```
type( [ a=1, b=2], ['name' = 'integer', 'name'= 'integer']);
```

true

Next we use the Maple type-checking routines in conjunction with procedure definition. First we perform the check at the top level and then it is carried out within the body of the procedure.

```
> A_TEST := proc(x: :numeric, y: :{posint, function})
  print('Parametrs check') end:

> A_TEST(1, cos(x));
```

Parameters check

```
> A_TEST(z, cos(x));
  Error, A_TEST expects its first argument x to be of type
   numeric but received z

> A_TEST_2 := proc()
  if type(args, numeric) then print('Ok')
  else print('Error')
    fi end:

> A_TEST_2(1234);
```

OK

```
> A_TEST_2(abcd);
```

Error

By comparing the two approaches, we can see that the first implementation is the simplest but allows no ability to recover from the error condition. The second implementation, on the other hand, is harder to program but it does give us the ability to recover gracefully from error conditions occurring as a result of the use of illegal arguments and to continue processing with default values for instance.

RETURN and ERROR

Loops and procedures can be terminated in one of two ways: the end condition becomes valid or a `RETURN` or an `ERROR` can be used. When a `RETURN` is encountered the current control statement is terminated and control is passed to the next level above. When an `ERROR` is encountered, on the other hand, control is passed to the top level and the current process chain is terminated.

```
FOR_TEST := proc(x, y) local n, nn;
 nn:=NULL:
 if x=0 and y=0 then ERROR('Non-zero entries required') fi;
 for n from x to 100 do
 if n>y then RETURN(nn) else nn:=nn,[n,n^3] fi;
 od;
nn;
end:
```

We can see how this works as follows:

```
FOR_TEST(90, 101);
```

$$[90, 729000], [91, 753571], [92, 778688], [93, 804357], [94, 830584], [95, 857375], [96, 884736], [97, 912673], [98, 941192], [99, 970299], [100, 1000000]$$

```
FOR_TEST(-1, 4);
```

$$[-1, -1], [0, 0,], [1, 1], [2, 8], [3, 27], [4, 64]$$

```
FOR_TEST(0, 0);
```

Error, (in FOR_TEST), Nonzero entries required.

Chapter 2

Active filter design and analysis

FILTERS ARE DESIGNED TO reject certain interference signals while allowing certain desired pieces of information to pass unattenuated. Filters can be realized in mechanical, electrical, chemical, and other applied science applications.

In the first application, we will concentrate on the analysis and synthesis of an electrical low-pass filter (LPF). In particular, the analog (as opposed to a digital hardware implementation) active LPF topology will have a Butterworth magnitude response. The Butterworth response is specifically characterized as having a maximally flat magnitude response. This means that while information is within the filter's passband, the signal's amplitude response will not change until it diminishes to exactly (at least mathematically) 0.7071 times its passband gain at the cutoff frequency. The LPF passes information below the cutoff, break, or 3-dB frequency while increasingly attenuating all frequencies above the cutoff frequency. The order of the LPF determines the rate (usually expressed in decibels per decade or octave) at which the higher frequencies are attenuated beyond the cutoff frequency.

The Butterworth LPF's characteristics are dependent on derived component values and tolerances and several test simulations are performed to show the reader how Maple can easily emulate certain electronic circuits.

In contrast, the second application is a switching bandpass filter (BPF) whose filter properties are dependent for the most part on a digital clocking signal and not on component values. Switching BPFs exhibit some very unique properties that are unavailable in an analog or continuous sense.

Case I: analog low-pass filter design and analysis

One of the most common electronic filters utilized in electronic design is the LPF. Today, the active resistor-capacitor (RC) filter is common due to its low cost and the extreme flexibility of its topology.

Figure 2.1 shows a typical second-order LPF whose Butterworth response is dictated by the passive component values R_1, R_2, C_1, C_2. This particular topology uses a noninverting unity gain configured operational amplifier with one positive feedback connection. Inverting topologies are also useful, but for maximum transfer bandwidth control with good input and output impedance across those frequencies, the topology in Figure 2.1 is about the best. The single application drawback is that the passband gain is fixed at unity.

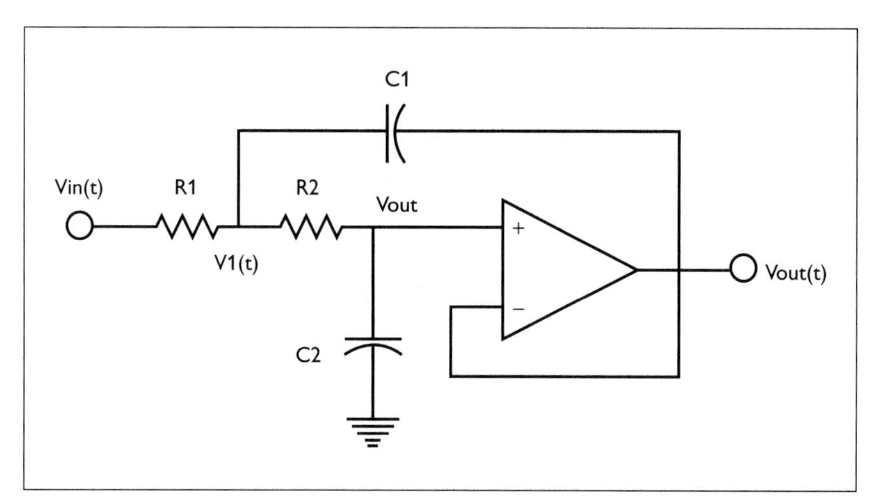

Figure 2.1
Second-order Butterworth LPF.

As designers, we are allowed to vary any of the passive components (R_1, R_2, C_1, C_2) to ensure that we obtain the Butterworth response with the desired cutoff frequency. Consequently, we should first determine the overall voltage transfer function associated with Figure 2.1 as

$$\text{Voltage transfer function} = \frac{V\text{out}(t)}{V\text{in}(t)}$$

To determine how each of the components relates to the cutoff frequency and Butterworth amplitude response, we use Maple to derive the necessary constituent relationships to form the symbolic design constraints. Finally, we take the symbolic expressions and substitute the desired filter specifications (i.e., cutoff frequency and Butterworth passband characteristics) to finalize the hardware design.

Use of Laplace transform explained

The analysis will be performed in the frequency domain using the Laplace transform. Many suitable engineering texts have been published that explain the use of the Laplace transform for performing nodal and mesh analyses for most linear time invariant (LTI) topologies [1–3]. LTI topologies—electrical, mechanical, chemical, etc.—exhibit constant coefficients associated with their respective dynamic equations. The dynamics could be with respect to any number of physical variables. In the case of the following LPF, all coefficients in the describing or constituent nodal equations have coefficients whose values are functions of the passive electronic components from which they are comprised.

Constituent relationships derived

The Laplace transform allows the designer to formulate the symbolic expressions describing the LPF's characteristics in the polynomial, s, defined by the following identity, which transforms a time function, f(t), into a frequency-domain function, F(s)

$$\text{Laplace}\{f(t)\} = F(s) \equiv \int_0^s f(t)e^{-st}\,dt$$

where

$$s = j\omega$$

Use of the Laplace transform allows the designer to use algebraic rather than differential equations to solve the LPF design problem.

First, to find the derivation of the constituent relationships via Kirchhoff's laws, we state the following two independent node expressions as shown in Figure 2.1 (using the Laplace transform):

currents about the node labeled $V_1(t)$

$$\frac{V_1 - V_{IN}}{R_1} + \frac{V_1 - V_{OUT}}{R_2} + sC_1(V_1 - V_{OUT}) = 0$$

currents about the node labeled $V_{out}(t)$

$$\frac{V_{OUT} - V_1}{R_2} + sC_2V_{OUT} = 0$$

To arrange the above expressions into matrix algebra form for the unknowns V_1, V_{OUT}, we rewrite the equations as

$$\left(\frac{1}{R_1} + \frac{1}{R_2} + sC_1\right)V_1 + \left(-\frac{1}{R_2} - sC_1\right)V_{OUT} = \frac{V_{IN}}{R_1}$$

$$\left(-\frac{1}{R_2}\right)V_1 + \left(\frac{1}{R_2} + sC_2\right)V_{OUT} = 0$$

or in matrix formulation

$$\begin{bmatrix} \left(\frac{1}{R_1} + \frac{1}{R_2} + sC_1\right) & \left(-\frac{1}{R_2} - sC_1\right) \\ \left(-\frac{1}{R_2}\right) & \left(\frac{1}{R_2} + sC_2\right) \end{bmatrix} \begin{bmatrix} V_1 \\ V_{OUT} \end{bmatrix} = \begin{bmatrix} \left(\frac{V_{IN}}{R_1}\right) \\ 0 \end{bmatrix}$$

or equivalently in the following vector equation [1]:

$$\mathbf{Ax} = \mathbf{y}$$

where

$$\mathbf{A} \rightarrow \begin{bmatrix} \left(\frac{1}{R_1} + \frac{1}{R_2} + sC_1\right) & \left(-\frac{1}{R_2} - sC_1\right) \\ \left(-\frac{1}{R_2}\right) & \left(\frac{1}{R_2} + sC_2\right) \end{bmatrix}$$

$$\mathbf{x} \rightarrow \begin{bmatrix} V_1 \\ V_{\mathrm{OUT}} \end{bmatrix}$$

$$\mathbf{y} \rightarrow \begin{bmatrix} \left(\frac{V_{\mathrm{IN}}}{R_1}\right) \\ 0 \end{bmatrix}$$

then solving for the variables V_1, V_{OUT} in vector form yields

$$\mathbf{x} = \mathbf{A}^{-1}\mathbf{y}$$

or if we expand it into a matrix formulation, we obtain

$$\begin{bmatrix} V_1 \\ V_{\mathrm{OUT}} \end{bmatrix} = \begin{bmatrix} \left(\frac{1}{R_1} + \frac{1}{R_2} + sC_1\right) & \left(-\frac{1}{R_2} - sC_1\right) \\ \left(-\frac{1}{R_2}\right) & \left(\frac{1}{R_2} + sC_2\right) \end{bmatrix}^{-1} \begin{bmatrix} \left(\frac{V_{\mathrm{IN}}}{R_1}\right) \\ 0 \end{bmatrix}$$

Solving this last expression in Maple is done by assigning the following expressions:

$$\mathbf{x} \rightarrow \begin{bmatrix} V_1 \\ V_{\mathrm{OUT}} \end{bmatrix} \xrightarrow{\text{Maple}} \text{Unknown\_Variable\_Matrix}$$

$$\mathbf{A} \rightarrow \begin{bmatrix} \left(\frac{1}{R_1} + \frac{1}{R_2} + sC_1\right) & \left(-\frac{1}{R_2} - sC_1\right) \\ \left(-\frac{1}{R_2}\right) & \left(\frac{1}{R_2} + sC_2\right) \end{bmatrix}^{-1} \xrightarrow{\text{Maple}} \text{A\_Matrix}$$

$$\mathbf{y} \rightarrow \begin{bmatrix} \left(\frac{V_{\text{IN}}}{R_1}\right) \\ 0 \end{bmatrix} \xrightarrow{\text{Maple}} \text{Y\_Matrix}$$

and then entering them into a Maple session:

```
with (linalg):
A_Matrix := array ([[1/R1+1/R2+s*C1,-1/R2-s*C1],
 [-1/R2,1/R2+s*C2]]);
Y_Matrix := array ([[Vin/R1],[0]]);
Unknown_Variable_Matrix := evalm(((A_Matrix)^(-1))&*
 (Y_Matrix));
```

$$\text{A\_Matrix} := \begin{bmatrix} \frac{1}{R1} + \frac{1}{R2} + sC1 & -\frac{1}{R2} - sC1 \\ -\frac{1}{R2} & \frac{1}{R2} + sC2 \end{bmatrix}$$

$$\text{Y\_Matrix} := \begin{bmatrix} \frac{V_{\text{IN}}}{R1} \\ 0 \end{bmatrix}$$

$$\text{Unknown\_Variable\_Matrix} := \begin{bmatrix} \frac{(1 + sC2\ R2)V_{\text{in}}}{1 + sC2\ R2 + R1\ sC2 + s^2C1\ R1\ R2\ C2} \\ \frac{V_{\text{in}}}{1 + sC2\ R2 + R1\ sC2 + s^2C1\ R1\ R2\ C2} \end{bmatrix}$$

Incidentally, since either column or row matrices are vectors, the user could set up the `Maple A_Matrix` and `Unknown_Variable_Matrix` variables as `A_Vector` and `Unknown_Variable_Vector` to more clearly represent these variables in a session. However, the single-order vector is only a special case of the multidimensional-order matrix.

Note that the matrix multiplication between `A_Matrix` inverse and `Y_Matrix` is performed with the "`&*`" operation rather than the "`*`" operator. Maple distinguishes between matrix and scalar multiplication with this notation.

Abstracting the variable's solutions,

```
V1 := Unknown_Variable_Matrix[1,1];
Vout := Unknown_Variable_Matrix[2,1];
```

$$V1 := \frac{(1 + sC2\ R2)V_{\text{in}}}{1 + sC2\ R2 + R1\ sC2 + s^2C1\ R1\ R2\ C2}$$

$$Vout := \frac{V_{\text{in}}}{1 + sC2\ R2 + R1\ sC2 + s^2C1\ R1\ R2\ C2}$$

Since the LPF's overall voltage transfer function was previously defined as

$$\text{LPF\_Transfer} = \frac{V_{\text{OUT}}}{V_{\text{IN}}}$$

we appear to have our transfer function with the second equation, thus

```
LPF_Transfer := Vout/Vin;
```

$$\text{LPF\_Transfer} := \frac{1}{1 + sC2\ R2 + R1\ sC2 + s^2C1\ R1\ R2\ C2}$$

We may rewrite the second-order `LPT_Transfer` expression as

$$\text{LPF\_Transfer} = \frac{1}{\dfrac{s^2}{\omega_N^2} + \dfrac{2\zeta}{\omega_N} + 1}$$

where ω_N and ζ are the natural frequency and damping factor, respectively, of the second-order system [1,2].

Now we must collect this coefficient information from the `LPF_Transfer` equation, hence

```
LPF_Transfer_Denominator := denom(LPF_Transfer):
Second_Order_Coeff := coeff(LPF_Transfer_Denominator,s,2);
First_Order_Coeff := coeff(LPF_Transfer_Denominator,s,1);
Zero_Order_Coeff := coeff(LPF_Transfer_Denominator,s,0);
```

$$\text{Second\_Order\_Coeff} := \text{C1 R1 R2 C2}$$
$$\text{First\_Order\_Coeff} := \text{C1 R2} + \text{R1 C2}$$
$$\text{Zero\_Order\_Coeff} := 1$$

Associating these Maple coefficient extraction values with the general `LPF_Transfer` results in the following second-order coefficients having the circuit component values shown in Figure 2.1:

```
Natural_Frequency_Radians := 1/sqrt(Second_Order_Coeff);
Damping_Factor := (Natural_Frequency_Radians/2)
 *First_Order_Coeff;
```

$$\text{Natural\_Frequency\_Radians} := \frac{1}{\sqrt{C1\ R1\ R2\ C2}}$$

$$\text{Damping\_Factor} := \frac{1}{2}\,\frac{C1\ R2 + R1\ C2}{\sqrt{C1\ R1\ R2\ C2}}$$

Comparing these expressions with any standard control theory text on second-order systems [1–3] shows the following for the natural frequency ω_N and damping factor ζ:

$$\omega_N = \frac{1}{\sqrt{R_1 R_2 C_1 C_2}}$$

$$\zeta = \frac{C_2(R_1 + R_2)}{2\sqrt{R_1 R_2 C_1 C_2}}$$

Hence, these Maple symbolically derived expressions agree with the previously cited texts.

If we want the `Natural_Frequency_Radians` in hertz, then

$$f_N = \frac{1}{2\pi\sqrt{R_1 R_2 C_1 C_2}}$$

To impose a Butterworth or maximally flat amplitude response, or flat passband gain, we need to impose the following on the damping factor:

$$\zeta = \frac{C_2(R_1 + R_2)}{2\sqrt{R_1 R_2 C_1 C_2}} \rightarrow \frac{1}{\sqrt{2}}$$

Hence, we have our first design constraint toward the design of a Butterworth LPF. This constraint comes from [1] if the reader is interested in discovering more about filter and network theory.

Now let's build ourselves a very common Butterworth filter with a 1-kHz cutoff frequency.

Designing a 1-kHz Butterworth LPF

From the previously derived `Natural_Frequency_Hertz` and ζ expressions, we state the following design constraints:

$$f_N = \frac{1}{2\pi\sqrt{R_1 R_2 C_1 C_2}} = 1{,}000 \text{ Hz}$$

$$\zeta = \frac{C_2(R_1 + R_2)}{2\sqrt{R_1 R_2 C_1 C_2}} = \frac{1}{\sqrt{2}}$$

Now we have four unknowns in two equations, hence we need to reduce the number of independent variables by two. The most direct way is to impose some initial constraints such as these:

$$R_1 = R_2 = R$$

$$C_1 = \beta C_2$$

Implementing this in a Maple session,

```
Natural_Frequency_Hertz := 1/(2*Pi*sqrt(R1*R2*C1*C2)):
Damping_Factor := C2*(R1+R2)/(2*sqrt(R1*R2*C1*C2)):
R1 := R:
R2 := R:
C1 := beta*C2:
Results :=
 solve({Natural_Frequency_Hertz=1000,
 Damping_Factor=1/sqrt(2)},{beta,R}):
Solutions := subs(Results,[beta,R]):
beta := op(1,Solutions);
R := op(2,Solutions);
```

$$\beta := 2$$

$$R := \frac{1}{4000}\frac{\sqrt{2}}{C2\ \pi}$$

Now we have two expressions that give us the nice result that the two capacitors are related by an integer and all resistors are defined once we pick a value for $C2$.

The fact that the capacitors are related by an integer is especially nice from the practical standpoint that capacitor values have the added problem of variable geometry (physical size) depending on type (polycarbonate, polypropylene, paper, electrolytic, etc.), voltage rating, and physical mounting configuration (axial or stand-up). Having an integer value for the capacitors allows the designer simply to purchase two, three, or whatever number of capacitors to complete the filter design. Matching capacitors for a design is not nearly as easy as it is for matching two or more resistors.

At this point we simply try some values to initiate the LPF design. Generally, a designer starts with some simple resistor and capacitor values. These values might be components that are readily available in the lab or from a vendor. Consequently, we will start by letting R = 10,000Ω, which forces C_2 = 11,253 pF, which we will call .01 μF. Both of these component values are common in most labs, electronic kits, or *gutted* electronic equipment. The complete component value set then becomes

$$R_1 = R_2 = 10{,}000\Omega$$
$$C_1 = 20{,}000 \text{ pF}$$
$$C_2 = 10{,}000 \text{ pF}$$

Let's use Maple to determine these coefficients' effect on `LPF_Transfer`:

```
LPF_Transfer_Complex := subs(s=I*w,LPF_Transfer):
LPF_Transfer_Mag := evalc(abs(LPF_Transfer_Complex));
LPF_Transfer_Phase :=
 arctan(evalc(Im(LPF_Transfer_Complex)/
 Re(LPF_Transfer_Complex)));
```

LPF_Transfer_Mag :=

$$1\Big/\left(1 - 2w^2\ C1\ R1\ R2\ C2 + w^4 C1^2\ R1^2\ R2^2\ C2^2 + w^2 C2^2\ R2^2 + 2w^2 C2^2\ R2\ R1 + R1^2\ w^2 C2^2\right)^{1/2}$$

$$\text{LPF\_Transfer\_Phase} := -\arctan\left(\frac{wC2\ R2 + R1\ wC2}{1 - w^2 C1\ R1\ R2\ C2}\right)$$

Bode magnitude and phase plots

Bode plots [1–3] are a convenient method for determining the magnitude and phase response of any linear (or nonlinear in special cases) system.

Magnitude response

The magnitude response, shown in Figure 2.2, depicts the gain of the Butterworth LPF over any particular frequency range. The Bode magnitude plot will indicate any part of the LPF's spectral response that is not "flat" or, in other words, that exhibits some kind of ripple or varying gain with frequency. The Butterworth criterion says this must not happen except when uniformly decreasing after the cutoff frequency (which we have decided is 1 kHz). The x-axis represents frequency units in hertz and the y-axis represents gain, which is measured in decibels (dB), which can be expressed as

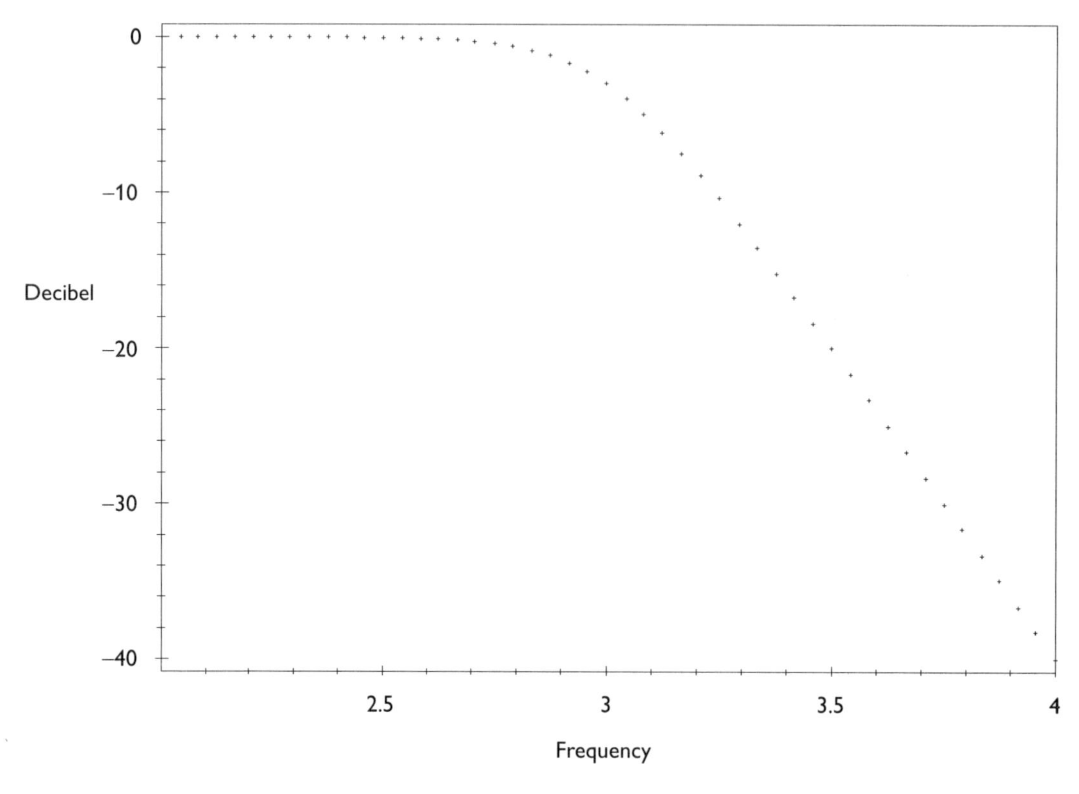

Figure 2.2 Bode plot for Butterworth LPF (cutoff frequency = 1 kHz, damping factor = 0.7071, y-axis is in decibels, x-axis is in 10^N Hz).

$$dB = 20 \log_{10}\left(\frac{\text{Output signal}}{\text{Input signal}}\right)$$

Technically speaking, gain is dimensionless; however, when expressed in decibels, what you are actually expressing on the y-axis is output response relative to the input excitation.

Phase response

The phase response, as shown in Figure 2.3, indicates how a sinusoidal output signal's phase will vary with respect to the input sinusoidal. Hence, if the input signal is defined as zero phase, then a negative phase response indicates that the output response will lag the input in time or, equivalently, phase. Conversely, a positive phase indicates the output signal anticipates the input signal. This last case is impossible (how can an output signal appear before the input signal was applied to the filter?) but can mathematically appear as a result. What this means is that the circuit or system has slipped a cycle, which might be 90 degrees, 180 degrees, 360 degrees, or other multiple value of the input sinusoidal phase. Such phase slips are sometimes called *Rice clicks* in FM (frequency modulated) receivers [4], which use a phase-lock loop (PLL) topology approach to detecting frequency changes from a central carrier frequency.

We plot the logarithmic form of `LPF_Transfer_Mag` versus frequency (in hertz) as follows:

```
with (plots):
readlib(log10):
LPF_Transfer_Mag_Subs := subs(w=2*Pi*f,R1=10000,
 R2=10000,C1=20000*10^(-12),C2=10000*10^(-12),
 LPF_Transfer_Mag):
LPF_Transfer_Mag_Decibel := 20*log10(LPF_Transfer_Mag_Subs):
f := 10^N:
plot (LPF_Transfer_Mag_Decibel,N=2..4,axes=boxed,
 style=point,
 symbol= cross ,color=black,labels=[Frequency,Decibel]);
```

The reader can see the nicely flat corner frequency at $N = 3$ (x-axis = 10^N Hz) or 1 kHz. Further, due to the second order, the response is around 35 to 40 dB below the passband, which is defined as the magnitude response below the cutoff frequency in a LPF.

Now let's look at the phase plot over the same frequency range:

```
LPF_Transfer_Phase_Subs :=
subs(w=2*Pi*f,R1=10000,R2=10000,C1=20000
 *10^(-12),C2=10000*10^(-12),LPF_ Transfer_Phase):
plot (180/Pi*LPF_Transfer_Phase_Subs,N=2..4,axes=boxed,
 color=black,labels=[Frequency,Phase]);
```

We can see a slight deviation from the ideal cutoff of 1 kHz in the Figure 2.3 plot, because the phase shift should be −90 degrees at $N = 3$ ($10^3 = 1$ kHz). Instead, we have a shift slightly less than the ideal −90 degrees. Making sure that we are close, let's numerically evaluate the `LPF_Transfer_Phase` expression at $\omega = 2\pi(1000)$ with Maple:

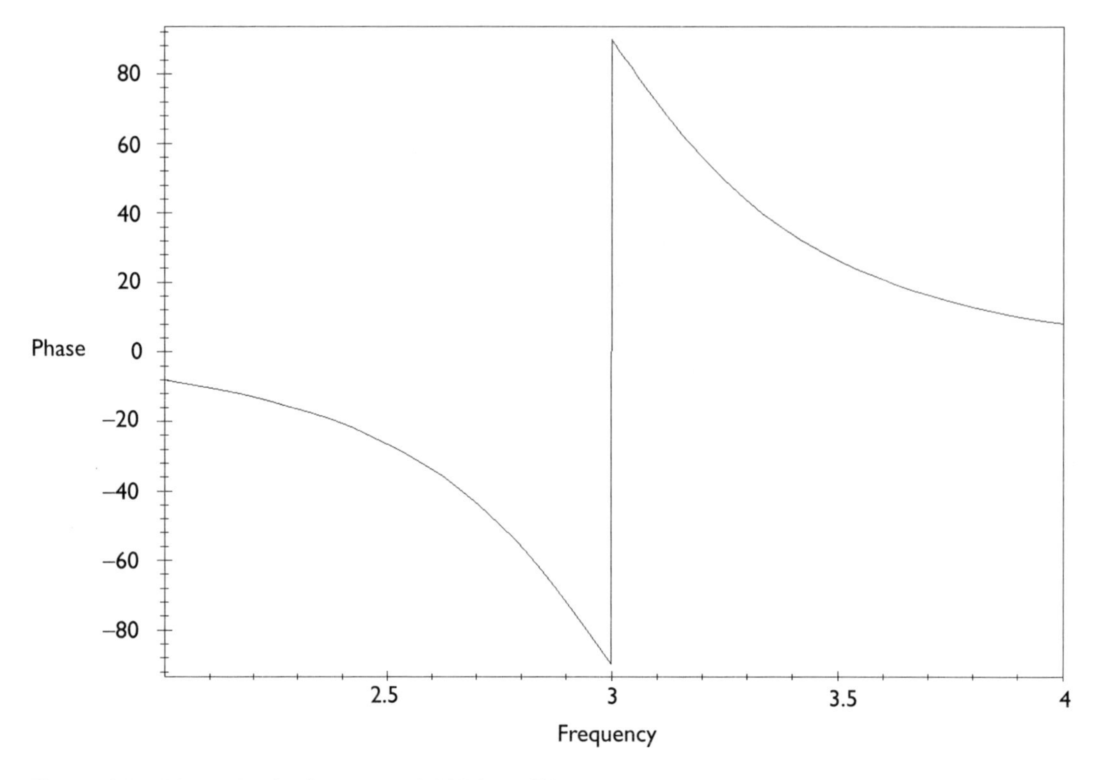

Figure 2.3 Phase plot for Butterworth LPF (cutoff frequency = 1 kHz, damping factor = 0.7071, y-axis is in degrees, x-axis is 10^N Hz).

```
Cutoff_Phase := evalf(subs(f=1000,180/Pi*
 LPF_Transfer_Phase_Subs));
```

$$\text{Cutoff\_Phase} := -80.49366995$$

This indicates that our cutoff frequency is higher (i.e., 1 kHz) than originally designed. In fact, using Maple, we can determine the actual half-power point as follows:

```
Phase_Cutoff_All := solve(LPF_Transfer_Phase_Subs=
 -Pi/2.00001,f);
Phase_Cutoff_Real_Positive := select(type,
 Phase_Cutoff_All],
 positive(numeric));
```

$$\text{Phase\_Cutoff\_All} := -1125.401644\,,\, 1125.389145$$
$$\text{Phase\_Cutoff\_Real\_Positive} := [1125.389145]$$

Obviously, the positive root is the one of interest, hence with the given component values, the LPF cutoff is 1125.389145 Hz.

We could have arrived at a similar (though not identical value) actual cutoff frequency by solving the `LPF_Transfer_Mag` expression for the half-power point or where

$$\text{LPF\_Transfer\_Mag} = \frac{1}{\sqrt{2}}$$

Therefore

```
Value := evalf(1/sqrt(2)):
Mag_Cutoff_All := solve(LPF_Transfer_Mag_Subs=Value,f);
Mag_Cutoff_Real_Positive := select(type,[Mag_Cutoff_All],
 positive(numeric));
```

$$\text{Mag\_Cutoff\_All} :=$$
$$1125.395395\,,\, -1125.395395\,,\, 1125.395395\ I, -1125.395395\ I$$
$$\text{Mag\_Cutoff\_Real\_Positive} := [1125.395395]$$

As expected, the cutoff frequency is slightly higher than designed, though by a small amount. Also, the numerical value is slightly different than by the phase computations, but well within numerical acceptability.

Improvement on the 1-kHz Butterworth LPF

We have seen that we were off the design cutoff frequency by around 10%. Consequently, for the sake of practicality, we will leave the two capacitors at their current values and adjust the resistive components to *trim* the cutoff frequency closer to the desired 1 kHz.

Starting with the following constraints:

$$C1 = .02\ \mu\text{F}$$
$$C2 = .01\ \mu\text{F}$$
$$R1 \neq R2$$

and further adhering to the Butterworth criterion,

$$\zeta = \frac{\sqrt{2}}{2}$$

We also use the previous equations describing the Butterworth's cutoff frequency and damping factor:

$$f_N = \frac{1}{2\pi\sqrt{R_1 R_2 C_1 C_2}} = 1000$$

$$\zeta = \frac{C_2(R_1 + R_2)}{2\sqrt{R_1 R_2 C_1 C_2}} = \frac{1}{\sqrt{2}}$$

We now solve for the resistive values in Maple:

```
C1 := .02*10^(-6):
C2 := .01*10^(-6):
Natural_Frequency_Hertz := 1/(2*Pi*sqrt(R1*R2*C1*C2)):
Damping_Factor := C2*(R1+R2)/(2*sqrt(R1*R2*C1*C2)):
Results :=
 fsolve({Natural_Frequency_Hertz=1000,
Damping_Factor=1/sqrt(2)},{R1,R2},co  mplex);
Solutions := subs(Results,[R1,R2]):
R1 := Re(op(1,Solutions));
R2 := Re(op(2,Solutions));
```

$$\text{Results} := \{R1 = 11253.95394 - .3785358168\ I,$$
$$R2 = 11253.95394 + .3785358168\ I\}$$
$$R1 := 11253.95394$$
$$R2 := 11253.95394$$

Notice that Maple found `complex` roots (after invoking the complex argument within the `fsolve` command) for the resistance values, hence we had to abstract the real values by invoking the `Re` command on the `Solutions` operands result. Further, note the relative magnitude difference between the real and imaginary parts of the $R1$, $R2$ `Result` solutions. What this indicates is that the solution roots (resistance values) are identical. However, due to round-off error within Maple's `fsolve` engine, the root-finding algorithm created some negative nonzero radicand value, which, when evaluated numerically, created an imaginary component. Consequently, users should always look at their results and evaluate these findings within the framework of the defining filter equations. To prove that the resistance values are valid, let's substitute these component values back into the Butterworth LPF's equations. You can easily do this by asking Maple to reiterate the `Natural_Frequency_Hertz` and `Damping_Factor` values, since all component values are still in Maple's *memory*. Hence,

```
Natural_Frequency_Check := evalf(Natural_Frequency_Hertz);
Damping_Factor_Check := evalf(Damping_Factor);
```

Natural_Frequency_Check := 1000.000001
Damping_Factor_Check := .7071067814

Not only are the component values numerically correct with easily found capacitive values, but we have the added bonus that finding 1% metal film resistors with the required 11.2-kΩ value is quite easily done. Now let's put the new 1% metal film resistor values into the Butterworth design and see what we obtain for a cutoff frequency and damping factor:

```
R1 := 11.2*10^3:
R2 := 11.2*10^3:
C1 := .02*10^(-6):
C2 := .01*10^(-6):
Natural_Frequency_Hertz :=
 evalf(1/(2*Pi*sqrt(R1*R2*C1*C2)));
Damping_Factor := C2*(R1+R2)/(2*sqrt(R1*R2*C1*C2));
```

Natural_Frequency_Hertz := 1004.817317
Damping_Factor := .7071067810

These component values are much better than our first attempt at realizing the desired Butterworth LPF design.

Butterworth LPF component sensitivity analysis

Now that we have seen (by both phase and magnitude computations) that our designed cutoff frequency is correct, we need to investigate the sensitivity of the LPF to component values. Consequently, let's define the sensitivity function, S_B^A, as

$$S_B^A \equiv \frac{B}{A}\frac{\partial A}{\partial B}$$

Simply stated, this function gives the percentage change of function A with respect to variable B. Consequently, the sensitivity of the cutoff frequency and damping factor to each of the component variables comprising the Butterworth topology is given in Table 2.1.

Table 2.1 Sensitivity cases of the Butterworth LPF

Component (variable)	**Cutoff frequency sensitivity** $\left(F_N = \frac{1}{2\pi\sqrt{R_1R_2C_1C_2}}\right)$	**Damping factor sensititivy** $\left(\zeta = \frac{1}{2}\frac{C_2(R_1 + R_2)}{\sqrt{R_1R_2C_1C_2}}\right)$
R1	$S_{R_1}^{f_N} \equiv$ CASE_1	$S_{R_1}^{\zeta} \equiv$ CASE_5
R2	$S_{R_2}^{f_N} \equiv$ CASE_2	$S_{R_2}^{\zeta} \equiv$ CASE_6
C1	$S_{C_1}^{f_N} \equiv$ CASE_3	$S_{C_1}^{\zeta} \equiv$ CASE_7
C2	$S_{C_2}^{f_N} \equiv$ CASE_4	$S_{C_2}^{\zeta} \equiv$ CASE_8

We let Maple handle the calculus and algebra for the eight cases:

```
N_F := 1/(2*Pi*sqrt(R1*R2*C1*C2)):
D_F := (C2*(R1+R2))/(2*sqrt(R1*R2*C1*C2)):
CASE_1 := simplify((R1/N_F)*diff(N_F,R1));
CASE_2 := simplify((R2/N_F)*diff(N_F,R2));
CASE_3 := simplify((C1/N_F)*diff(N_F,C1));
CASE_4 := simplify((C2/N_F)*diff(N_F,C2));
CASE_5 := simplify((R1/D_F)*diff(D_F,R1));
CASE_6 := simplify((R2/D_F)*diff(D_F,R2));
CASE_7 := simplify((C1/D_F)*diff(D_F,C1));
CASE_8 := simplify((C2/D_F)*diff(D_F,C2));
```

Then we put the computed results into a tabular form, as shown in Table 2.2. From Table 2.2, we see that for six out of eight possible cases, the cutoff frequency and/or damping factor can exhibit a negative 50% change or shift in cutoff frequency for a unit change in a particular component's value. However, cases 5 and 6 indicate that if $R1 = R2$, then there is a zero component sensitivity for the LPF's damping factor. This was the initial design exercise for creating the Butterworth LPF. However, after constraining the two capacitors to be integer multiples of each other (a practicality issue), we found that the resistive values still came out equal. Consequently, this final design result has the added bonus of having the lowest damping factor change for any change between the two resistor values.

Table 2.2
Maple computer sensitivies

Component (variable)	**Cutoff frequency sensitivity**	**Damping factor sensitivity**
R1	$S_{R_1}^{f_N} \equiv \text{CASE\_1} = -\frac{1}{2}$	$S_{R_1}^{\zeta} \equiv \text{CASE\_5} = \frac{1}{2}\frac{R_1 - R_2}{R_1 + R_2}$
R2	$S_{R_2}^{f_N} \equiv \text{CASE\_2} = -\frac{1}{2}$	$S_{R_2}^{\zeta} \equiv \text{CASE\_6} = \frac{1}{2}\frac{R_2 - R_1}{R_1 + R_2}$
C1	$S_{C_1}^{f_N} \equiv \text{CASE\_3} = -\frac{1}{2}$	$S_{C_1}^{\zeta} \equiv \text{CASE\_7} = -\frac{1}{2}$
C2	$S_{C_2}^{f_N} \equiv \text{CASE\_4} = -\frac{1}{2}$	$S_{C_2}^{\zeta} \equiv \text{CASE\_8} = -\frac{1}{2}$

What about situations when the values of the resistors are not equal? Could this happen by design? Is it desirable to have unbalanced resistor values in the Butterworth LPF topology? Let's take a quick look at this prospect.

Unequal resistance values in the Butterworth LPF topology

From Table 2.2, we saw that unequal resistance values certainly increase the cutoff frequency's sensitivity to resistor component value change. That is certainly not a desirable result. However, for the sake of design interest, let's force the following constraint:

$$R_1 = \alpha R_2$$
$$C_1 = \beta C_2$$

where α, β are real scalars. Remember, we want β to retain its integer value for practical reasons (that is, because of the difficulty of finding arbitrarily related capacitor values). Now the Butterworth design equations become

$$f_N = \frac{1}{2\pi\sqrt{R_1 R_2 C_1 C_2}} = \frac{1}{2\pi R_2 C_2 \sqrt{\alpha\beta}} = 1000$$
$$\zeta = \frac{C_2(R_1 + R_2)}{2\sqrt{R_1 R_2 C_1 C_2}} = \frac{R_2 C_2(\alpha + 1)}{2 R_2 C_2 \sqrt{\alpha\beta}} = \frac{1}{\sqrt{2}}$$

Again, we have four unknowns and two equations; therefore, let's impose the following criteria:

$$R_2 = 12\ \text{k}\Omega$$
$$C_2 = .01\ \mu\text{F}$$

Solving for α, β we obtain

```
R2 := 12*10^3:
C2 := .01*10^(-6):
Natural_Frequency_Hertz := 1/(2*Pi*R2*C2*sqrt(alpha*beta)):
Damping_Factor :=
 R2*C2*(alpha+1)/(2*R2*C2*sqrt(alpha*beta)):
Results := solve ({Natural_Frequency_Hertz=1000,
 Damping_Factor=1/sqrt(2)},{alpha,beta});
Solutions := subs(Results,[alpha,beta]):
alpha := op(1,Solutions);
beta := op(2,Solutions);
```

$$\text{Results} := \{\beta = 2.008828030\ ,\ \alpha = .8756589909\}$$
$$\alpha := .8756589909$$
$$\beta := 2.008828030$$

Consequently, under these constraints there was a unique solution and we did retain a strongly integer relationship between $C1, C2$ while obtaining different resistor values. Therefore, the new component values of the Butterworth LPF are as follows:

$$R_1 = \alpha R_2 = 10.50\ \text{k}\Omega$$
$$R_2 = 12\ \text{k}\Omega$$
$$C_1 = \beta C_2 = .02\ \mu\text{F}$$
$$C_2 = .01\ \mu\text{F}$$

Again, the R1s are easily obtained in a 1% metal thin-film-type resistor as are the R2s. However, as shown before in the sensitivity discussion, the damping factor now has a greater resistance component sensitivity. In fact, the sensitivity functions numerically compute to be

$$S_{R_1}^{\zeta} = \frac{1}{2}\frac{R_1 - R_2}{R_1 + R_2} = \frac{1}{2}\frac{(10.5 - 12)}{(10.5 + 12)} = -.033$$
$$S_{R_2}^{\zeta} = \frac{1}{2}\frac{R_1 - R_2}{R_1 + R_2} = \frac{1}{2}\frac{(12 - 10.5)}{(10.5 + 12)} = +.033$$

which indicates an approximate 3% variation in the damping factor for every unit variation in the resistance values. Further, since the signs are opposite and the resistors will have identical temperature coefficients (provided they are the same type and physically close), any resistance change due to temperature variation will be compensated for, causing little to no significant effect on the Butterworth's damping factor.

The capacitors' sensitivity functions were not changed by this analysis because the sensitivity function for them was independent of their specific or relative value to each other.

One final cautionary note about the solve *and* fsolve *commands*

Always view your results (regardless of what type of engineering problem you are analyzing) to see if you have come up with any unexpected imaginary terms as a result of Maple's evaluation. In the previous component evaluation case, we had imaginary terms, hence we needed to use the `Re` command to abstract the real part of the solution. If we had not bothered to look at the Results evaluation and simply went to abstract them, no output would have been produced. This would have led the user to incorrectly believe there are no solutions (which Maple can do if no solutions exist to a set of simultaneous equations). So make sure you always observe the `solve` or `fsolve` commands' results before continuing with your analysis.

Now that we have a realistic 1-kHz Butterworth LPF design, let's examine some simulations of output responses to some sinusoidal and step functions.

The intent of a LPF is to attenuate frequencies beyond the designed filter's cutoff frequency. The question becomes one of how strongly we want to attenuate and, therefore, how close can an interfering signal be to the desired bandwidth before it becomes a problem for a given design.

Butterworth LPF test setup

Figure 2.4 represents our test setup for evaluating the Butterworth LPF under the following conditions:

- Out-of-band signal attenuation;
- Step response.

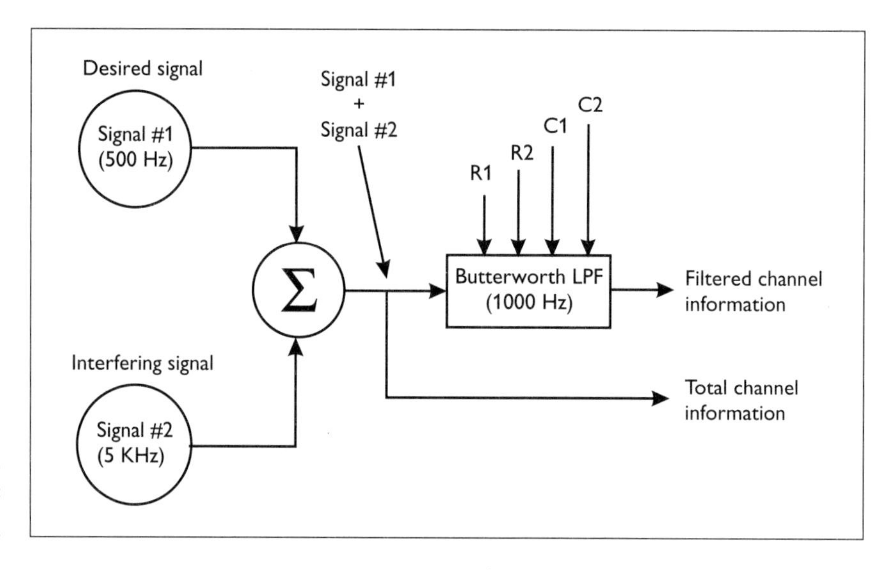

Figure 2.4 Test setup for LPF measurements.

Creating the signal sources in Maple, we get

```
F1 := 500:
F2 := 5000:
Signal_1 := sin(2*Pi*F1*t);
```

```
Signal_2 := sin(2*Pi*F2*t);
Channel_Signal := Signal_1 + Signal_2;
```

$$\text{Signal\_1} := \sin(1000\ \pi\ t)$$
$$\text{Signal\_2} := \sin(10{,}000\ \pi\ t)$$
$$\text{Channel\_Signal} := \sin(1000\ \pi\ t) + \sin(10{,}000\ \pi\ t)$$

then taking the Laplace transform of the total channel signal, which is the summed input of the test signals shown in Figure 2.4, we have

```
with(inttrans):
Channel_Signal_Laplace := expand(Laplace
 (Channel_Signal,t,s));
```

$$\text{Channel\_Signal\_Laplace} :=$$
$$11{,}000\ \frac{\pi s^2}{\left(s^2 + 1{,}000{,}000\ \pi^2\right)\left(s^2 + 100{,}000{,}000\ \pi^2\right)} + 110{,}000{,}000{,}000\ \frac{\pi^3}{\left(s^2 + 1{,}000{,}000\ \pi^2\right)\left(s^2 + 100{,}000{,}000\ \pi^2\right)}$$

substituting the component values into the `LPF_Transfer` function previously derived results in the following:

$$\text{LPF\_Transfer} = \frac{1}{R_1 R_2 C_1 C_2 s^2 + (R_1 C_2 + R_2 C_2)\, s + 1}$$

and multiplying the channel information by the Butterworth `LPF_Transfer` function yields the filter's output:

```
LPF_Transfer := 1/(R1*R2*C1*C2*s^2+(R1*C2+R2*C2)*s+1):
LPF_Transfer_Subs := subs(R1=10.5*10^3,R2=12*10^3,
 C1=.02*10^(-6),C2=.01*10^(-6),LPF_Transfer):
Filter_Output_Laplace := Channel_Signal_Laplace*
 LPF_Transfer_Subs;
```

Filter_Output_Laplace :=

$$\left(\pi \frac{s^2}{\left(s^2 + 1{,}000{,}000\ \pi^2\right)\left(s^2 + 1{,}000{,}000\ \pi^2\right)} + 110{,}000{,}000{,}000 \frac{\pi^3}{\left(s^2 + 1{,}000{,}000\ \pi^2\right)\left(s^2 + 100{,}000{,}000\ \pi^2\right)}\right) \Big/ \left(.2520000000\ 10^{-7}\ s^2 + .0002250000000s + 1\right)$$

Performing the inverse Laplace transform yields the time-domain response, therefore,

```
Filter_Output_Time := evalf(simplify(invlaplace
 (Filter_Output_Laplace,s,t)));
```

$$\begin{aligned}\text{Filter\_Output\_Time} := {} & .7060431942\ \sin(3141.592654\ t) \\ & - .6642910232\ \cos(3141.592654\ t) \\ & - .03851414293\ \sin(31415.92654\ t) \\ & - .01140445884\ \cos(31415.92654\ t) \\ & + .4518843917\ e^{(-4464.285714\ t)}\ \sin(4444.400154\ t) \\ & + .6756954821\ e^{(-4464.285714\ t)}\ \cos(4444.400154\ t)\end{aligned}$$

The `Filter_Output_Time` expression exhibits a steady-state and transient output response. The transient aspects are easy to spot. They are the output terms with the decaying exponential coefficients.

Now let's plot the unfiltered and filtered information as shown in Figure 2.4 to compare how well our Butterworth LPF has attenuated the 5-kHz interference signal:

```
with(plots):
Filtered_Plot :=
 plot(Filter_Output_Time,t=0..2*(1/500),style=line,
 color=black):
Unfiltered_Plot :=
 plot(Channel_Signal,t=0..2*(1/500),
 linestyle=2,color=black,numpoints=512):
display({Filtered_Plot,Unfiltered_Plot},axes=boxed,
 labels=[Time,Output]);
```

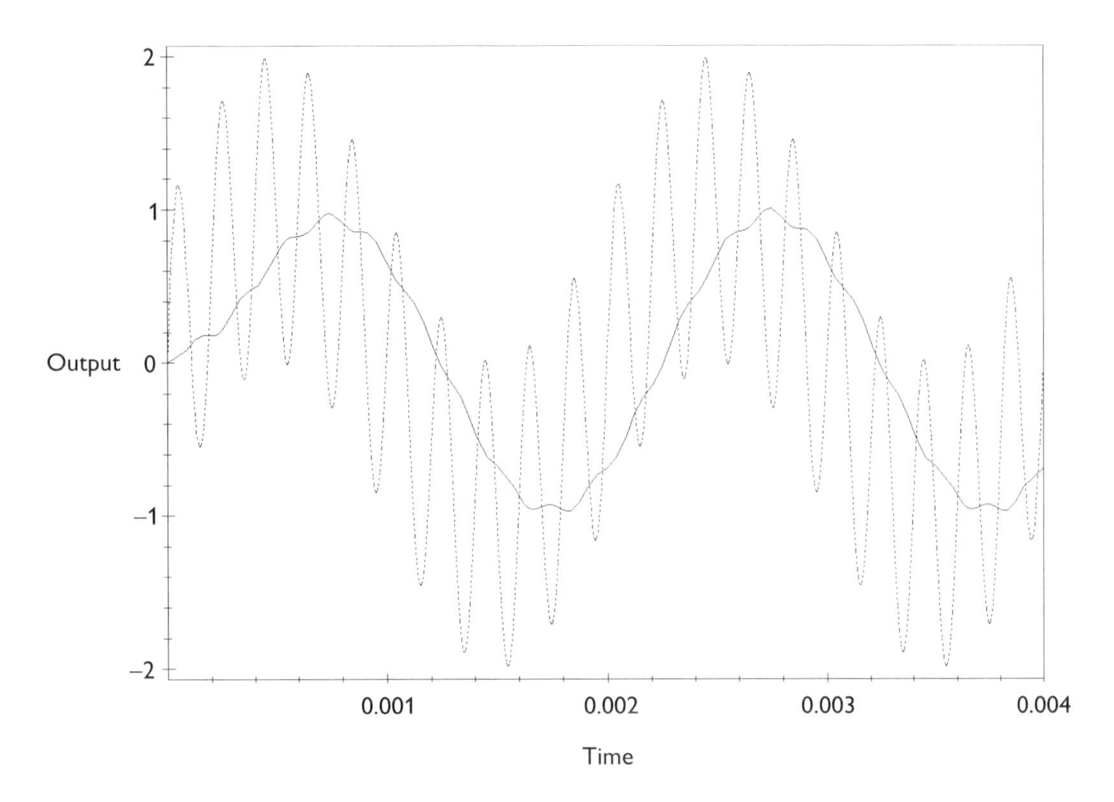

Figure 2.5 Comparison between filtered and unfiltered output, both at 500 Hz (filtered signal, solid line; unfiltered signal, dashed line).

Figure 2.5 indicates that we have greatly attenuated the 5-kHz interference, which was strongly corrupting our 500-Hz information. To increase filtering, we would need to either lower the cutoff frequency toward the desired frequency and/or increase the order of the Butterworth LPF. The current design is attenuating the 5-kHz signal by a factor of

$$\text{Attenuation} \approx \left(\frac{f_{\text{Noise}}}{f_{-3\text{dB}}}\right)^2 = \left(\frac{5000 \text{ Hz}}{1000 \text{ Hz}}\right)^2 = 25$$

Hence, we have improved signal to noise (SNR) by $20 \log_{10}(25) \approx 28$ dB.

Design iteration of LPFs for newer filtering requirements

The question now arises: Is this sufficient attenuation for the given filtering application? The answer is that it depends on the specification given to the designer. Let's say the customer looks at our design and realizes the final design requires another 5 dB of 5-kHz signal attenuation before the product is viable. There are two immediate approaches for obtaining the increased 5-kHz signal attenuation:

1. Decrease the cutoff frequency;
2. Increase the order of the Butterworth LPF.

Solution 2 is viable when most of the passband up to 1 kHz is required. Using a higher order LPF than the currently designed second-order LPF will more rapidly attenuate signals beyond the cutoff frequency without causing excessive attenuation within the 1-kHz passband. However, the higher order filter will require more hardware, design time, and a higher cost. If, on the other hand, we only need to see the 500-Hz signal, then following solution 1 by reducing the cutoff frequency in the original topology should be sufficient to fulfill the customer's requirement.

Consequently, we reiterate the capacitive component restriction,

$$C_1 = .02\ \mu\text{F}$$
$$C_2 = .01\ \mu\text{F}$$

also knowing that we now need a total 5-kHz attenuation of

$$28\ \text{dB} + 5\ \text{dB} = 33\ \text{dB} \rightarrow 5\ \text{kHz}$$

or about a 45-to-1 attenuation at 5 kHz. Further, we also want the following criterion to hold true as well:

$$\zeta = \frac{C_2(R_1 + R_2)}{2\sqrt{R_1 R_2 C_1 C_2}} = \frac{1}{\sqrt{2}}$$

in order for the LPF topology to retain the Butterworth response. Therefore, entering these constraints and constants into Maple,

```
readlib(unassign):
unassign ('R1','R2'):
C1 := .02*10^(-6):
C2 := .01*10^(-6):
freq := 5000:
LPF_Transfer := 1/(R1*R2*C1*C2*s^2+(R1*C2+R2*C2)*s+1):
LPF_Transfer_Mag := evalc(abs(subs(s=2*Pi*I*freq,
 LPF_Transfer)));
Damping_Factor := C2*(R1+R2)/(2*sqrt(R1*R2*C1*C2));
```

$$\text{LPF\_Transfer\_Mag} := 1 \Bigg/ \left(\left(-.2000000000\ 10^{-7}\ R1\ R2\ \pi^2 + 1 \right)^2 + (.0001000000000\ \ \pi R1 + .0001000000000\ \ \pi R2)^2 \right)^{1/2}$$

$$\text{Damping\_Factor} := .3535533907\ \frac{R1 + R2}{\sqrt{R1\ R2}}$$

Now, working backwards, we solve the `LPF_Transfer_Mag` equation for the unknown resistors, $R1$,$R2$, by invoking the computed filter's magnitude response at 5 -kHz and the `Damping_Factor` constraint for ensuring a Butterworth response:

```
Solutions_45db := solve({LPF_Transfer_Mag=1/45,
 Damping_Factor=1/sqrt(2)},{R1, R2});
Results := subs(Solutions_45db,[R1,R2]):
R1 := op(1,Results);
R2 := op(2,Results);
```

$$\text{Solutions\_45db} := \{R1 = 19042.39265\ ,\ R2 = 12694.92848\}$$

$$R1 := 19042.39265$$

$$R2 := 12694.92848$$

These resistor values correspond to a new cutoff frequency (in hertz) of

```
Natural_Frequency_Hertz :=
 evalf(1/(2*Pi*sqrt(R1*R2*C1*C2)));
```

$$\text{Natural\_Frequency\_Hertz} := 723.8177791$$

which is acceptable, since this does not attenuate the 500-Hz signal, retains the Butterworth response, and produces 5 dB more attenuation at the 5-kHz interference spectral location.

The final Butterworth LPF filter design that incorporates all of the previous customer requirements is as follows:

$$
\begin{aligned}
R_1 &= 19\text{K}\Omega \ \rightarrow 1\% \ \text{metal film} \\
R_2 &= 12.7\text{K}\Omega \ \rightarrow 1\% \ \text{metal film} \\
C_1 &= .02\mu\text{F} \\
C_2 &= .01\mu\text{F}
\end{aligned}
$$

Now, let's use the test setup with these new design values and see what the output appears as.

```
with(plots):
with(inttrans):
F1 := 500:
F2 := 5000:
Signal_1 := sin(2*Pi*F1*t):
Signal_2 := sin(2*Pi*F2*t):
Channel_Signal := Signal_1 + Signal_2:
Channel_Signal_Laplace := expand(laplace
 (Channel_Signal,t,s)):
LPF_Transfer := 1/(R1*R2*C1*C2*s^2+(R1*C2+R2*C2)*s+1):
LPF_Transfer_Subs := subs(R1=19*10^3,R2=12.7*10^3,
 C1=.02*10^(-6),C2=.01*10^(-6),LPF_Transfer):
Filter_Output_Laplace := Channel_Signal_Laplace*
 LPF_Transfer_Subs:
Filter_Output_Time :=
 evalf(simplify(invlaplace
 (Filter_Output_Laplace,s,t))):
Filtered_Plot :=
 plot(Filter_Output_Time,t=0..2*(1/500),
 style=line,color=black):
Unfiltered_Plot :=
plot(Channel_Signal,t=0..2*(1/500),
 linestyle=2,color=black,numpoints=512):
display({Filtered_Plot,Unfiltered_Plot},axes=boxed,
 labels=[Time,Output]);
```

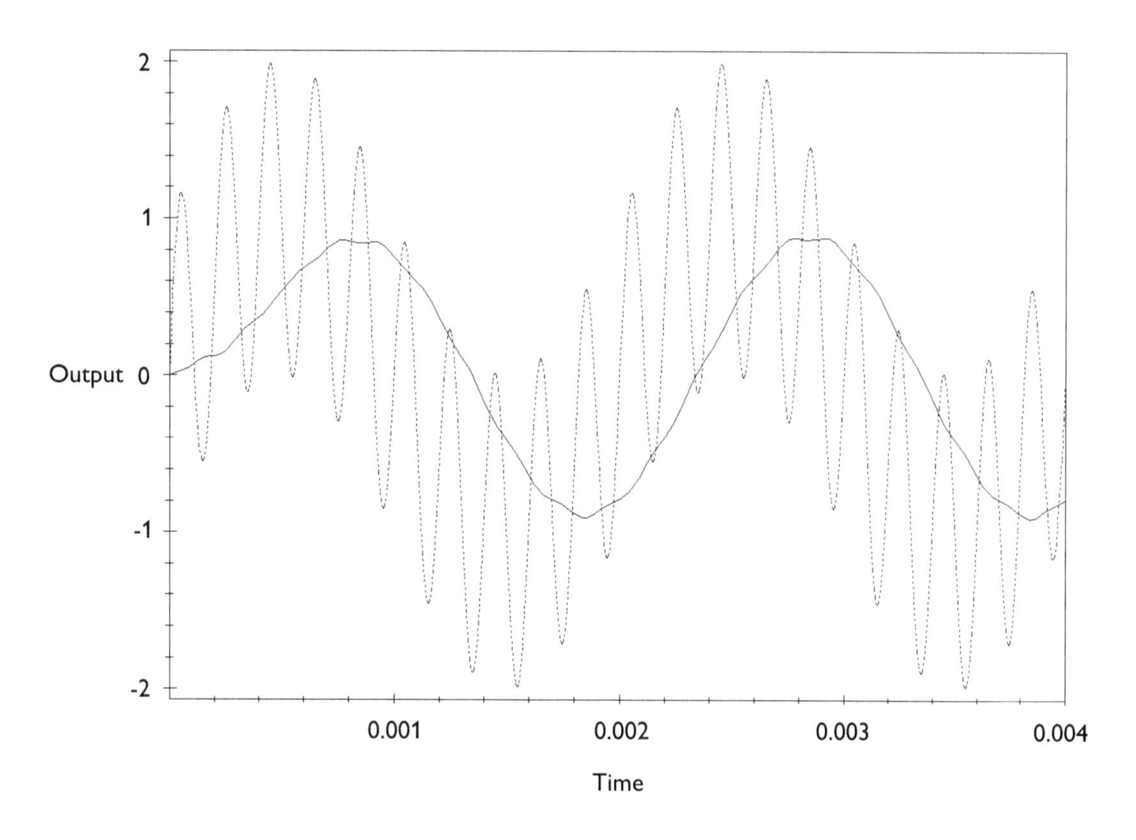

Figure 2.6 Final Butterworth LPF design (filtered signal [500 Hz], solid line; unfiltered signal [500 Hz + 5000 Hz], dashed line).

Compare Figure 2.6 to Figure 2.5 and there are two immediate observations. First, the filtered output (solid line) is somewhat smoother in Figure 2.6 and, second, there is a slightly greater phase shift between the filtered and unfiltered signals in Figure 2.6. The extra smoothness in Figure 2.6 is directly attributable to the added filtering (5 dB more) of the 5-kHz interference signal. The added phase shift is again caused by the extra low-pass filtering and should not be a surprise.

In brief, we have gone through two iterations of the Butterworth LPF to improve on the required electronic components and a third iteration to accommodate the added 5 dB at 5-kHz attenuation required in the customer's updated requirement.

Unit step response

Another important aspect of filtering is the filter's step response. This measure indicates how well a filter allows rapidly changing signals to be reconstructed after filtering. Obviously, a low-pass filter is not going to pass a fast transition very well by definition; however, with that said, certain step responses are still required of LPFs even when one is trying to suppress high-frequency noise.

Using our previously described Butterworth 1-kHz LPF, let's introduce the step function:

$$V_{IN} = \begin{Bmatrix} 0 & t < 0 \\ 1 & t \geq 0 \end{Bmatrix}$$

The Laplace transform of this is simply

$$L\,[V_{IN} = \text{Unit step}] = \frac{1}{s}$$

then recomputing the Butterworth LPF's output with the step input, we get

```
with(plots):
readlib(laplace):
R1 := 19*10^3:
R2 := 12.7*10^3:
C1 := .02*10^(-6):
C2 := .01*10^(-6):
LPF_Transfer := 1/(R1*R2*C1*C2*s^2+(R1*C2+R2*C2)*s+1):
```

Multiplying by the input step transform yields the output Laplace expression

```
Output_Laplace := LPF_Transfer*(1/s);
```

$$\text{Output\_Laplace} := \frac{1}{\left(.4826000000\ 10^{-7}\ s^2 + .0003170000000\ s + 1\right) s}$$

and taking the inverse transform gives the final equation for the output's time domain response, `Output_Time`,

```
Output_Time := invlaplace(Output_Laplace,s,t);
```

$$\text{Output\_Time} := 1. - 1.042002364\ e^{(-3284.293411\ t)} \sin(3151.905911\ t) - 1.\ e^{(-3284.293411\ t)} \cos(3151.905911\ t)$$

Not surprisingly, we have the unit output with two exponentially decaying sinusoidals. Plotting this result shows one of the reasons why Butterworth filters are not the best for maximally flat transient response.

```
with(plots):
Unit_Step := 1:
Step_Plot := plot(Unit_Step,t=0..+.002,linestyle=3,
 color=black):
Response_Plot :=
 plot(Output_Time,t=0..+.002,style=line,
 color=black):
display({Step_Plot,Response_Plot},axes= boxed,labels=
 [Time,Output]);
```

Figure 2.7 shows a substantial overshoot at around 0.001 second, which is caused by the group phase delay associated with any LPF topology. Butterworth filters are maximally flat passband magnitude, not phase, filters. Thompson or Bessel filters, which have a small amount of passband magnitude ripple, will give much smoother phase response to fast or transient signals. In fact, these filter topologies are known as maximally flat phase filters [3,4].

Conclusion

In this section, Maple has allowed the us to characterize, analyze, and design the dynamics of the Butterworth LPF in terms of:

1. The overall transfer function;
2. Optimized component values that are physically realizable;
3. LPF sensitivity to component values;
4. Altered design values in order to change the LPF's response;
5. Viewing the LPF's time-domain unit step response.

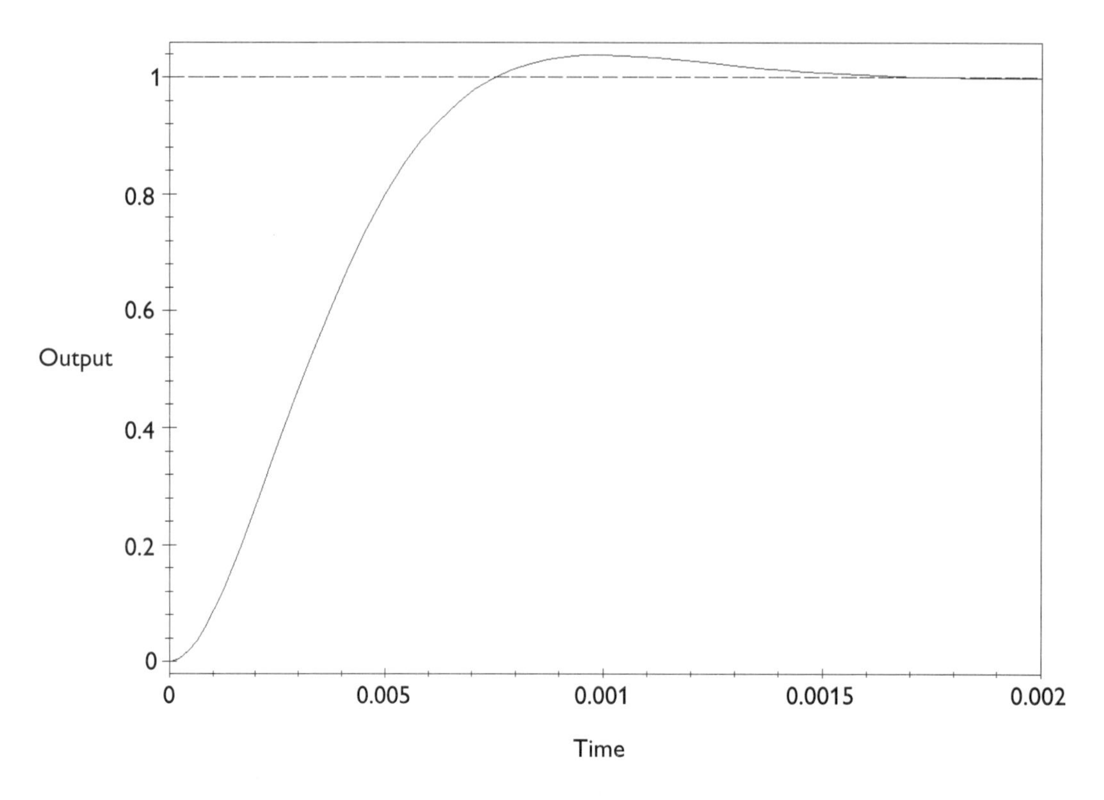

Figure 2.7 Butterworth LPF step response (cutoff = 723.81 Hz).

Maple has the great ability to exercise any dynamic topology (in this case, a linear electronic circuit) under any number of interesting conditions. In the case of the Butterworth LPF, we chose a rather common application of interference signal attenuation. We arbitrarily created a desired and undesired signal channel and simulated the Butterworth LPF's function. Then with Maple's symbolic, numeric, and graphic capabilities, we were able to see and test the results of our design and how sensitive the design was to real component values. This, in turn, helps the designer realize the efficacy of the chosen LPF, the component values, and the topology's relative sensitivity to component variations.

Admittedly, we reused many Maple command lines repetitively throughout this and other sections, but with one specific purpose in mind. The more the user sees certain command structures, the more those basic structures should become instinctive when used for more advanced and general-purpose applications.

Case II: comb filter analysis and design

A very interesting set of filters has been derived using switched capacitor circuits. These filters have the desirable property of programmable attributes such as passband gain, cutoff frequency, and other features completely specified by a simple digital clocking signal. One of the more common implementations of this idea uses a charge-coupled device (CCD) technology. This approach is fabricated in silicon and is functionally equivalent to a series of capacitors that moves their respective signal charge via a discrete clocking signal from one end of a serial register to the opposing end as shown in Figure 2.8. The reader should note that Figure 2.8 is only a crude model of how real CCD technology is implemented, but this simple model will suffice for our analysis purposes.

Table 2.3 shows the basic description of the switches (S1, S2, S3, ..., S*n*) as a function of the clocking period (1/Fclock).

Upon the nth clock period, the signal is entered into the output buffer by switching to S*n*'s "0" state. Consequently, after *n* clock periods, the CCD register is completely updated with new values introduced by the input buffer (provided we have a one translation per clock signal). In essence, you have an analog *first-in/first-out* register or FIFO. The CCD shift registers come in a variety of sizes starting with as few as 64 cells and increase all the way up to around 16,000 cells as well as some serial-parallel combinations for producing serial-in/parallel-out protocols.

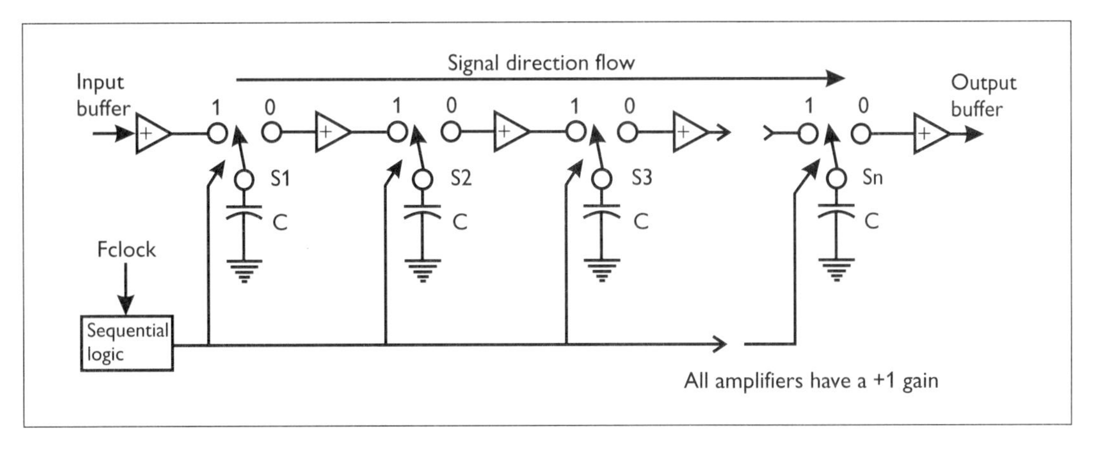

Figure 2.8 Basic circuit model of a serial CCD.

	Switch states				
Clock period	S1	S2	S3	...	S*n*
0	1	0	0		0
1	0	1	0		0
2	0	0	1		0
3	0	0	0		0
*n*_1	0	0	0		1
n	1	0	0		0

Table 2.3 Switch states for CCD model

Filter derivation and analysis

A simple analog model of this delay element can be stated as

$$f(t - T)$$

where

$$T = \frac{\text{No. of CCD cells}}{F_{clk}}$$

Taking the Laplace transform with Maple of this delay element yields

```
with(inttrans):
Time_Delay_Integrate := simplify(int(f(t-T)*exp(-s*t),
 t=0..infinity));
Time_Delay_Laplac  e := expand(laplace(F(t-T),t,s));
```

$$\text{Time\_Delay\_Integrate} := \int_0^{\infty} \mathrm{f}(t - T)\, e^{(-st)}\, dt$$

$$\text{Time\_Delay\_Laplace} := \text{Laplace}\,(\mathrm{F}(t - T),\ t,\ s)$$

Unfortunately, Maple cannot evaluate this transform with either the definition form, `Time_Delay_Integrate`, or through the internal library, `Time_Delay_Laplace`. Consequently, we need to look it up in a mathematics table, which will show you that the following is a Laplace transform of the general delay function:

$$\mathcal{L}(f(t - T)) = e^{-sT}$$

The comb filter uses this delay element in a classic feedback topology that is used to create a periodic passband transfer function. Graphically, this control diagram appears as shown in Figure 2.9. In Figure 2.9, the feedback gain, α, has the adjustable range

$$0 \leq \alpha < 1$$

The feedback cannot be allowed to reach unity due to the generation of a sustained output with no input (oscillation). However, in some designs, this aspect is desired.

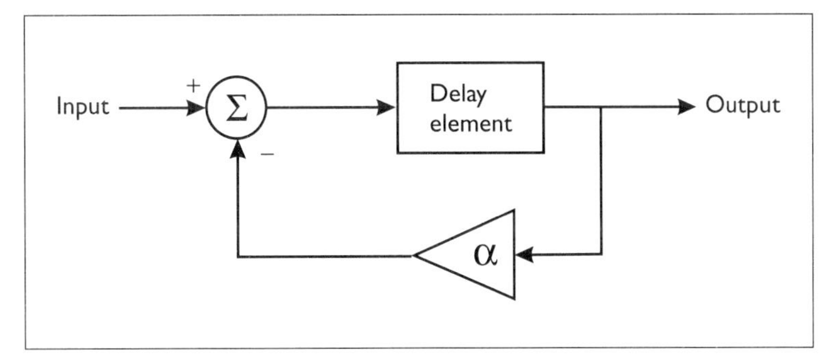

Figure 2.9 Control diagram of comb filter.

The overall transfer function of Figure 2.9 then becomes

$$H(s) = \frac{e^{-sT}}{1 + \alpha e^{-sT}}$$

Entering this expression into Maple $\alpha \rightarrow$ alfa

```
H_Laplace := (exp(-s*T))/(1+alfa*exp(-s*T));
```

$$\text{H\_Laplace} := \frac{e^{(-sT)}}{1 + \text{alfa}\ e^{(-sT)}}$$

upon substituting $s = I\omega$ into the transfer expression, we obtain

```
H_JW := subs(s=I*w,H_Laplace);
```

$$\text{H\_JW} := \frac{e^{(-I\,w\,T)}}{1 + \text{alfa}\ e^{(-I\,w\,T)}}$$

Taking advantage of Euler's identity, which states the following:

$$e^{\pm j\omega T} = \cos(\omega T) \pm j\ \sin(\omega T)$$

and substituting this identity into the `H_JW` expressions and simplifying, we get these results:

```
H_Euler := subs(exp(-I*w*T)=cos(w*T)-I*sin(w*T),H_JW);
```

$$\text{H\_Euler} := \frac{\cos(wT) - I\ \sin(wT)}{1 + \text{alfa}\ (\cos(wT) - I\ \sin(wT))}$$

Remembering that Maple uses "I" instead of "j" to represent an imaginary quantity, we evaluate the transfer function's magnitude response:

```
H_Mag := simplify(evalc(abs(H_Euler)));
```

$$\text{H\_Mag} := \sqrt{\frac{1}{1 + 2\ \text{alfa}\ \cos(wT) + \text{alfa}^2}}$$

Determining the maximal responses requires us to substitute a dummy variable for the wT product in the `H_Mag` expression, then take the derivative with respect to that dummy variable:

$$wT = A$$

$$\frac{d(\text{H\_Mag})}{d(A)} = 0$$

and then substitute the value of wT into the `H_Mag` expression. Therefore,

```
H_Mag_A := subs(w*T=A,H_Mag):
H_Mag_Derivative := diff(H_Mag_A,A);
```

$$\text{H\_Mag\_Derivative} := \frac{\text{alfa}\ \sin(A)}{\sqrt{\frac{1}{\text{alfa}^2 + 2\ \text{alfa}\ \cos(A) + 1}}\left(\text{alfa}^2 + 2\ \text{alfa}\ \cos(A) + 1\right)^2}$$

and solving for the variable A that will gives us zero derivative,

```
Solution := solve(H_Mag_Derivative=0,A);
```

$$\text{Solution} := \pi,\ 0$$

However, we do not know which value corresponds to the maximum rather than the minimum associated with the valleys between peaks. Therefore, we should take a second derivative and substitute the $[0,\ \pi]$ values to determine which value is referring to the maximum. To facilitate this computation, we will arbitrarily assign `alfa` =.50 so the derivative test functions evaluate numerically. This is acceptable, because `alfa` affects not the location but the heights of the maximal peaks.

```
H_Mag_Derivative_2 := diff(H_Mag_Derivative,A):
H_Mag_D1 := evalf(subs(A=0,alfa=.5, H_Mag_Derivative_2));
H_Mag_D2 := evalf(subs(A=Pi,alfa=.5, H_Mag_Derivative_2));
```

$$\text{H\_Mag\_D1} := .1481481482$$
$$\text{H\_Mag\_D2} := -4.000000000$$

Clearly, the second computation, `H_Mag_D2`, indicates the negative change when evaluated at π, hence the maximal peaks are at the odd integer multiples of π or

$$A = \pi,\ 3\pi,\ 5\pi, \ldots \quad \text{or}$$
$$A = (2N + 1)\pi \text{ where } \quad N = 0,\ 1,\ 2,\ 3, \ldots$$

So now, let's redefine the `H_Mag` expression by substituting this periodic function for the dummy variable, A,

```
H_Mag := subs(A=(2*N+1)*Pi,H_Mag_A);
```

$$\text{H\_Mag} := \sqrt{\frac{1}{\text{alfa}^2 + 2\ \text{alfa}\ \cos((2\ N + 1)\ \pi) + 1}}$$

To get a sense of how this filter's transfer function behaves, let's plot several graphs onto one common plot with various values of `alfa`:

```
with(plots):
alfa := .25:
Plot_1 := plot(H_Mag,N=0..2,color=black,style=point,
 symbol=cross):
alfa := .50:
Plot_2 := plot(H_Mag,N=0..2,color=black,style=point,
 symbol=diamond):
alfa := .75:
Plot_3 := plot(H_Mag,N=0..2,color=black,style=point,
 symbol=circle):
display({ Plot_1,Plot_2,Plot_3},axes=boxed,
 labels=[N_Value,MAG]);
```

Figure 2.10 shows each discrete *resonant* peak whose amplitude or gain function within any peak region is greatly affected by the `alfa` variable. In fact, the peak value is completely determined by this variable.

To get an even larger picture of the comb filter's behavior let's produce a 3-D plot of the transfer function's response versus `alfa` and `N`. First we must redefine `alfa` as a variable, since this Maple session remembers it as a numerical value (namely, `alfa`=.75 from Figure 2.10):

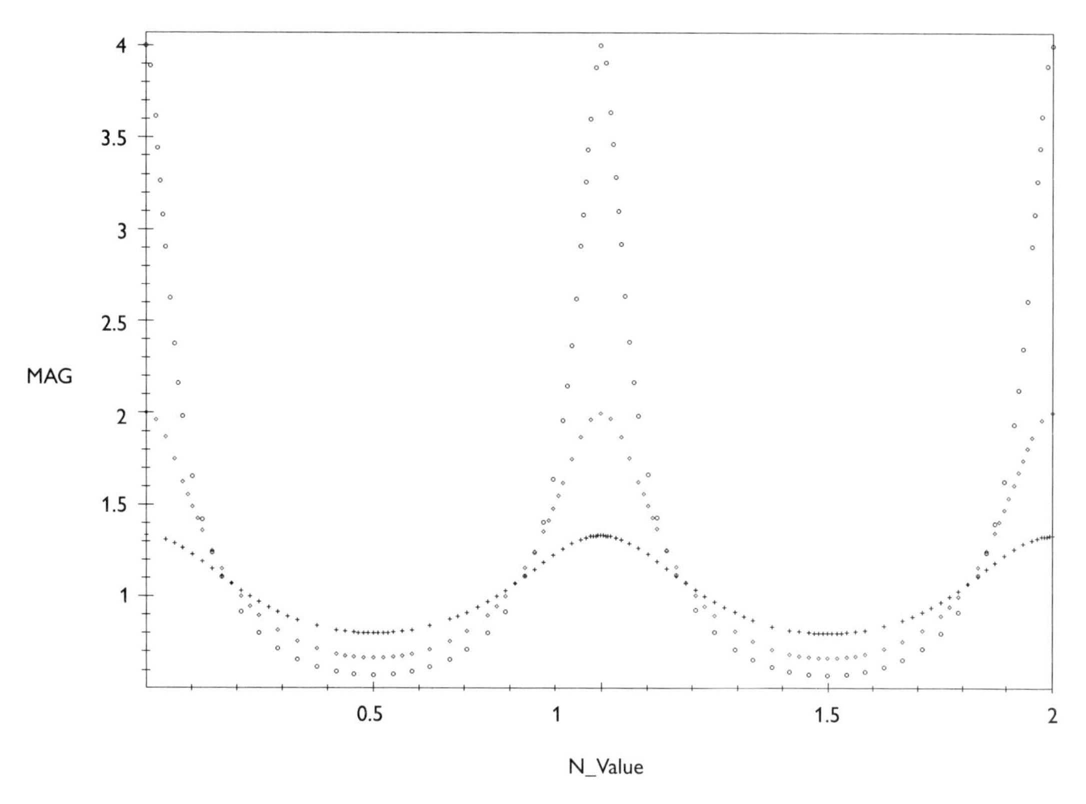

Figure 2.10 General comb filter response for various values of Alfa (alfa =.75, circles; alfa =.50, diamonds; alfa = .25, crosses).

```
alfa := 'alfa':
plot3d(H_Mag,alfa=.25..+.75,N=0..2,color=black,
 style=hidden,axes=boxed,labels=[alfa,N_Value,MAG]);
```

Notice that the peak locations in Figure 2.11 do not change with alfa; however, as stated, the magnitude of the peak responses is greatly affected. Also, the general shape about any peak is identical and not a function of location, but, again, of the variable `alfa`.

Determining the peak response value for any peak is simply `H_Mag` evaluated by any integer value of `N`, hence,

```
H_Mag_Peak_Value := subs(cos((2*N+1)*Pi)=-1,H_Mag);
```

$$\text{H\_Mag\_Peak\_Value} := \sqrt{\frac{1}{\text{alfa}^2 - 2\ \text{alfa} + 1}}$$

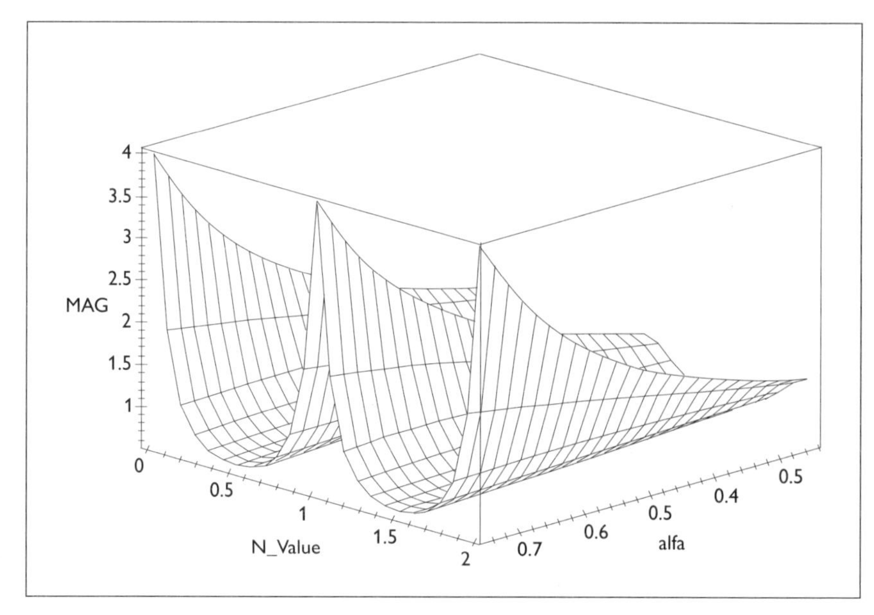

Figure 2.11 3-D plot of the comb filter's response.

Warning: a common pitfall associated with plotting

At this point, the author would like to point out a common error associated with deriving graphic results without having some *a priori* knowledge about the problem.

Let's replot Figure 2.11, but give it many more peaks, say, `N` = 10, and produce the corresponding 3-D plot over the same range of `alfa`.

```
plot3d(H_Mag,alfa=.25..+.75,N=0..10,color=black,
 style=hidden,axes=boxed,labels=[alfa,N_Value,MAG]);
```

Figure 2.12 shows that we have 11 peaks, which is correct (`N`=0,1, 2, 3, ..., 10), but the peaks are badly distorted. So what happened? The mathematics are correct, but the graphics are undersampled for the function being plotted. What this means is when you have 11 peaks, we should have a resolution (at least in the `N_Value` axis) of twice this value (the sampling theorem strikes again!). Maple has a default 3-D plotting grid of 25 × 25, which, in this case, was sufficient to show the peaks' existence, but not of sufficient quality to depict the equally high and spaced profiles we expected. Hence, replotting Figure 2.12 with an increase in the resolution using the `grid` plotting option should increase the quality of the distorted peaks.

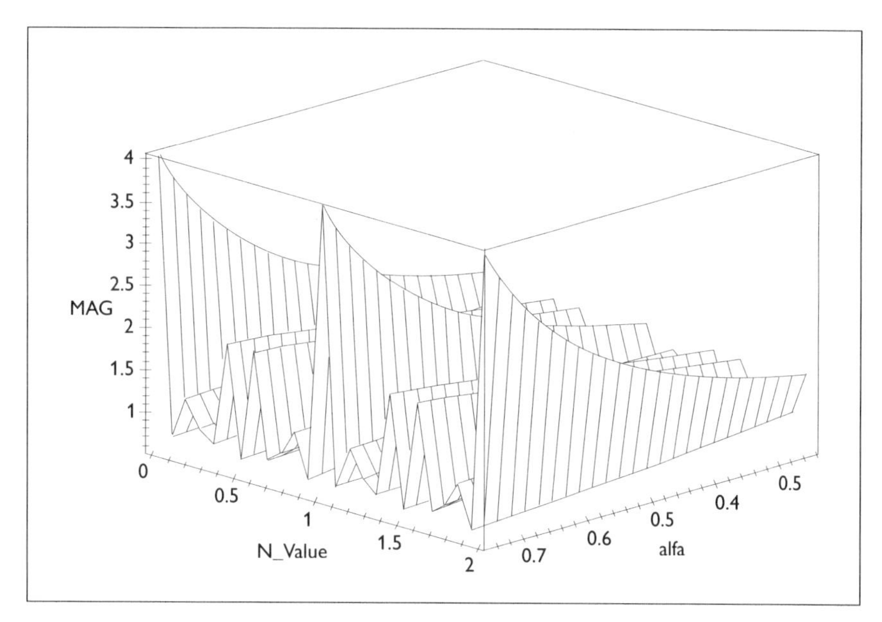

Figure 2.12
3-D plot of comb filter response.

```
plot3d(H_Mag,alfa=.25..+.75,N=0..10,color=black,
 style=hidden,axes=boxed,labels=[alfa,N_Value,MAG],
 grid=[50,50]);
```

Figure 2.13 definitely shows higher quality information, but the peaks are still badly distorted because we are not seeing the equally peaked resonant peaks associated with the comb filter. It appears that simply doubling the resolution is not quite good enough resolution to show graphically the 11 peak responses. Hence, we increase the resolution to `GRID=[100,100]`:

```
plot3d(H_Mag,alfa=.25..+.75,N=0..10,color=black,
 style=hidden,axes=boxed,labels=[alfa,N_Value,MAG],
 grid=[100,100]);
```

Figure 2.14 is much better, but, again, one could come to a false conclusion about the filter's response over this resonance range.

If we were to increase the 3-D resolution to 200 × 200, we would see pretty much what is shown in Figure 2.14, but at a much heavier (four times heavier) computational cost to our computer's resources. The lesson here is simple: Display only what you need to display and have a little *a priori* knowledge about your problem. These simple rules will keep you from getting erroneous results.

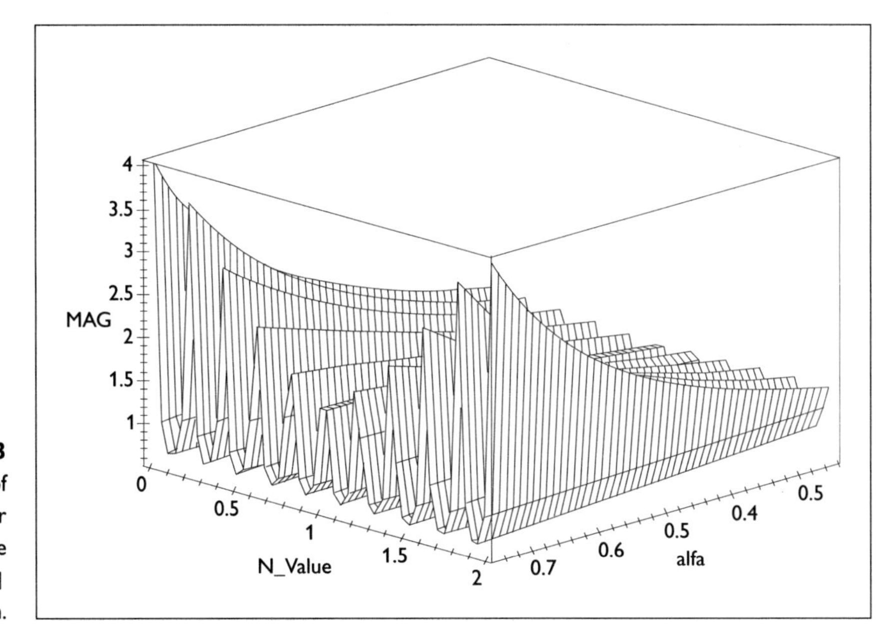

Figure 2.13
3-D plot of comb filter response with the GRID=[50,50] option.

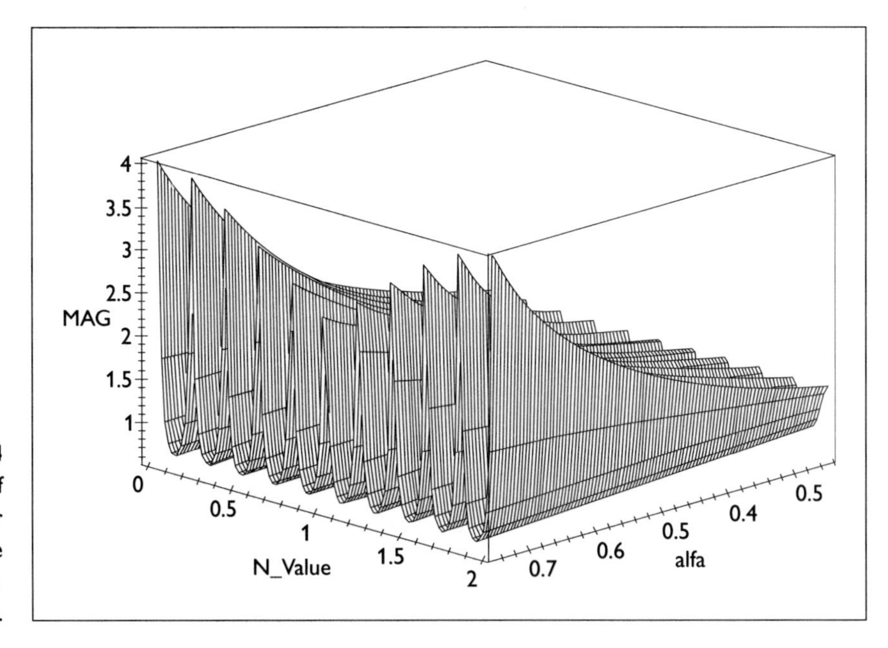

Figure 2.14
3-D plot of comb filter response with the GRID=[100,100] option.

Another approach that is extremely useful is that, because we do not need high resolution along the alfa axis, we could have specified a GRID=[200,25] plotting option. This would greatly reduce computational resources (one-eighth that of the 200 × 200 grid plot) and still allow us to visualize graphically the comb filter's response.

Separating a known signal from an interfering neighboring background design

Now returning to our comb filter analysis, suppose we want to examine how much filtering of a signal from an interfering neighboring signal we can derive by using this type of filter.

Figure 2.15 shows the experimental test setup we are going to emulate with Maple. Further, let's define the input signals as follows:

$$\text{Signal } 1 = \sin(2\pi \times 100t)$$
$$\text{Signal } 2 = \sin(2\pi \times 200t)$$

Since we want to recover the signal #1 sinusoidal, the comb filter should have its peak centered at signal #1's value. At this point, we need to define the wT argument as

$$wT = 2\pi f T$$

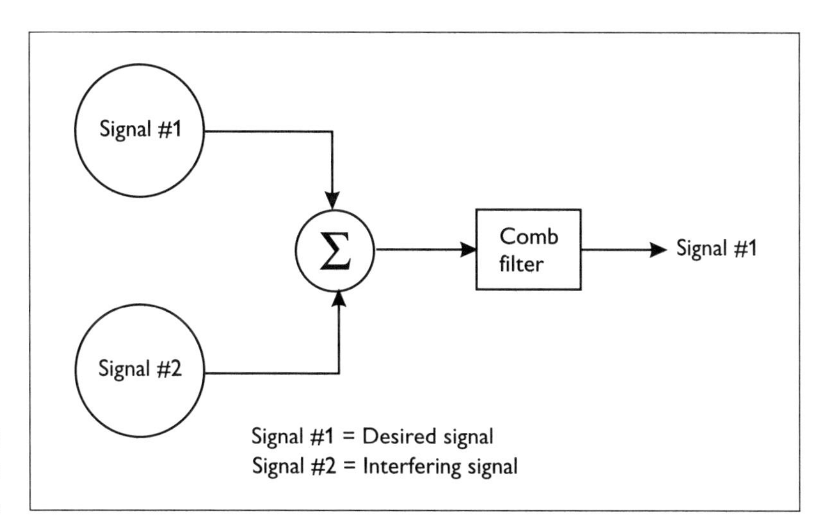

Figure 2.15 Signal separation application.

where, in particular,

$$w = 2\pi f_i$$
$$T = \frac{\text{No. of CCD cells}}{f_{clk}} \rightarrow \text{Effective CCD time delay}$$
$$f_i \rightarrow \text{Information frequency}$$
$$f_{clk} \rightarrow \text{CCD clocking frequency}$$

We defined this at the peak response when the sinusoidal argument was defined by

$$wT = (2N + 1)\ \pi$$

where $N = 0, 1, 2, 3, \ldots$

but now we need to implement the variables associated with the physical device.

The variable T is defined as the total delay with the CCD, which is a function of both clocking speed, f_{clk}, and number of delay cells, (number of CCD cells). Substituting this knowledge into the wT expression results in

$$wT = 2\pi f_N \left(\frac{\text{Cells}}{f_{clk}}\right) + (2N + 1)\pi$$

Therefore, solving for f_N yields

$$f_N = \frac{2N + 1}{2T}$$

Further, since the number of CCD delay cells can be made arbitrary without affecting the quality of the analysis, let's give it an arbitrary value of 1000 (actually, most real units have 1024 as a normal cell count). Substituting this fact into the f_N expression gives the location (in hertz) of the comb filter's peak responses as a function of the clocking rate and peak harmonic, N:

$$f_N(\text{Hz}) = \frac{2N + 1}{2}\left(\frac{f_{clk}}{\text{Cells}}\right)$$

$$= \frac{2N + 1}{2}\left(\frac{f_{clk}}{1000}\right)$$

$$= \frac{(2N + 1)\ f_{clk}}{2000}$$

Finally, we give the fundamental and harmonic peak centers and the associated frequencies for this physical design in Table 2.4.

Table 2.4 Relationship between fundamental and harmonic peak centers and associated passed signals (based on 1024 cells)

N	f_N (Hz)	Frequency passed (Hz)
0 (fundamental)	$\frac{f_{clk}}{2000}$	100
1 (third harmonic)	$\frac{3f_{clk}}{2000}$	300
2 (fifth harmonic)	$\frac{5f_{clk}}{2000}$	500

From Table 2.4, we see that the following is true if the CCD's clocking speed is set to acquire the fundamental (i.e., N = 0):

$$\frac{f_{clk}}{2000} = 100 \quad \text{or}$$
$$f_{clk} = 200 \ \text{kHz}$$

Now let's set up the signals from before and compute the magnitude response of the comb filter:

```
with (inttrans):
H := (cos(w*T)-I*sin(w*T)+alfa)/(alfa^2+2*alfa*cos(w*T)+1):
H_Mag := simplify(evalc(abs(H)));
```

$$\text{H\_Mag} := \sqrt{\frac{1}{\text{alfa}^2 + 2 \ \text{alfa} \ \cos(w \ T) + 1}}$$

Now we know from Table 2.4 that we need a clocking frequency of 200 kHz to acquire the desired signal (100 Hz) and maximally reject the interfering neighbor (200 Hz). Hence, we substitute this value, along with the CCD cell count to arrive at the effective T value or

$$T = \frac{\text{CCD cell count}}{f_{clk}} = \frac{100}{200\ \text{kHz}} = .005$$

also substituting

$$w = 2\pi f_1$$

Hence, we put these values into the Maple `H_Mag` expression:

```
> H_Mag_General := subs(T=.005,w=2*Pi*Fi,H_Mag);
```

$$\text{H\_Mag\_General} := \sqrt{\frac{1}{\text{alfa}^2 + 2\ \text{alfa}\ \cos(.010\ \pi\ Fi) + 1}}$$

We then compute the comb filter's rejection ratio as defined by

$$\text{Rejection ratio} = \frac{\text{Passband response}}{\text{Stopband response}} = \frac{\text{H\_Mag\_General}\ (Fi = 100\ \text{Hz})}{\text{H\_Mag\_General}\ (Fi = 200\ \text{Hz})}$$

for a given `alfa` and implement this definition in Maple:

```
> H_Mag_General_100 := subs(Fi=100,H_Mag_General):
  H_Mag_General_200 := subs(Fi=200,H_Mag_General):
  Rejection_Ratio :=
  simplify(H_Mag_General_100/H_Mag_General_200);
```

$$\text{Rejection\_Ratio} = \frac{\sqrt{\frac{1}{\text{alfa}^2 - 2.\ \text{alfa} + 1}}}{\sqrt{\frac{1}{\text{alfa}^2 + 2.\ \text{alfa} + 1.}}}$$

Let's plot this function as a function of `alfa` to see what kind of values we need to consider for sufficient attenuation of the neighboring signal (200 Hz).

```
plot(Rejection_Ratio,alfa=0..+.9,color=black,axes=boxed,
 labels=[alfa,RR]);
```

It might be easier to replot Figure 2.16 with a log plot, since engineers usually discuss ratios in terms of decibels (see Figure 2.17). Hence,

```
readlib(log10):
Rejection_Ratio_DB := 20*log10(Rejection_Ratio):
plot(Rejection_Ratio_DB,alfa=0..+.9,color=black,axes=boxed
 labels=[alfa,DB]);
```

Therefore, for a given rejection ratio, we can decide how much the neighboring signal needs to be attenuated. If the interfering signal were closer, say, at 120 Hz, we would have to operate much higher values of `alfa` (i.e., closer to unity), but with the disadvantage of amplitude stability at the desired signal.

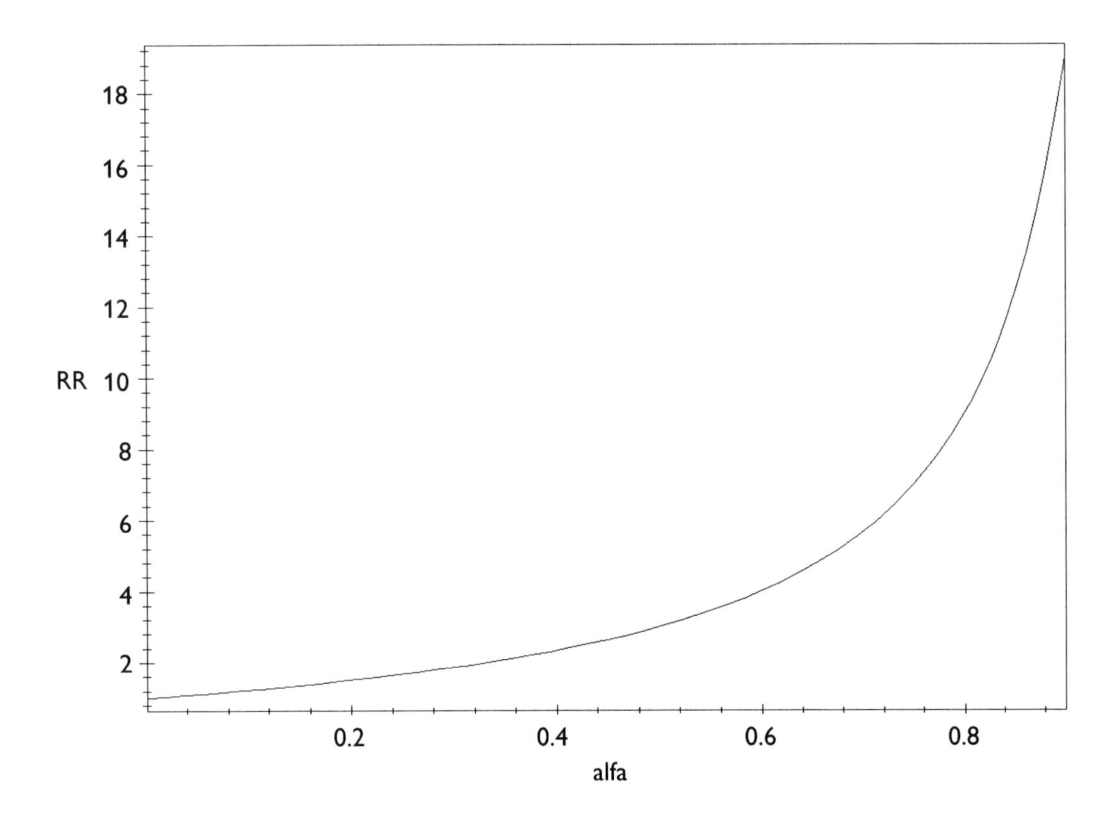

Figure 2.16 Rejection ratio (RR) between 100- and 200-Hz signals.

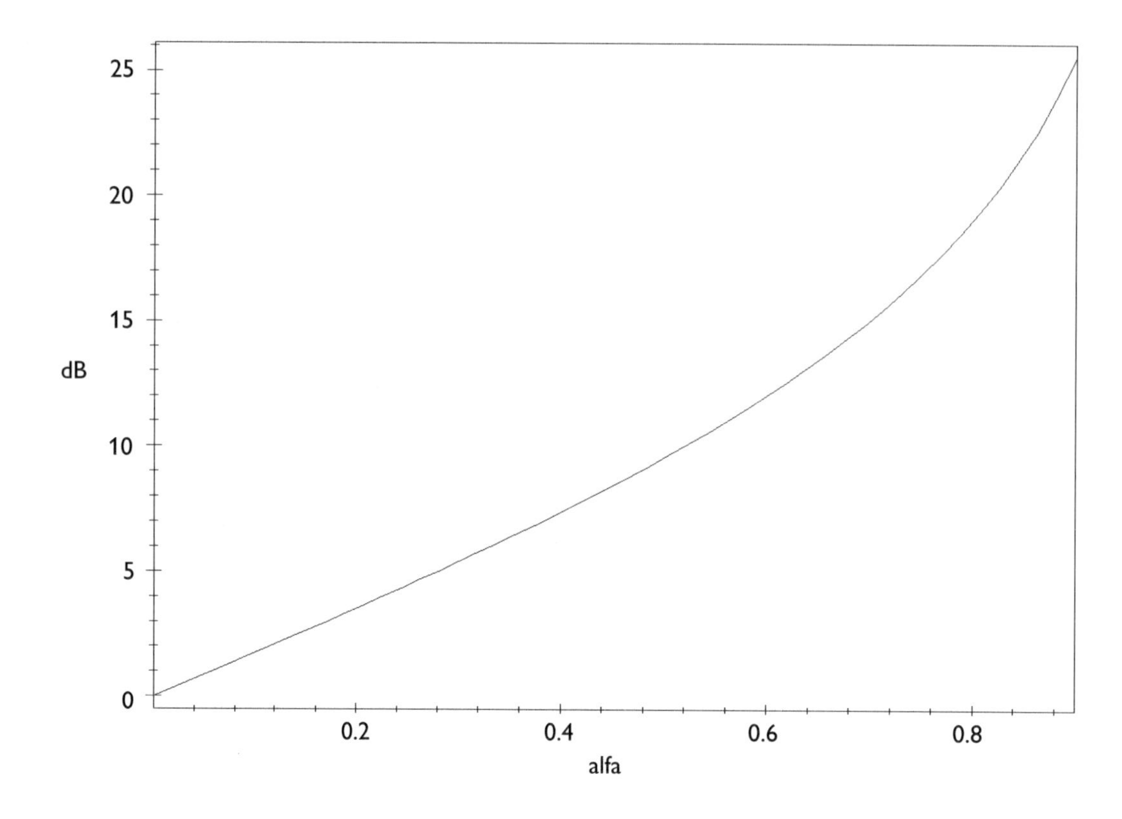

Figure 2.17 Log plot of Figure 2.16.

An easier and more often used method for increasing the attenuation of very close signals is by cascading two or more comb filters. This method allows the use of more stable `alfa` values on a per-stage basis, while greatly increasing the attenuation of the unwanted neighboring frequency.

Cascading comb filters

We can quickly look at the passband profile result associated with two filters that are cascaded, because all of the constituent CCD switching filter equations are in Maple's memory, and there is no need to derive any previous relationships because that has already been done for the single filter case. Therefore, assuming no loading of the filters (i.e., the output of the first filter does not "see" the input impedance of the second filter ... a valid approximation for real hardware), and identical filter topologies, the cascaded transfer function can be simply stated from Figure 2.18.

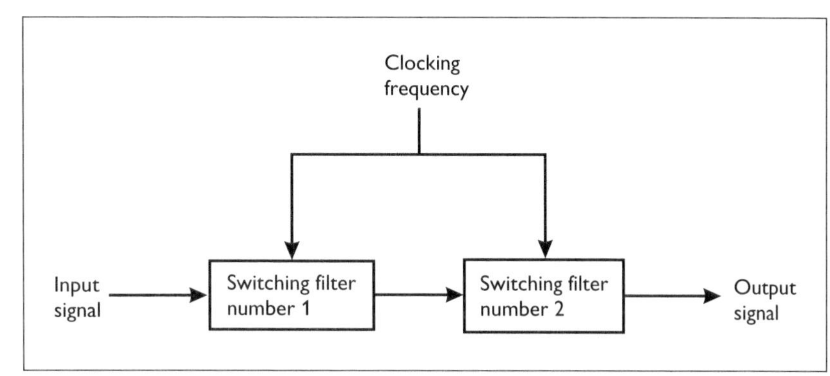

Figure 2.18 Cascaded switching filter topology.

Enter into Maple the transfer function magnitude of the original filter transfer function

```
with(inttrans):
H_Laplace := (exp(-s*T))/(1+alfa*exp(-s*T)):
```

then obtain the magnitude

```
H_JW := subs(s=I*w,H_Laplace):
H_Euler := subs(exp(-I*w*T)=cos(w*T)-I*sin(w*T),H_JW):
H_Mag := simplify(evalc(abs(H_Euler)));
```

$$\text{H\_Mag} := \sqrt{\frac{1}{1 + 2\ \text{alfa}\ \cos(wT) + \text{alfa}^2}}$$

Because this is the magnitude, we need only square the `H_Mag` and substitute the sinusoidal argument function, $wT = (2N + 1)\,\pi$, into both expressions, to obtain the two periodic transfer functions. Hence,

```
H_Mag := subs(w*T=(2*N+1)*Pi,H_Mag);
H_Mag_Cascade := subs(w*T=(2*N+1)*Pi,H_Mag^2);
```

$$\text{H\_Mag} := \sqrt{\frac{1}{1 + 2\ \text{alfa}\ \cos((2\ N + 1)\pi) + \text{alfa}^2}}$$

$$\text{H\_Mag\_Cascade} := \frac{1}{1 + 2\ \text{alfa}\ \cos((2\ N + 1)\ \pi) + \text{alfa}^2}$$

We will use the same three alfa values as computed and displayed in Figure 10, hence

```
with(plots):
alfa := .25:
Plot_1 := plot(H_Mag_Cas-
cade,N=0..2,color=black,style=point,symbol=cross):
alfa := .50:
Plot_2 := plot(H_Mag_Cascade,N=0..2,color=black,style=point,
 symbol=diamond):
alfa := .75:
Plot_3 := plot(H_Mag_Cascade,N=0..2,color=black,style=point,
 symbol=c ircle):
display({Plot_1,Plot_2,Plot_3},axes=boxed,
 labels=[N_Value,MAG]);
```

Note the gain goes as the square, hence the selectivity is greater than the single filter stage. This is evident since the rate of change as you move along the x-axis in Figure 2.19 is greater than that in Figure 2.10. To prove this, let's evalute the ratio of the two normalized (i.e., both transfers maximized for unity) filter functions and plot the result as a function of different alfas:

```
H_Mag_Norm := H_Mag*abs(alfa-1):
H_Mag_Cascade_Norm := H_Mag_Norm^2:
H_Mag_Ratio := simplify
 (subs(w*T=(2*N+1)*Pi,H_Mag_Norm/H_Mag_Cascade_Norm));
```

$$\text{H\_Mag\_Ratio} := \sqrt{\frac{1}{1 - 2\ \text{alfa}\ \cos(2\ \pi\ \mathcal{N}) + \text{alfa}^2}}\ \left(1 - 2\ \text{alfa}\ \cos(2\ \pi\ \mathcal{N}) + \text{alfa}^2\right) / (|\ \text{alfa} - 1\ |)$$

Now plotting H_Mag_Ratio,

```
with(plots):
alfa := .25:
Plot_1 := plot(H_Mag_Ratio,N=0..2,color=black,style=point,
 symbol=cross
```

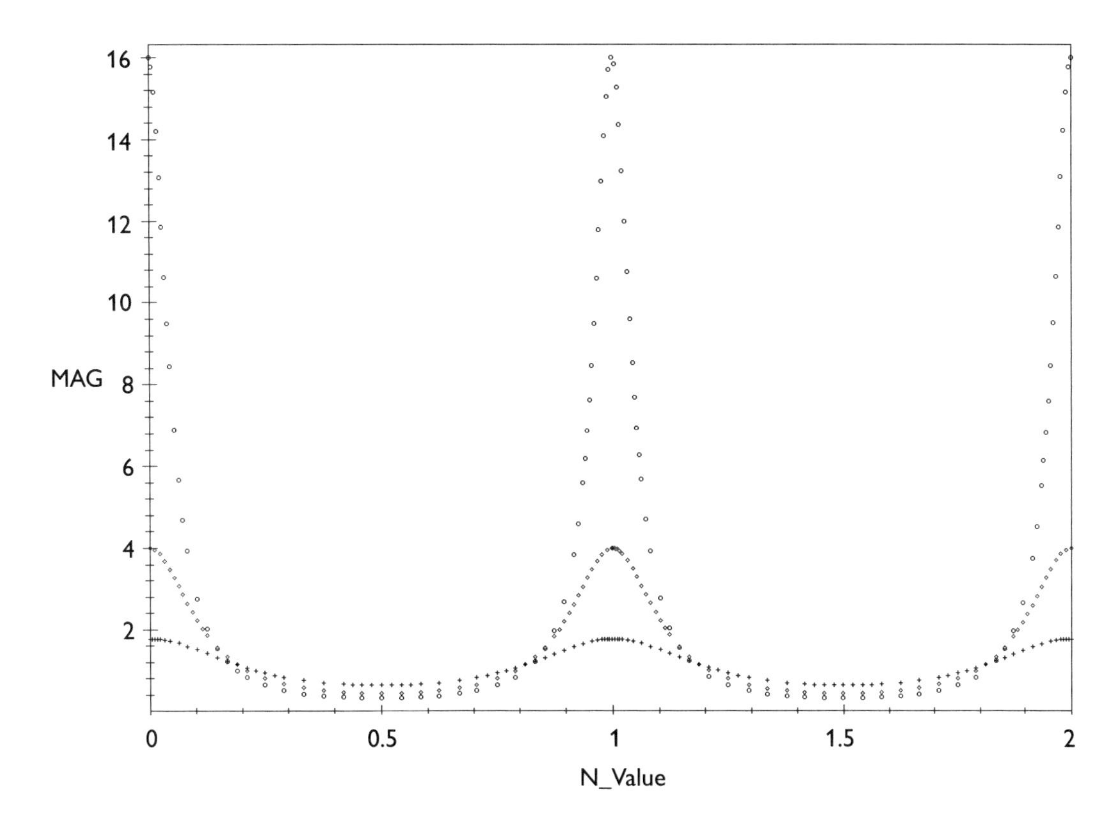

Figure 2.19 General cascaded comb filter response for various alfa (alfa = .75, circles; alfa = .50, diamonds; alfa = .25, crosses).

```
Plot_2 := plot(H_Mag_Ratio,N=0..2,color=black,style=point,
 symbol=diamo    nd):
alfa := .75:
Plot_3 := plot(H_Mag_Ratio,N=0..2,color=black,style=point,
 symbol=circle):
display({Plot_1,Plot_2,Plot_3},axes=boxed,
 labels=[N_Value,MAG]);
```

Figure 2.20 shows that as we vary the `alfa` value, the ratio is always greater than unity, indicating that the denominator function (normalized cascaded transfer) is less than the numerator (normalized noncascaded transfer). Note further that at the passband peaks (`N_Value`=0, 1, 2, ...), the ratio must be unity because both functions were normalized for their maximums. Consequently, all values outside the peaks are ratiometrically greater than unity, indicating the cascaded switching filter has higher selectivity for any given `alfa`.

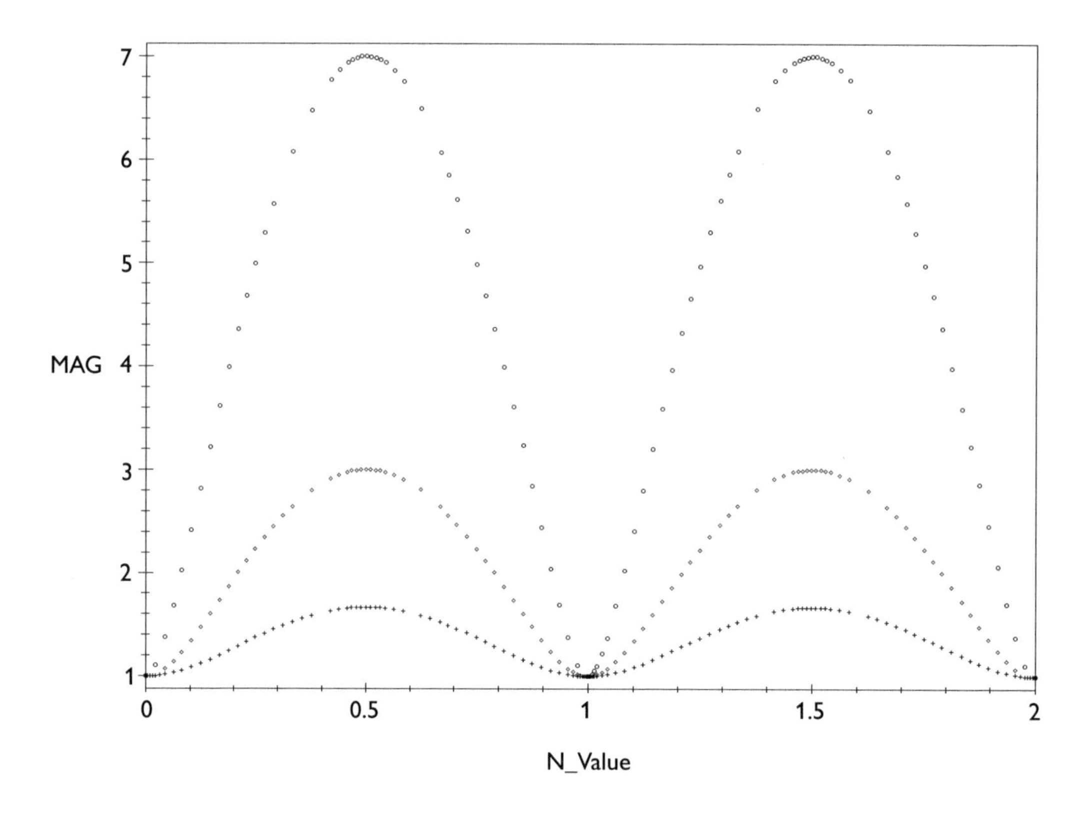

Figure 2.20 Ratio of single-to-cascaded transfer functions versus various values of alfa (alfa = .75, circles; alfa = .50, diamonds; alfa = .25, crosses).

Conclusion

The idea of controlling the center or pass frequency by adjusting a digital clocking signal has great appeal for those of us involved with computer-controlled processes. Certainly, such a filter is extremely useful in the telecommunications business where sharp neighborhood signal rejection is important when a company is trying to squeeze as many voice channels onto one line a possible without crosstalk or aliasing taking place within any one isolated information band. Other immediate applications of the switching bandpass filter can be used in a tracking filter scheme where an input signal is *tracked* as the clocking frequency changes to maximize the received input signal. The correction or error signal generated would be fed to a voltage-controlled oscillator (VCO) that clocks the CCD. This, in effect, is what your frequency-modulated (FM) receiver does, though not using CCD technology (primarily due to cost).

Maple allowed us to delve into some of the dynamics of the equations that describe CCD operation. From the manipulation of these equations, we were able to ascertain operating parameters necessary to acquire a chosen frequency, while rejecting a close neighboring interfering signal. We were also able to see graphically what was happening in the neighboring spectrum (i.e., the rate of increasing attenuation around the centrally passed frequency).

As to the issue of cascading switching filters to enhance neighboring signal attenuation, the basic premise of multiplying transfer function magnitudes and Maple gave us a quick look into what improvements we can expect without getting the user reinvolved with the constituent relationships. This saved a tremendous amount of time and effort by not having to revisit the basic mathematical structures of the CCD switching filter.

REFERENCES:

[1] Tietze, U., and C. Schenk, *Advanced Electronic Circuits*, New York: Springer-Verlag, 1978.

[2] Roberge, J. K., *Operational Amplifiers: Theory and Practice*, New York: John Wiley & Sons, 1975.

[3] Weinberg, L., *Network Analysis and Synthesis*, New York: McGraw-Hill Book Co., 1962.

[4] Krauss, H. L., C. W. Bostian, and F. H. Raab, *Solid State Radio Engineering*, New York: John Wiley & Sons, 1980.

Chapter 3

Curve fitting

Introduction

THE EXAMPLE WE examine in this chapter is a typical situation that scientists and engineers come in contact with during routine analysis of a new piece of equipment, phenomena, or an algorithm development. There are two general types of data curve fitting. The first is involved with determining the best fit given very few data points (a data-starved problem). The second type is used when there are thousands of data points (a data-rich problem). Generally, abstracting the desired information from each of these categories can be tedious and full of pitfalls in terms of what the observer believes he or she is getting. Beyond these two classifications of data fitting, the reader should realize that we are going to talk about interpolators (i.e., estimators) as opposed to extrapolators (i.e., predictors), which could generate several books of information by themselves.

Another important aspect about regression estimators is the cost function used to determine "goodness of fit" of the observed data to the model. Perhaps the most common is called the least squares error (LSE), which defines the error as

$$\text{LSE} = \min \sum_{i=1}^{N} \left(y_i^{\text{observed}} - y_i^{\text{estimate}} \right)^2$$

Conceptually, an algorithm (and there are many) attempts to minimize differential or error between the observed (y_{observed}) and estimated ($y_{\text{estimated}}$) data at successive points along the modeled function. Another common cost or penalty function is defined as the least median squares error (LMSE), which is defined as

$$\text{LMSE} = \min \left[\frac{1}{N-1} \sum_{i=1}^{N} \left(y_i^{\text{observed}} - y_i^{\text{estimate}} \right)^2 \right]$$

In the LMSE cost function, the algorithm is always trying to minimize the median or average of the squared error. Contrast this with the LSE, which minimizes the square error. The LMSE cost function gives much better robustness (less variance of error) when a few data points lie relatively far away from the majority of the rest. However, in this brief expose, we will use the more common LSE function without any loss of generality as it relates to curve-fitting data. If the reader wants to use a more rigorous cost function, then the following Maple code needs only slight modification to implement a more aggressive cost function. Several excellent references exist that give practical and theoretical explanations of what we will investigate in this section [1–3]. These references give typical and advanced examples that the reader can try out on his or her own with Maple.

Regardless of which cost function is used, the reader should be able to use the following Maple procedures to apply these principles to whatever he/she is required to analyze or use for data design.

Case study: Gaussian peak estimator filter example with regressive curve fitting

Our first example is of a data-starved variety in which we are trying to get the best precision possible for a peak intensity reading. Precision means the instrument has the ability to reproduce specific readings within certain tolerances given a set of conditions. This aspect is of particular importance when absolutes are not necessary (except when calibrating the instrument at the beginning of an experiment). However, the instrument's ability to reproduce readings (in this case, peak light intensity at a given wavelength) is crucial to the machine's functional and economic viability.

The following data are taken from an atomic spectroscopy unit that can spectrally detect very close wavelengths (excellent optical dispersion) for the purposes of determining what atomic elements exist in a sample mixture. The instrument's ability to identify what elements are present is a direct function of how well the instrument can decipher and reproduce peak intensity readings.

For the first "go through," let's take one data set from our new instrument. In this situation, our instrument will only be allowed to obtain 16 peak intensities (it could be less). From these readings, we can assume that the intensity profile is Lorentzian [1], i.e.,

$$\text{Intensity profile} \rightarrow I_0[\text{sinc}(x)]^2$$

but a Gaussian profile

$$\text{Intensity profile} \rightarrow I_0 e^{k^1(k2 - \mu)^2}$$

will be close enough to suit our requirements.

By assuming an intensity profile, we can eliminate much of the decision space that must be searched to arrive at the optimal regression coefficients for our estimator. This statement says that the more we bias the intensity profile to a known form, the less chance we have of getting final results that are far afield of the observed intensity profile function.

Hence the strategy for us to follow is to determine what specific Gaussian profile best fits the observed intensity data. We do this by applying the general Gaussian profile function:

$$\text{Estimated intensity} = k1 e^{k2(x - k3)^2}$$

After finding the $k1$, $k2$, $k3$ coefficients, via a Maple session, we will compare the estimated and actual intensity data to see how well the curve fit worked.

Along with this approach, we will also approximate the Gaussian profile function with the same data using a nonlinear regression technique called the Levenberg-Marquardt algorithm (LMA). Finally, we will show the reader the more dangerous approach of using a high-order general polynomial, i.e.,

$$\text{Estimated intensity} = a_0 + a_1x + a_2x^2 + a_3x^3 + \ldots + a_Nx^N$$

for approximating the intensity profile from the observed data. This is the most dangerous approach because the results can easily mislead the investigator as to the estimation obtained from the observations.

Starting the Maple regression session

The very first thing we need to do is to initialize Maple with the appropriate mathematics libraries required to perform the regression analysis. The libraries are `[STATS]` (statistics) and `[PLOTS]` (graphical output plotting). Remember that the user only needs to start these libraries once per Maple session. However, during this case study, as in other application chapters, the authors will reiterate the libraries so that the reader will associate certain Maple libraries with certain approaches used in the template applications.

Linear regression using a logarithmic representation of the Gaussian model

```
with (stats):
with (plots):
```

Next, enter the observable intensity data (`Yvalues_raw`) at uniform and discrete window step positions (`Xvalues`) into Maple:

```
Yvalues_raw :=
[25059,34459,56923,109885,152544,198619,256505,
 289850,295849,273272,225068,171780,126180,70684,43297,
 25515];
Xvalues := [seq(i,i=0..nops(Yvalues_raw)-1)];
```

Yvalues_raw := [25059 , 34459 , 56923 , 109885 , 152544 , 198619 , 256505 , 289850 , 295849 , 273272 , 225068 , 171780 , 126180 , 70684 , 43297 , 25515]
Xvalues := [0, 1, 2, 3, 4, 5, 6, 7, 8, 9, 10, 11, 12, 13, 14, 15]

Since the linear regression algorithm cannot not directly deal with the Gaussian exponential form, we need to convert the model into a linear combination of increasing polynomials by converting the raw intensity values (`Yvalues_raw`) to the natural logarithm form (`Yvalues`) or

Estimated Intensity =

$$\text{Yvalues\_raw} = k1e^{k2(x-k3)^2} \xrightarrow{\text{Transform}} a_0 + a_1x + a_2x^2$$

By taking the natural logarithm of the both sides

$$\begin{aligned}\ln(\text{Yvalues\_raw}) &= \ln\left(k1e^{k2(x-k3)^2}\right)\\ &= \ln(k1) + k2(x-k3)^2\\ &= \ln(k1) + k2x^2 - 2k2k3x + k2k3^2\end{aligned}$$

Equating coefficients from the general quadratic polynomial form gives

$$\begin{aligned}a_2 &= (k2)\\ a_1 &= 2(k_2k_3)\\ a_0 &= \left(k2k3^3 + \ln(\text{k1})\right)\end{aligned}$$

or grouping like powers of x, we obtain

$$\ln(\text{Yvalues\_raw}) = \text{Yvalues} = (k2)x^2 + 2(k2k3)x + \left(k2k3^2 + \ln(k1)\right)$$

Therefore, by taking the natural logarithm of the raw intensity data, we perform regression on this data to the aforementioned quadratic polynomial. Let's start by getting Maple to take the logarithm of the intensity data:

```
Yvalues := evalf(transform[multiapply[(y)-ln(y)]]
  ([Yvalues_raw]));
```

Yvalues := [10.12898832 , 10.44752549 , 10.94945476 , 11.60718964 , 11.93520836 , 12.19914370 , 12.45490344 , 12.57711883 , 12.59760447 , 12.51822292 , 12.32415786 , 12.05396987 , 11.74546474 , 11.16597452 , 10.67583863 , 10.14702179]

Now, to perform the least squares regression with the assumed quadratic polynomial function (`y=a+b*x+c*x^2`) and abstract the appropriate {*a*, *b*, *c*} regression coefficients, we use the following code:

```
eq_fit := fit[leastsquare[[x,y],y=a+b*x+c*x^2]]
  ([Xvalues,Yvalues]);
a := coeff(rhs(eq_fit),x,0);
b := coeff(rhs(eq_fit),x,1);
c := coeff(rhs(eq_fit),x,2);
```

$$eq\_fit := y = 9.920667962 + .689997947\, x - .04516343047\, x2$$

$$a := 9.920667962$$

$$b := .689997947$$

$$c := -.04516343047$$

Remembering how we obtained the logarithmic regression form, we must now substitute these coefficients back into the original Gaussian function. This is accomplished by doing a little "back-of-the-envelope" algebra, which results in the following:

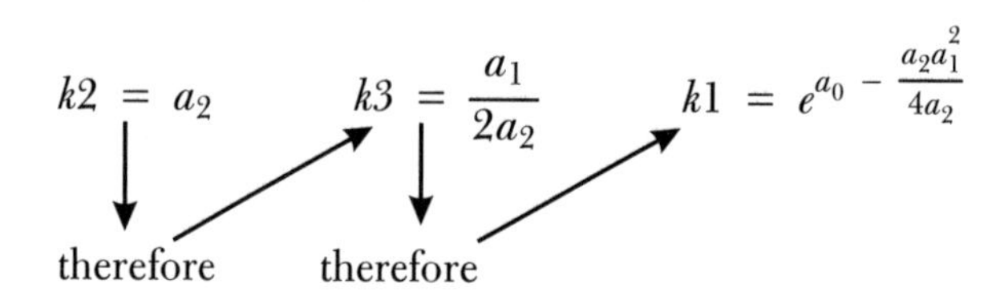

Substituting the regression coefficients back into the original Gaussian form, we obtain the estimator function (`Y_Estimated`):

```
Y_Estimated :=
 (exp(a-(b^2/(4*c))))*((exp(c*(x+b/(2*c))^2)));
```

$$Y\_Estimated := 283815.8695\ e^{(-.04516343047\ (x\ -\ 7.638900985)^2\ )}$$

We can determine the x-value (`X_Max_Value`) by solving where the derivative of the estimated intensity function (`Y_Estimated_Derivative`) equals zero. Then, knowing this allows us to determine the approximate step position where the peak intensity (`Y_Max_Value`) exists and its value:

```
Y_Estimated_Derivative := diff(Y_Estimated,x);
X_Max_Value := solve(Y_Estimated_Derivative=0,x);
Y_Max_Value := evalf(subs(x=X_Max_Value,Y_Estimated));
```

$$Y\_Estimated\_Derivative := 283815.8695$$

$$(-.09032686094x + .6899979470\ )\ e^{(-.04516343047(x\ -\ 7.638900985)^2)}$$

$$X\_Max\_Value := 7.638900985 Y\_Max\_Value := 283815.8695$$

Plotting the curve fit and residual error

Plotting the actual and estimated intensities versus window step position subjectively indicates how close an estimate or "fit" we have accomplished (see Figure 3.1):

```
Data_Set := zip((a,b)->[a,b],Xvalues,Yvalues_raw):
Data_Plot := pointplot(Data_Set,style=point,
 symbol= circle,color=black):
Estimated_Plot := plot(Y_Estimated,x=0..nops(Yvalues_raw)-1,
 style=line,color=black):
display({Estimated_Plot,Data_Plot},axes=boxed,
 labels= [Window_Step_Position,Intensity]);
```

Now let's code and plot the residual error between the estimated and the actual intensity data as shown in Figure 3.2:

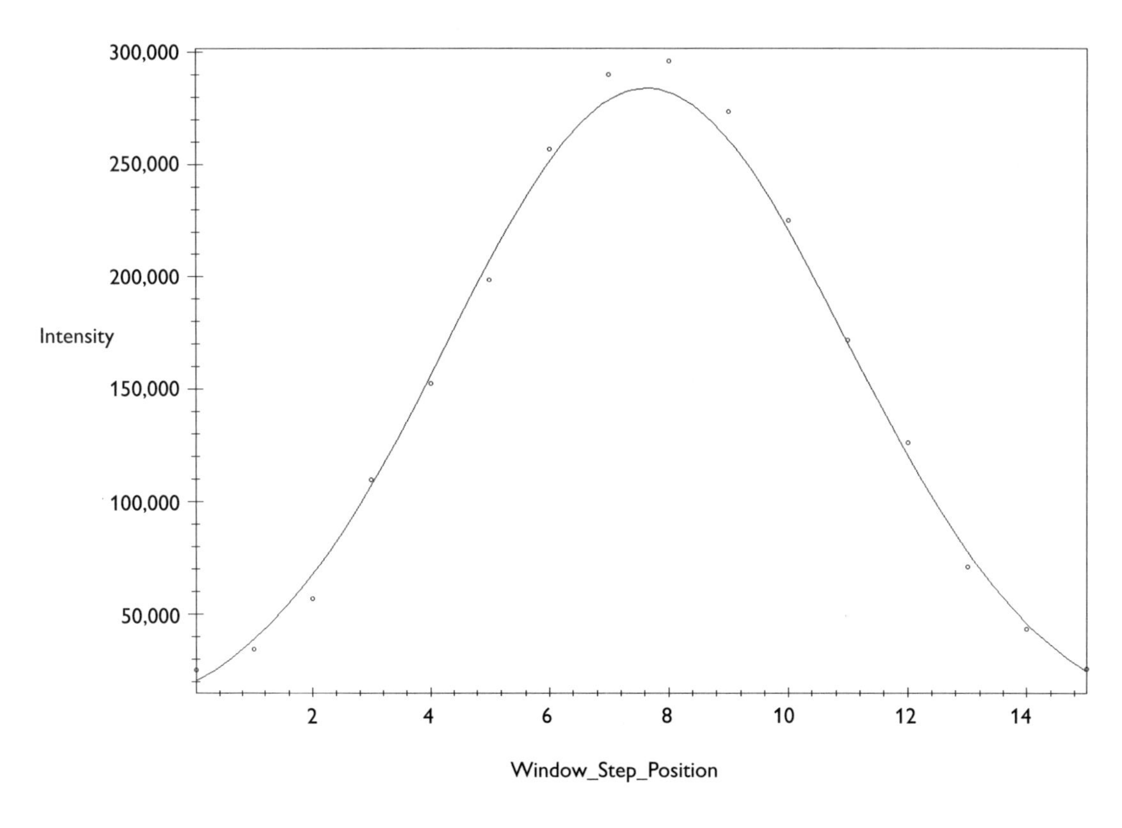

Figure 3.1 Estimated (solid line) and actual (open circles) intensity versus window step position (second-order polynomial Ln form).

```
Estimator := y=Y_Estimated:
Transform_Equation := unapply(rhs(Estimator),x):
Estimated_Data :=
 transform[apply[Transform_Equation]](Xvalues):
Residual_Error := transform[multiapply[(x,y)->x-y]]
 ([Yvalues_raw,Estimated_Data]):
Error_Set := zip((a,b)->[a,b],Xvalues,Residual_Error):
Error_1 := pointplot(Error_Set,style=line,color=black):
Error_2 := pointplot(Error_Set,style=point,symbol= circle,
 color=black):
display({Error_1,Error_2},labels=[Window_Step_Position,
 Intensity],axes=boxed);
```

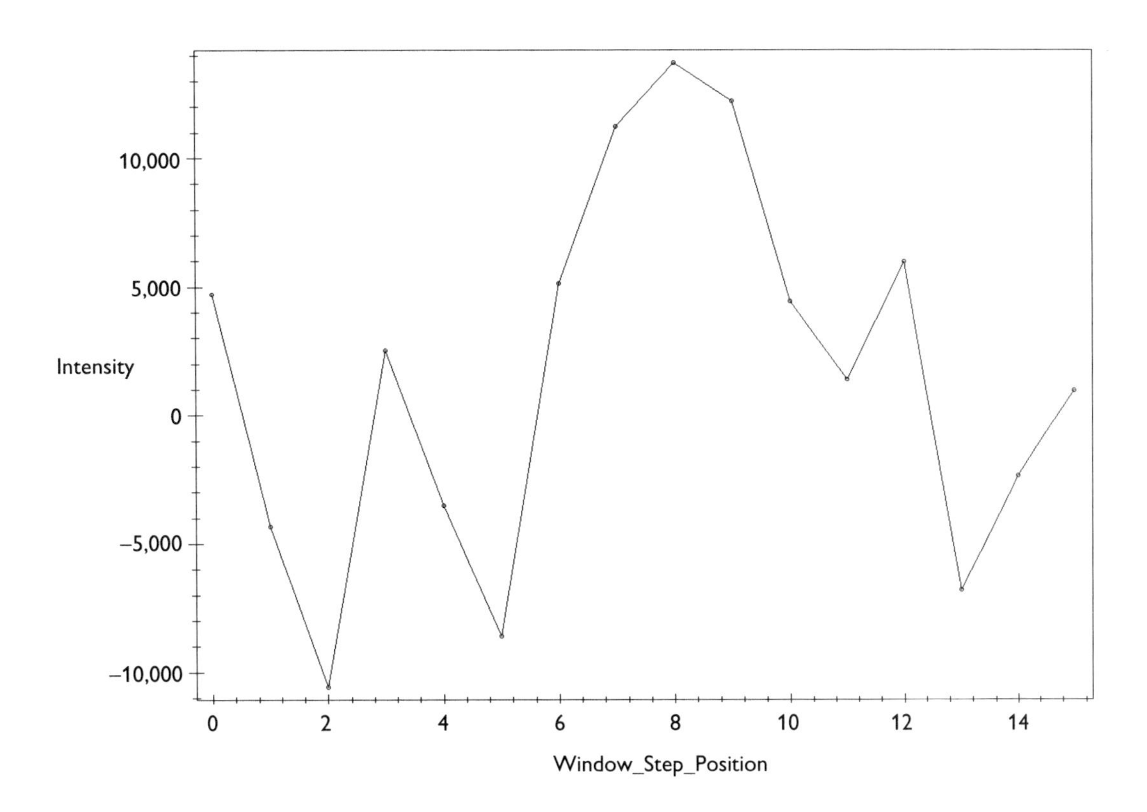

Figure 3.2 Residual error between estimated and actual intensity versus window step position.

Now plotting residual error versus window step position as a relative percentage error at each step, we generate Figure 3.3:

```
Percent_Residual_Error := transform[multiapply[(x,y)
 -> ((x-y)/x)*100]]([Yvalues_raw,Estimated_Data]):
Percent_Error_Set :=
 zip((a,b)->[a,b],Xvalues,Percent_Residual_Error):
Percent_Error_1 := pointplot(Percent_Error_Set,style=line):
Percent_Error_2 :=
 pointplot(Percent_Error_Set,style=point,symbol=circle,
 color=black):
display({Percent_Error_1,Percent_Error_2},labels=
 [Window_Step_Position,Percent_Error],axes=boxed);
```

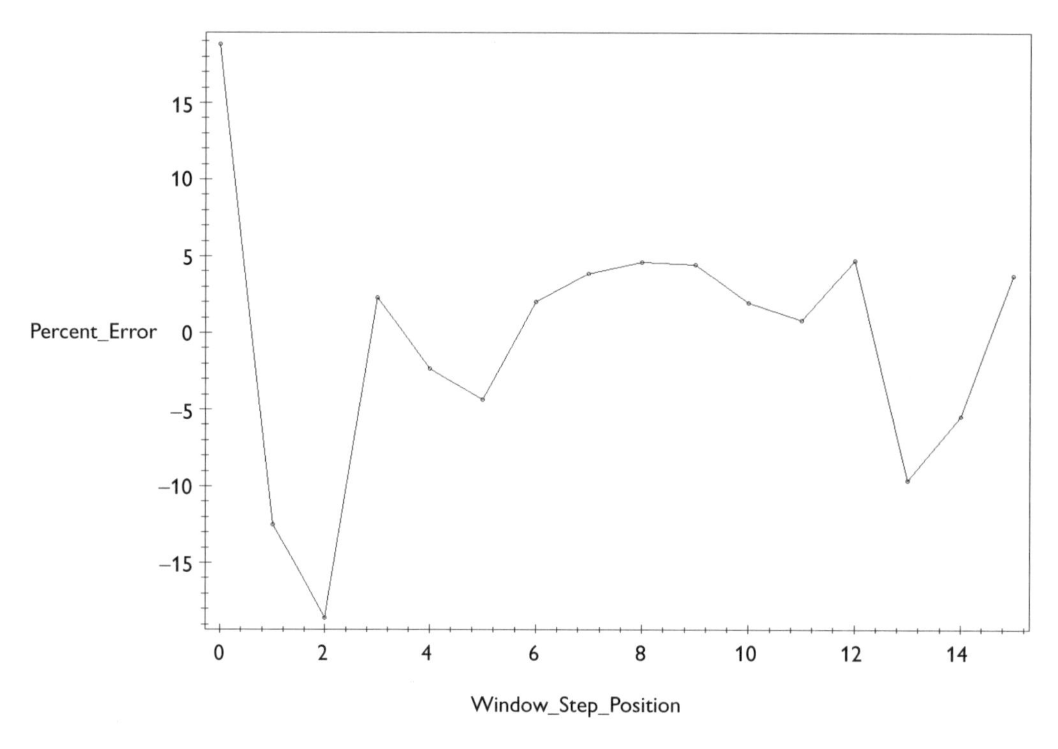

Figure 3.3 Percent relative residual error versus window step position (second-order fit).

Figure 3.3 indicates a fairly well behaved regression in terms of relative positional error. Notice that residual errors generally show up at the beginning and end, while retaining a fairly smooth error variance in the middle of the data set. Simply stated, the regression process is trying to "smooth" the data behavior, but until it gets a few starting pieces of data, it may wildly fly around with a few data points in the beginning. Once the regression has some extra data or *trend momentum*, it begins to calm down and settle about some mean or averaging process dictated by the cost function. The end of the estimation window exhibits a large excursion because the quadratic function has a certain amount of information bandwidth. Hence, when the data suddenly stop, the quadratic information filter has a transient response. This fact becomes extremely evident with higher order polynomials and small data sets. The reader will experience this situation when we perform a regression with a high-order polynomial later in this section.

Obtaining peak intensity RSD by linear regression on multiple data sets

If we have a number of *runs* or data sets of intensity data, we can determine the relative standard deviation (RSD) of the peak intensity associated with the regressive curve fitter. Hence, let's use a loop to abstract the data from 10 runs and perform the appropriate averaging of peak levels or maxima of the Gaussian-shaped intensity profiles.

Again, we initialize Maple for the libraries that are needed to perform the peak estimator's synthesis:

```
with (plots):
with (stats):
```

To determine the estimator's RSD capability, we first start by entering the 10 strings of window raw intensity data:

```
Yvalues_raw(1) :=
[25059,34459,56923,109885,152544,198619,256505,289850,295849,
 273272,225068,171780,126180,70684,43297,25515]:
Yvalues_raw(2) :=
[16142,23101,37309,82903,123089,161534,223857,267502,294544,
 296350,2          64693,226965,181950,119709,79059,52353]:
Yvalues_raw(3) :=
[23853,34409,54547,107772,147408,190335,253463,284703,293676,
 284066,233891,184085,137663,82013,47451,28750]:
Yvalues_raw(4) :=
[42339,59722,87024,155752,204287,245568,290199,289464,272198,
 246212,187953,141533,98621,56186,33778,21203]:
Yvalues_raw(5) :=
[86172,108518,147092,219739,259478,287404,301906,271885,
 232533,18794 8,125513,83811,58396,29801,16991,11355]:
Yvalues_raw(6) :=
[72548,94368,133371,197988,249126,279774,292374,268231,
 243549,200351,138546,94043,62918,32122,18877,12192]:
Yvalues_raw(7) :=
[48609,65256,94868,151736,203552,244670,289439,295627,282365,
 253035,19195        5,140082,102895,54583,31780,19525]:
Yvalues_raw(8) :=
[16907,24572,39079,79189,121543,165090,225141,272662,295347,
 296574,268700,216644,169936,106075,68572,43873]:
Yvalues_raw(9) :=
[6628,8489,13492,29828,50541,76777,130350,189591,224634,
```

```
 255294,288537,280874,270477,210871,163214,119500]:
Yvalues_raw(10) :=
[6337,8282,12323,27675,46061,71486,125680,173236,215764,
 256192,290531,286705,277525,223899,177035,136611]:
```

Next, we generate a procedure called `GaussianEstimator`, which will produce the following quantities (on a per-intensity-run basis) for us once we give the procedure a list of intensity values, `Yvalues_raw`.

- The peak value for the intensity data set;
- The data set variance between the estimation and actual data;
- The data set standard deviation between the estimation and actual data;
- The residual error for the data set given in list form.

If you look closely, the following procedure is similar to the first go through of the estimator's program structure. The difference now is the implementation of previous effort into a Maple procedure. This approach is more efficient with Maple's (and your computer's) resources than using a subscripted session where each Maple line of code generates a large array.

Then by performing a do loop after the data set has been processed by the procedure, we can abstract the appropriate statistical information. This do loop will keep a record of each do loop pass in order to accumulate results for each individual data set. This is necessary to compute averages, standard deviations, variances, and the like for the individual data sets as well as to generate the required database for interdata set analysis.

```
GaussianEstimator := proc(Yvalues_raw)
# Declaration of local variables used in algorithm...
local Xvalues,              # the sequenced window step positions
 Yvalues,                   # the logarithmic raw intensity values
 eq_fit,                    # least squares equation fit
 a,                         # zero order regression coefficient
 b,                         # first order regression coefficient
 c,                         # second order regression coefficient
 Y_Estimated,               # estimated intensity function
 Y_Estimated_Derivative,    # derivative of estimated function
 X_Max_Value,               # window step position at estimated peak intensity
 Y_Max_Value,               # estimated peak intensity
 Estimator,                 # interim equation
 Transform_Equation,        # interim equation
 Estimated_Data,            # estimated intensity for given window step position
 Residual_Error,            # residual error between estimated and actual intensities
 Estimator_Variance,        # variance of the residual error
```

```
 Estimator_Standard_Deviation, # standard deviation of the estimator error
 Residual_Error_Set,           # residual error data set
 indexer;                      # indexing variable
 options 'Copyright 1995 by Chris Tocci';
# Main Algorithm...
 Xvalues := [seq(indexer,indexer=0..nops(Yvalues_raw)-1)]:
 Yvalues := evalf(transform[multiapply[(y)->ln(y)]]([Yvalues_raw])):
 eq_fit := fit[leastsquare[[x,y],y=a+b*x+c*x^2]]([Xvalues,Yvalues]):
  a := coeff(rhs(eq_fit),x,0):
  b := coeff(rhs(eq_fit),x,1):
  c := coeff(rhs(eq_fit),x,2):
 Y_Estimated := (exp(a-(b^2/(4*c))))*((exp(c*(x+b/(2*c))^2))):
 Y_Estimated_Derivative := diff (Y_estimated,x):
 X_Max_Value := solve(Y_Estimated_Derivative=0,x):
 Y_Max_Value := evalf(subs(x=X_Max_Value,Y_Estimated)):
 Estimator := y=Y_Estimated:
 Transform_Equation := unapply(rhs(Estimator),x):
 Estimated_Data := transform[apply[Transform_Equation]](Xvalues):
 Residual_Error := transform[multiapply[(x,y)->x-y]]([Yvalues_raw,Estimated_Data]):
 Estimator_Variance     := describe[variance](Residual_Error):
 Estimator_Standard_Deviation := describe[standarddeviation](Residual_Error):
 Residual_Error_Set := zip((a,b)-[a,b],Xvalues,Residual_Error):
 [Y_Max_Value,Estimator_Variance,Estimator_Standard_Deviation,Residual_Error_Set]:
end:
```

NOTE that in the last line of the Maple procedure, all of the important information is put together in a single list:

```
[Y_Max_Value,Estimator_Variance,
 Estimator_Standard_Deviation,Residual_Error_Set]:
```

where `Y_Max_Value` is the peak intensity within any one run, `Estimator_Variance` is the intensity variance between estimation and actual intensity data, `Estimator_Standard_Deviation` is the intensity standard deviation between estimation and actual intensity data, and `Residual_Error_Set` is the ordered pairs representing the windows step position and the corresponding intensity value differential between the estimation and actual.

Now, the `do` loop is used for abstracting and accumulating (indexing, if you like) the required statistical information from the `GaussianEstimator` procedure for each data set. We have incorporated some *line print* commands (`lprint`) to report these previously mentioned variables as the program goes through each loop iteration. These line prints can be useful when the user needs to see what is going on inside the loop during execution. Also, line printing, in this case, specifically gives the user important numerical or symbolic insight associated with each of the intensity data sets. This insight might be necessary if things do not make sense at the end of a Maple session. Get in the habit of line printing your values in Maple procedures and do loops until you are sure of the veracity of your results.

```
for n from 1 to 10 do
Estimated_Yvalues_Peak(n) := op(1,GaussianEstimator(Yvalues_raw(n))):
lprint('Intensity Data Set',n):
lprint('Estimated Peak is',op(1,GaussianEstimator(Yvalues_raw(n)))):
lprint('Estimator Variance is',op(2,GaussianEstimator(Yvalues_raw(n)))):
lprint('Estimator Standard Deviation is',op(3,GaussianEstimator(Yvalues_raw(n)))):
lprint(''):
lprint('Residual Error Data Set for plotting purposes is as follows...'):
lprint(''):
lprint(op(4,GaussianEstimator(Yvalues_raw(n)))):
lprint('    ______________________________________________'):
od:
                    Intensity Data Set  1
Estimated Peak is  283815.8723
Estimator Variance is  50163044.16
Estimator Standard Deviation is  7082.587208
Residual Error Data Set for plotting purposes is as follows...
[[0, 4712.42278], [1, -4314.89519], [2, -10585.59134], [3, 2498.3726], [4,
-3523.6542], [5, -8608.3845], [6, 5112.1444], [7, 11218.4699], [8, 13699.5998], [9,
12236.5170], [10, 4424.3144], [11, 1385.7233], [12, 5956.3119], [13, -6815.09532],
[14, -2346.06094], [15, 955.22218]]
        ______________________________________________
                    Intensity Data Set  2
Estimated Peak is  288260.8159
Estimator Variance is  35235193.39
Estimator Standard Deviation is  5935.926132
Residual Error Data Set for plotting purposes is as follows...
[[0, 3011.29849], [1, -2785.20830], [2, -9607.34324], [3, 4730.30757], [4, 3342.9527],
[5, -7098.9760], [6, 5534.2348], [7, 7647.1998], [8, 10204.1923], [9, 10314.5978],
[10, 161.6716], [11, 2054.5480], [12, 6150.6093], [13, -6619.1958], [14, -4397.16492],
[15, 1666.56408]]
        ______________________________________________
                              .
                              .
                              .
                    Intensity Data Set  10
Estimated Peak is  269799.2195
Estimator Variance is  151521708.2
Estimator Standard Deviation is  12309.41440
Residual Error Data Set for plotting purposes is as follows...
[[0, 2064.904548], [1, -566.056479], [2, -4759.16049], [3, -3066.54453], [4,
-5508.93017], [5, -9154.90192], [6, 8135.5349], [7, 13524.2144], [8, 13480.8444], [9,
17372.1348], [10, 27704.5001], [11, 17082.6882], [12, 19696.7633], [13, -5923.7953],
```

```
[14, -13925.5693], [15, -11293.8014]]
```

Finally, we compute the statistical information from Maple's `do` loop and computing the overall intensity peak average, `Y_Max_Value_Average`, the peak value standard deviation, `Y_Max_Value_StanDev`, and the desired RSD, `Y_Max_Value_RSD`, for all 10 data sets:

```
Digits := 4:
Y_Max_Average_Seq := [seq(Estimated_Yvalues_Peak(n),n=1..10)]:
Y_Max_Value_Average := describe[mean](Y_Max_Average_Seq);
Y_Max_Value_StanDev := describe[standarddeviation](Y_Max_Average_Seq);
Y_Max_Value_RSD := (Y_Max_Value_StanDev/Y_Max_Value_Average)*100;
```

Y_Max_Value_Average := 280900.
Y_Max_Value_StanDev := 6855
Y_Max_Value_RSD := 2.440

Next, for comparison purposes, we abstract the maximum raw intensity values (`Yvalues_raw_Max`) from the individual data sets and compute the data's RSD over 10 data sets in the same manner. Writing a quick Maple `do` loop for abstracting the peak intensity values from each of the 10 data sets yields the desired results:

```
for m from 1 to 10 do
Yvalues_raw_Peak(m) := op(nops(Yvalues_raw(m)),
 sort(Yvalues_raw(m))):
od:
Yvalues_raw_Peak_Seq := [seq(Yvalues_raw_Peak(m),m=1..10)]:
Yvalues_raw_Peak_Average := evalf(describe[mean]
 (Yvalues_raw_Peak_Seq));
Yvalues_raw_Peak_StanDev :=
 evalf(describe[standarddeviation]
 (Yvalues_raw_Peak_Seq));
Yvalues_raw_Peak_RSD := evalf((Yvalues_raw_Peak_StanDev/
 Yvalues_raw_Peak_Average)*100);
```

and the output is:

Yvalues_raw_Peak_Average := 294200.
Yvalues_raw_Peak_StanDev := 3735
Yvalues_raw_Peak_RSD := 1.270

As we can see, the RSD of the curve estimator (`Y_Max_Value_RSD`) is about half as good as the straightforward *peak picking* method or approach (`Yvalues_raw_Peak_RSD`) of 10 successive windows. The reason is that the estimator depends on *all* data presented to the filter, whereas the simple peak detector approach, depends on only one value within the window.

The drawback to the peak detector method is if any one peak is bad it can significantly affect the overall RSD measurement. On the other hand, if any one, two, or even three measured values are *bad*, they are averaged against others in the window, hence their influence is reduced significantly. Therefore, even though the estimator might not have the best RSD for one time, it will be much more robust under a much wider range of noisy window data than the peak method.

Comparison of peak method versus regression for robustness against outlier data

To prove this statement, let's alter some of the previous data's peak values to see the comparison of the estimator against the peak method (in other words, we artificially create outlier data).

First, we alter the first two window data sets' peak values by 10%, say, make them 10% higher than actually measured, i.e.,

Yvalues_raw(1)(peak) = 295849 to 325433.9(+10%)
Yvalues_raw(2)(peak) = 296350 to 325985.0(+10%)

```
Yvalues_raw(1) :=
[25059,34459,56923,109885,152544,198619,256505,289850,
 325433.9,273272,225068,171780,126180,70684,43297,25515]:
Yvalues_raw(2) :=
[16142,23101,37309,82903,123089,161534,223857,267502,294544,
 325985.0,264693,226965,181950,119709,79059,52353]:
Yvalues_raw(3) :=
[23853,34409,54547,107772,147408,190335,253463,284703,293676,
 284066,233891,184085,137663,82013,47451,28750]:
Yvalues_raw(4) :=
[42339,59722,87024,155752,204287,245568,290199,289464,272198,
 246212,187953, 141533,98621,56186,33778,21203]:
Yvalues_raw(5) :=
[86172,108518,147092,219739,259478,287404,301906,271885,
 232533,187948,125513,83811,58396,29801,16991,11355]:
Yvalues_raw(6) :=
```

```
[72548,94368,133371,197988,249126,279774,292374,268231,
 243549,200351,138546,94043,62918,32122,18877,12192]:
Yvalues_raw(7) :=
[48609,65256,94868,151736,203552,244670,289439,295627,282365,
253035,191955,140082,102895,54583,31780,19525]:
Yvalues_raw(8) :=
[16907,24572,39079,79189,121543,165090,225141,272662,295347,
 296574,268700,216644,169936,106075,68572,43873]:
Yvalues_raw(9) :=
[6628,8489,13492,29828,50541,76777,130350,189591,224634,
 255294,288537,280874,270477,210871,163214,119500]:
Yvalues_raw(10) :=
[6337,8282,12323,27675,46061,71486,125680,173236,215764,
 256192,290531,286705,277525,223899,177035,136611]:
```

Performing the same algorithm via the `GaussianEstimator` procedure for the 10 intensity data sets, we get

```
for n from 1 to 10 do
Estimated_Yvalues_Peak(n) := op(1,GaussianEstimator
 (Yvalues_raw(n))):
od:
```

computing the overall intensity average, standard deviation, and RSD of the intensity peaks over the 10 new data sets:

```
Digits := 4:
Y_Max_Average_Seq := [seq(Estimated_Yvalues_Peak(n),
 n=1..10)]:
Y _Max_Value_Average := describe[mean](Y_Max_Average_Seq);
Y_Max_Value_StanDev := describe[standarddeviation]
 (Y_Max_Average_Seq);
Y_Max_Value_RSD := (Y_Max_Value_StanDev/
 Y_Max_Value_Average)*100;
```

and the output is:

Y_Max_Value_Average := 281600.
Y_Max_Value_StanDev := 7537.
Y_Max_Value_RSD := 2.676

The RSD value is only slightly higher this time (2.676% versus 2.440%) using the Gaussian estimator. This increase corresponds to an RSD increase of around 9.67%.

Now let's see what happens when deriving the same measures with only the peak intensity from each data set:

```
for m from 1 to 10 do
Yvalues_raw_Peak(m)  := op(nops(Yvalues_raw(m)),sort
 (Yvalues_raw(m))):
od:
Yvalues_raw_P eak_Seq := [seq(Yvalues_raw_Peak(m),m=1..10)]:
Yvalues_raw_Peak_Average := evalf(describe[mean]
 (Yvalues_raw_Peak_Seq));
Yvalues_raw_Peak_StanDev := evalf(describe
 [standarddeviation] (Yvalues_raw_Peak_Seq));
Yvalues_raw_Peak_RSD := evalf((Yvalues_raw_Peak_StanDev/
 Yvalues_raw_Peak_Average)*100);
```

and the result is:

Yvalues_raw_Peak_Average := 300200.
Yvalues_raw_Peak_StanDev := 13310.
Yvalues_raw_Peak_RSD := 4.434

While the estimator's RSD changed about +9.67%, the peak picked RSD changed by about +249.13%! Consequently, the estimator is clearly more dependent on the collective data set, whereas the peak method is dependent on one-out-of-N points.

Obviously, had we altered any of the data points around the peak, the peak method would have shown no RSD change, whereas the estimator would have reflected some change. Therefore, the efficacy of an estimator depends on whether the user has a very precise way of obtaining the peak value and how important nonpeak values are to one's final analysis. If the user has limited resources and an excellent peak detection hardware system, then the much simpler and faster peak detection method should be used. If, on the other hand, one requires peak data under a volatile data acquisition environment, then use the estimator approach for a more uniform RSD behavior under a much noisier set of conditions.

Problem data set for linear regression

Now let's look at a data set that can "confuse" an estimator based on the previous approach of taking the logarithm of the raw data values in order to implement a linear polynomial combination.

```
with (plots):
with (stats):
Yvalues_raw :=
[569322,647595,871287,904318,820139,700099,434252,216687,150058,81671,57118,32746,25717
 17639,13063,11527]:
Xvalues := [seq(i,i=0..nops(Yvalues_raw)-1)]:
Yvalues := evalf(transform[multiapply[(y)->ln(y)]]([Yvalues_raw])):
eq_fit := fit[leastsquare[[x,y],y=a+b*x+c*x^2]]([Xvalues,Yvalues]):
a := coeff(rhs(eq_fit),x,0);
b := coeff(rhs(eq_fit),x,1);
c := coeff(rhs(eq_fit),x,2);
`———————————————————————————————';
'Original Coefficients prior to performing the Log Conversion';
`———————————————————————————————';
k1 := exp(a-(b^2)/(4*c));
k2 := c;
k3 := b/(2*c);
`———————————————————————————————';
Y_Estimated := (exp(a-(b^2/(4*c))))*((exp(c*(x+b/(2*c))^2))):
Y_Estimated_Derivative := diff(Y_Estimated,x):
X_Max_Value := solve(Y_Estimated_Derivative=0,x);
Y_Max_Value := evalf(subs(x=X_Max_Value,Y_Estimated));
Data_Set := zip((a,b)->[a,b],Xvalues,Yvalues_raw):
Data_Plot := pointplot(Data_Set,style=point,symbol=circle,color=black):
Estimated_Plot := plot(Y_Estimated,x=0..nops(Yvalues_raw)-1,style=line,color=black):
display({Estimated Plot,Data_Plot},axes=boxed,labels-[Window_Step_Position,Intensity],
 title= 'Estimated (line) & Actual (point) Intensity versus Window Step Position');
```

Looking at only the numerical results of the LMS estimator's output:

$$a := 13.76022887$$
$$b := -.084211727$$
$$c := -.01640900803$$

$$k1 := .1054179619\ 10^7$$
$$k2 := -.01640900803$$
$$k3 := 2.566021263$$

$$X\_Max\_Value := -2.566021263$$
$$Y\_Max\_Value := .1054179619\ 10^7$$

The reason for this grossly inaccurate estimation is due to the fact that taking the logarithm of the raw data creates increasingly larger errors for the window extremum. For instance, we can reverse the data sequence and get the same grossly inaccurate estimation, i.e.,

```
Yvalues_raw :=
[11527,13063,17639,25717,32746,57118,81671,150058,216687,
 43425 2,700099,820139,904318,871287,647595,569322]:
```

and the output (see Figure 3.4)

$$a := 8.805026173$$
$$b := .576481963$$
$$c := -.01640900774$$
$$X\_Max\_Value := 17.56602143$$
$$Y\_Max\_Value := .1054179640\ 10^7$$

Notice that the estimator puts the peak value exactly the same amount outside the window (2.566021 = 17.566021 – 15) in this case as it did during the previous estimation (– 2.566021 = 0 – 2.566021) when it estimated the peak before the window. Hence, one may deduce that the estimation error mechanism is the same and indeed it is.

The reason for the error is twofold: First, by taking the logarithm, you have compressed information, thereby forfeiting resolution of that information; and, second, after compressing the information, you are only obtaining one or two more data points before there is no more window data. This means that the regression mechanism does not have enough or sufficiently new or updated information about phenomena information before it runs into the end of the data string. Consequently, with lower resolution and fewer pieces of information to compensate against the lower resolution, the regression process begins to become unstable. This type of regression error is especially true of data-starved acquisition systems where the amount of data is so low that "recovering" from a few "wild" points can be nearly impossible. This effect leads to the grossly inaccurate estimates shown previously.

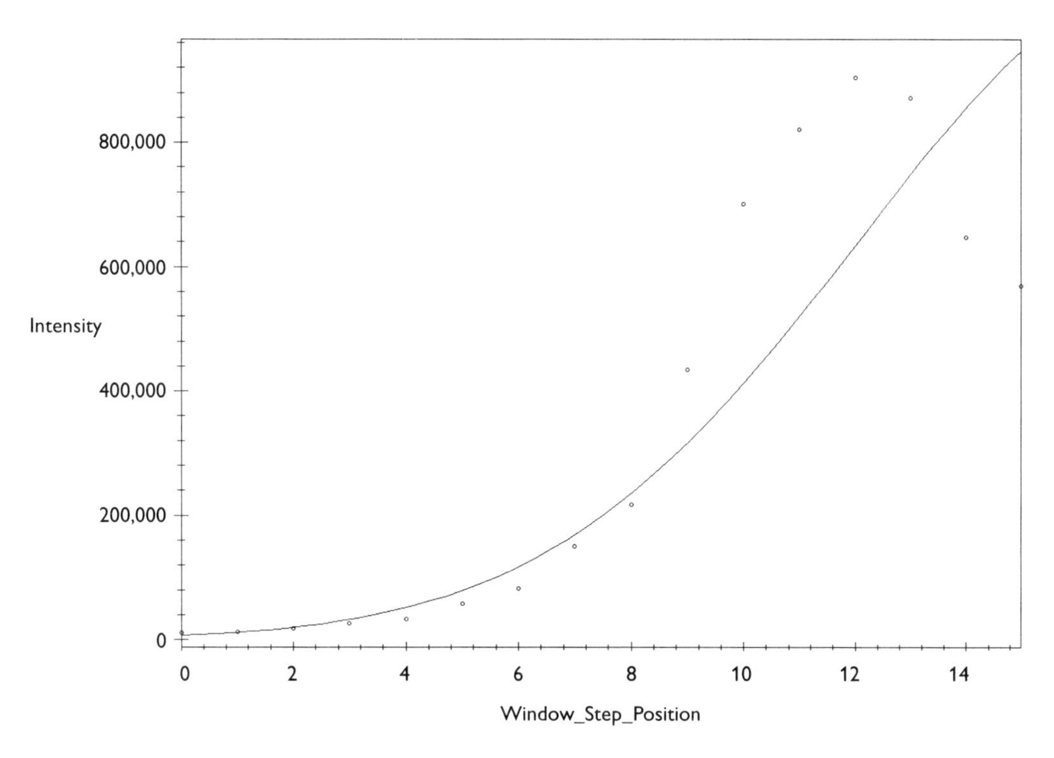

Figure 3.4 Estimated (solid line) and actual (open circles) intensity versus window step position.

Regression improvement by cheating or data stuffing

Let's prove the preceding statement by artificially increasing or *stuffing* the amount of information at the end of the window by attempting to follow the Gaussian profile to see if our estimator performance improves.

```
with (plots):
with (stats):
Yvalues_raw_new :=
[25717,32746,57118,81671,150058,216687,434252,700099,820139,
 904318,887802,871287,759441,647595,608458,569322]:
Xvalues := [seq(i,i=0..nops(Yvalues_raw_new)-1)]:
Yvalues := evalf(transform[multiapply[(y)-ln(y)]]
 ([Yvalues_raw_new])):
eq_fit := fit[leastsquare[[x,y],y=a+b*x+c*x^2]]([Xvalues,
 Yvalues]):
a := coeff(rhs(eq_fit),x,0);
b := coeff(rhs(eq_fit),x,1);
c := coeff(rhs(eq_fit),x,2);
```

```
Y_Estimated :=
 (exp(a-(b^2/(4*c))))*((exp(c*(x+b/(2*c))^2))):
Y_Estimated_Derivative := diff(Y_Estimated,x):
X_Max_Value := solve(Y_Estimated_Derivative=0,x);
Y_Max_Value := evalf(subs(x=X_Max_Value,Y_Estimated));
Data_Set := zip((a,b)->[a,b],Xvalues,Yvalues_raw_new):
Data_Plot := pointplot(Data_Set,style=point,symbol=circle,
 color=black):
Estimated_Plot := plot(Y_Estimated,
 x=0..nops(Yvalues_raw_new)-1,
 style=line,color=black):
display({Estimated_Plot,Data_Plot},axes=boxed,labels=
 [Window_Step_Position,Intensity]);
```

$$a := 9.786117301$$
$$b := .709475575$$
$$c := -.03230337911$$
$$\text{X\_Max\_Value} := 10.98144520$$
$$\text{Y\_Max\_Value} := 874712.7975$$

Figure 3.5 indicates a better fit; however, it also appears that we have extended the window. Unfortunately, this method is only useful for exemplifying how sensitive the linear regression is to the initial and final data in a data-starved situation. Obviously, one cannot artificially add data where there are no data without causing an immediate outcry from one's professional peers.

Fortunately, to circumvent this type of problem with certain data sets, one can use a nonlinear regression technique that does not rely on taking the logarithm or any other trick to convert the regression process into a linear polynomial combination curve fit.

Nonlinear regression: the Levenberg-Marquardt algorithm

Rather than altering the original model to fit into a function that can be represented by a linear combinatorial polynomial, nonlinear regression uses the modeled equation (in our case, a Gaussian profile) directly via gradient and difference equations. Nonlinear regressions move incrementally, constantly looking for changes in the residual errors (difference between the actual and estimated values) and compensating accordingly to keep them at a minimum as defined by some cost function and independent variable(s).

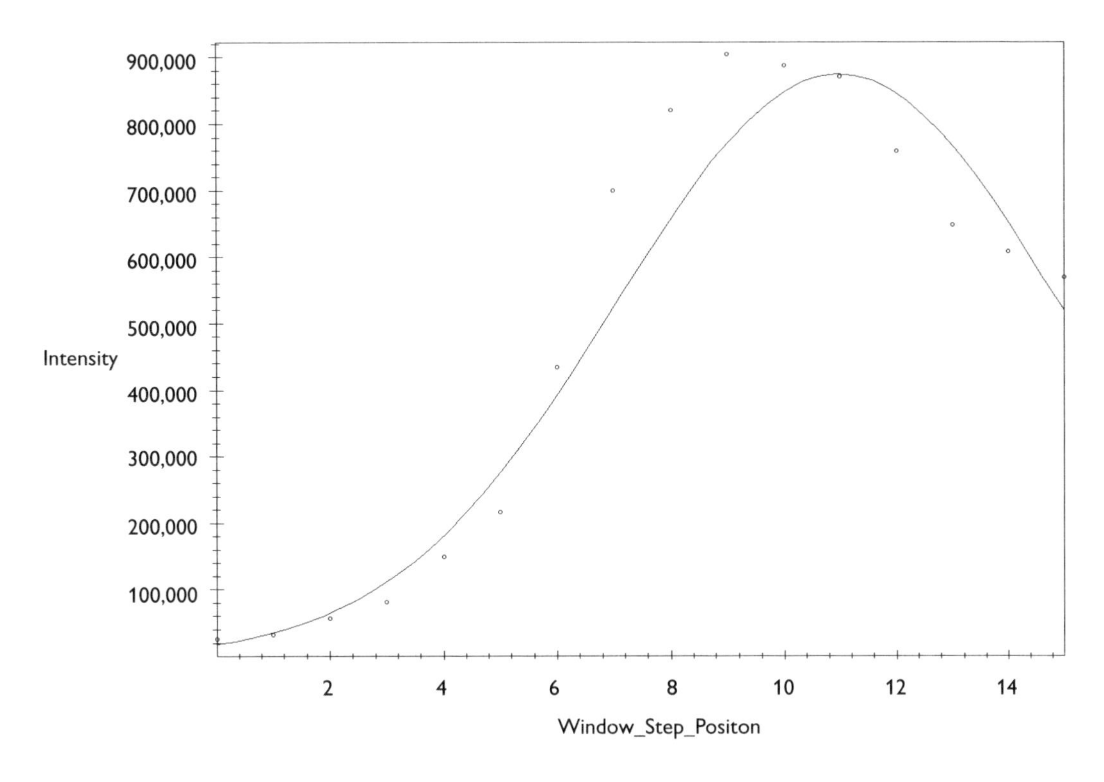

Figure 3.5 Estimated (solid line) and actual (open circles) intensity versus window step position (data shifting to improve regression curve fit).

In particular, the Levenberg-Marquardt algorithm (LMA) is one of the more commonly used nonlinear regression techniques today. The following LMA was developed by Dr. Jerome M. Lang at Waterloo Maple Software, Inc., and incorporates some initial Newtonian line searching to facilitate rapid location of the global optimization region associated with the solution. (For further reading about nonlinear optimization techniques, [1,2] are recommended.)

```
with (linalg):
with (plots):
with (stats):
Digits := 5:
LevenbergMarquardt:=proc(expr,vars,initial,epsilon)
local
d,   # dimension
delta_f_k, # change in function
```

```
delta_k,    # correction vector
delta_q_k, # change in quadratic approximation to the func-
tion
f,  # function to minimize
f_k, # function at current point
f_k_delta, # function at trial point
g,  # gradient function
g_k, # grad function at current point
G,   # hessian function
G_k, # hessian at current point
H_k, # coeff matrix to solve for correction
i,  # counter
Id, # identity matrix
loop_no,    # loop number
nu_k,       # line-search parameter
r_k, # accuracy ratio measuring how close is the quadratic
       appx to the exact function
var_subs,  # to go from user vars to ours
x,  # name of indexed variables to replace vars
x_k; # current position
options 'Copyright 1994 by Waterloo Maple Software';
d:=nops(vars);
var_subs:={seq(vars[i]=x[i], i=1..d)};
f:=unapply(subs(var_subs,expr), x);
g:=subs('_BODY'=convert
linalg[grad](f(x),[x[1],x[2],x[3]]),'list'),
'_PARMS'=x,_PARMS->array(_BODY) );
G:=subs('_BODY'=convert(linalg
 [hessian](f(x),[x[1],x[2],x[3]]),
 'listlist'),'_PARMS'=x,_PARMS-array(_BODY) );
# INITIALIZATION
x_k:=linalg[vector](evalf(initial));
nu_k:=1.0;
Id:=array(identity,1..d,1..d);
r_k:=NULL; # for printing first time through
delta_f_k:=infinity;
if not(nu_k>0) then ERROR('nu_k must be positive') fi;
userinfo(2, 'stats', 'loop_no, f_k, delta_f_k, x_k, nu_k,
r_k');
# MAIN LOOP
for loop_no do
#  step 1
 f_k:=f(x_k); g_k:=g(x_k); G_k:=G(x_k);
```

```
 userinfo(2, 'stats', loop_no, f_k, delta_f_k, con-
vert(x_k,'list'),
 nu_k, r_k);
#  step 2
 do
 H_k:=linalg[matadd](G_k, Id, 1, nu_k);
 if linalg[definite](H_k,'positive_def') then break; fi;
 nu_k:=4*nu_k;
 od;
#  step 3
 delta_k:=linalg[linsolve](H_k, linalg[scalarmul](g_k, -1));
# STOPPING CONDITION
 if linalg[dotprod](delta_k,delta_k)<epsilon^2 then break
fi;
#   step 4
 f_k_delta:=f(linalg[matadd](x_k,delta_k));
 delta_f_k:=f_k_delta-f_k;
delta_q_k:=linalg[dotprod](g_k,delta_k)+1/2*linalg[innerprod]
 ( delta_k, G_k, delta_k );
 r_k:=delta_f_k/delta_q_k;
#  step 5
 if r_k</4 then nu_k:=4*nu_k
 elif r_k3/4 then nu_k:=nu_k/2
 else # nu_k is unchanged
 fi;
#  step 6
 if r_k>0 then x_k:=linalg[matadd](x_k, delta_k);
 else  # x_k is unchanged
 fi;
od;
[seq( vars[i]=x_k[i], i = 1..d)];
end:
infolevel[stats] := 2; # This line gives the LMA's output
iteration toward the solution.
```

$$\text{infolevel}_{\text{stats}} := 2$$

Now with the LMA defined, let's recast the problem in a slightly different way. The LMA simply requires that we define the difference between the actual and estimated to be minimal, hence, the cost function to be minimized will be:

$$\texttt{Gaussian\_Fit\_Error\_Squared} = (k1e^{k2(x\ +\ k3)^2} - \text{Yvalues\_raw})^2$$

The variables to be varied are:

$$[k1, k2, k3]$$

and the initial starting point for these variables can be obtained from the previously performed regression; hence,

$$\begin{aligned}
&k1(\text{initial}) = .1054179587 \cdot 10^7\\
&k2(\text{initial}) = -.01640900845\\
&k3(\text{initial}) = 2.566021006
\end{aligned}$$

or

$$[.1054179587 \cdot 10^7, -.01640900845, 2.566021006]$$

and an arbitrary error criterion of, say 1% or

$$.01$$

But we must first get the original estimation form into a compatible form for the LMA optimization. As stated, we defined the cost function to be minimized as `Gaussian_Fit_Error_Squared`, which is equal to the difference squared between the actual and estimated values. The squaring prohibits any algebraically added values to create zero error (cost); i.e., only zero differentials can contribute zero error.

```
Yvalues_raw :=
[569322,647595,871287,904318,820139,700099,434252,216687,
 150058,81671,57118,32746,25717,17639,13063,11527]:
Xvalues := [seq(i,i=0..nops(Yvalues_raw)-1)]:
k1_initial := .1054179587*10^7:
k2_initial := -.01640900845:
k3_initial := 2.566021006:
Yvalues_est := transform[multiapply[(x)->(k1*exp(-k2*
 (x-k3)^2))]]([Xvalues]):
Gaussian_Fit_Error_Squared := transform[multiapply
```

```
 (x,y)-(x-y)^2]]([Yvalues_raw,Yvalues_est]):
Sum_Gaussian_Fit_Error_Squared := evalf
 (add(i,i=Gaussian_Fit_Error_Squared)):
LMA_Result := LevenbergMarquardt(Sum_Gaussian_Fit_Error_
 Squared,[k1,k2,k3],[k1_initial,k2_initial,k3_initial],.01);
Solutions := subs(LMA_Result,[k1,k2,k3]):
k1 := Solutions[1];
k2 := Solutions[2];
k3 := Solutions[3];
```

The output iteration associated with the LMA's convergence toward the minimal least squares error of the model with the data is

```
LevenbergMarquardt:  loop_no,  f_k,        delta_f_k,   x_k,                              nu_k,        r_k
LevenbergMarquardt:    1      .34523e15    infinity     [.10542e7, -.16409e-1, 2.5660]  1.0
LevenbergMarquardt:    2.     .12727e15   -.21796e15    [.10542e7, -.11988e-1, 2.3218]  .14075e15    1.2206
LevenbergMarquardt:    3.     .50500e14   -.76770e14    [.10542e7, -.79964e-2, 2.1712]  .70375e14    1.2449
LevenbergMarquardt:    4.     .21452e14   -.29048e14    [.10542e7, -.40671e-2, 2.0611]  .35188e14    1.2569
LevenbergMarquardt: +  5.     .98006e13   -.11651e14    [.10542e7,  .150e-4,   1.9686]  .17594e14    1.2653
LevenbergMarquardt:    6.     .47928e13   -.50078e13    [.10542e7,  .44776e-2, 1.8785]  .87970e13    1.2726
LevenbergMarquardt: +  7      .24521e13   -.23407e13    [.10542e7,  .96326e-2, 1.7784]  .43985e13    1.2774
LevenbergMarquardt: +  8.     .12708e13   -.11813e13    [.10542e7,  .15893e-1, 1.6624]  .21993e13    1.2751
LevenbergMarquardt:    9.     .68161e12   -.58919e12    [.10542e7,  .23536e-1, 1.5475]  .10997e13    1.2611
LevenbergMarquardt:   10.     .43731e12   -.24430e12    [.10542e7,  .32202e-1, 1.4993]  .54985e12    1.2504
LevenbergMarquardt:   11.     .34121e12   -.9610e11     [.10542e7,  .42014e-1, 1.6618]  .27493e12    1.2472
LevenbergMarquardt: + 12.     .13940e12   -.20181e12    [.10542e7,  .65746e-1, 2.5901]  .13747e12     .92799
LevenbergMarquardt:   13.     .80053e11   -.59347e11    [.10542e7,  .86241e-1, 2.9358]  .68735e11     .97217
LevenbergMarquardt:   14.     .77059e11   -.2994e10     [.10542e7,  .91583e-1, 2.9134]  .34368e11    1.0506
```

$$\text{LMA\_Result} := \left[k1 = .10542\ 10^7,\ k2 = .091583,\ k3 = 2.9134\right]$$

$$k1 := .10542\ 10^7$$

$$k2 := .091583$$

$$k3 := 2.9134$$

Now let's plot this LMA regressed estimation model along with the original data to see how the fit appears:

```
Estimated := k1*exp(-k2*(x-k3)^2):
Actual := zip((a,b)-[a,b],Xvalues,Yvalues_raw):
Estimated_1 :=
 plot(Estimated,x=0..nops(Yvalues_raw)-1,color=black,
 style=line):
```

```
Actual_1 := pointplot(Actual,style=point,symbol=circle,
 color=black):
display({Estimated_1,Actual_1},labels=[Window_Step_Position,
 Intensity],axes=boxed);
```

Figure 3.6 shows that the LMA did a reasonable job at modeling the data profile with the Gaussian function. The primary danger with most nonlinear regression algorithms is one of initiation. Some nonlinear regression techniques take a very long time to converge if they are initialized too far away from the solution region. Other nonlinear regression approaches can yield completely erroneous results if the initial starting points are too far from the solution space. That is why the reader should use a prefilter regression, even if it is wrong, as we did in this example. In that way, the nonlinear regression has an excellent start due to close proximity to the solution space. This will reward the user with reasonably fast convergence and good results.

General polynomial regression

Another approach simply and blindly assumes the estimation can follow a general polynomial form:

$$y_{\text{estimate}} = a + bx + cx^2 + dx^3 + ex^4 + \ldots$$

To show how easy this method is, let's use the previously difficult data set that we had to curve fit using the LMA:

```
with (plots):
with (stats):
Yvalues :=
[569322,647595,871287,904318,820139,700099,434252,216687,
 150058,81 671,57118,32746,25717,17639,13063,11527]:
Xvalues := [seq(i,i=0..nops(Yvalues)-1)]:
eq_fit := fit[leastsquare[[x,y],
 y=a+b*x+c*x^2+d*x^3+e*x^4+f*x^5+g*x^6]]
 ([Xvalues, Yvalues]):
a := evalf(coeff(rhs(eq_fit),x,0)):
b := evalf(coeff(rhs(eq_fit),x,1)):
c := evalf(coeff(rhs(eq_fit),x,2)):
```

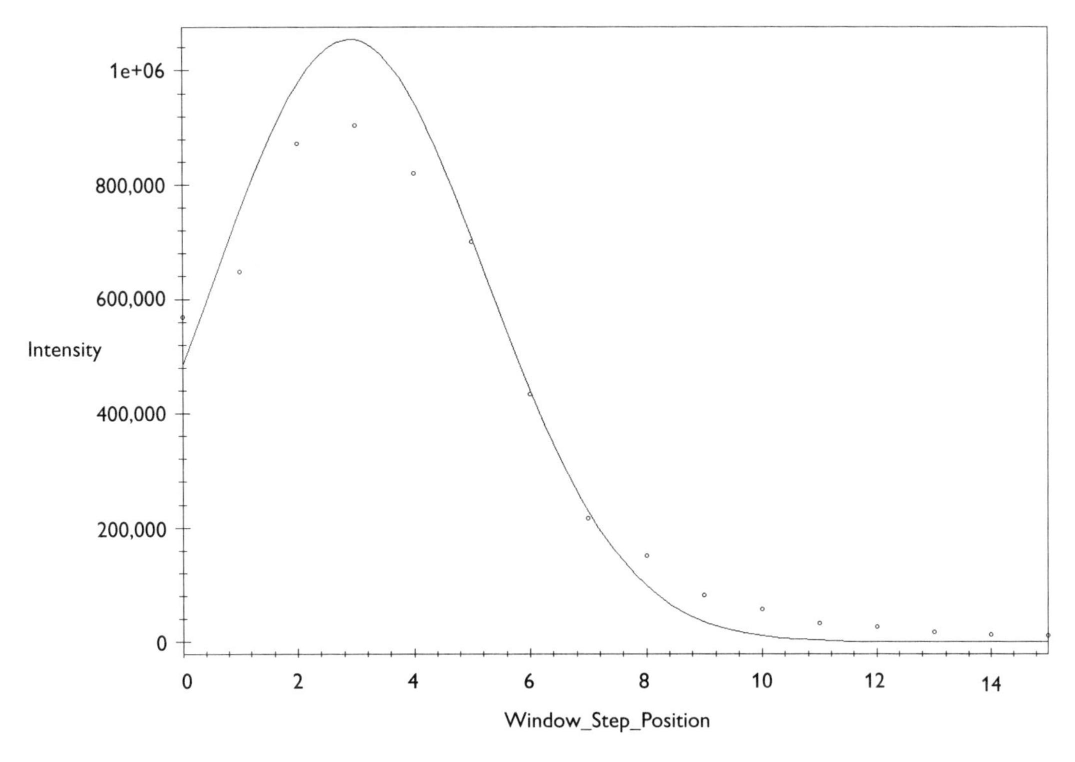

Figure 3.6 LMA estimated (solid line) and actual (open circles) intensities versus window step position (residual square error constrained to 1%).

```
d := evalf(coeff(rhs(eq_fit),x,3)):
e := evalf(coeff(rhs(eq_fit),x,4)):
f := evalf(coeff(rhs(eq_fit),x,5)):
g := evalf(coeff(rhs(eq_fit),x,6)):
```

Now that we have the regression coefficients, solve for the multiple peak values (`Y_Max_Values`) and their corresponding window step positions (`X_Max_Values`) associated with the polynomial model:

```
Y_Estimated := a+b*x+c*x^2+d*x^3+e*x^4+f*x^5+g*x^6:
Y_Estimated_Derivative := diff(Y_Estimated,x):
X_Peak_Values := [fsolve(Y_Estimated_Derivative=0,x)];
for i from 1 to nops(X_Peak_Values)
 do Y_Estimated_Peak(i) := subs(x=X_Peak_Values[(i)],
 Y_Estimated):
od:
```

```
Y_Peak_Values := [seq(Y_Estimated_Peak(i),i=1..nops
 (X_Peak_Values))];
```

X_Peak_Values := [−.1370943937 , 2.951874268 , 9.828839319 , 11.64172962 , 14.25935997]

Y_Peak_Values := [554151.9313 , 899447.9633 , 36962.397 , 55348.95 , −22785.75]

Note that some solutions to this sixth order are pure filter artifact from the regression process. The `X_Peak_Values` value equal to −.13709... and the `Y_Max_Values` value equal to −22785.75 clearly do not represent anything associated with the data set. The `X_Max_Values` artifact is *noise* because this filter is not a predictor (extrapolation), but an estimator (interpolation). The negative `Y_Max_Values` artifact is noise because there is no such thing as "negative" light (make sure during your modeling that you experience "reality checks" frequently!).

Now that we have the x-y values associated with the polynomial's extrema, let's abstract the polynomial's result for the maximal intensity (`Y_Max_Value`) and its corresponding window step position (`X_Max_Value`):

```
Zipped_Pairs :=
 zip((x,y)->[y,x],X_Peak_Values,Y_Peak_Values):
Boolean_ Condition := (a,b)->if op(1,a)<op(1,b) then true
else false fi:
Sorted_Ordered_Pairs :=
 sort(Zipped_Pairs,Boolean_Condition):
XY_Max_Coordinate := op(nops(Sorted_Ordered_Pairs),
 Sorted_Ordered_Pairs):
Y_Max_Value := op(1,XY_Max_Coordinate);
X_Max_Value := op(2,XY_Max_Coordinate);
```

Y_Max_Value := 899447.9633
X_Max_Value := 2.951874268

These x-y values graphically correspond to the data set peak fairly well as shown in Figure 3.7 when we generate a composite curve fit and data plot:

```
Data_Set := zip((a,b)->[a,b],Xvalues,Yvalues):
Data_Plot := pointplot(Data_Set,style=point,symbol=circle,
 color=black):
Estimated_Plot := plot(Y_Estimated,x=0..nops(Yvalues)-1,
 style=line,color=black):
display({Es timated_ Plot,Data_Plot},axes=boxed,labels=
 [Window_Step_Position,Intensity]);
```

In Figure 3.7, note the rippling of the curve fit due to the polynomial. Remember, a curve has $N-1$ extrema, where N is the order of the function. In this case, $N=6$, hence there are 5 extrema of which 4 are clearly visible within the confines of the window and the fifth is just starting to show at the window's entry (i.e., `Window_Step_Position =x=0`).

By plotting the residual error between the actual and estimated data, we can see where the interpolation errors are greatest. Also, when the user uses very high order polynomials, the residual error plot (Figure 3.8) will exhibit a large amount of "zigzag" behavior.

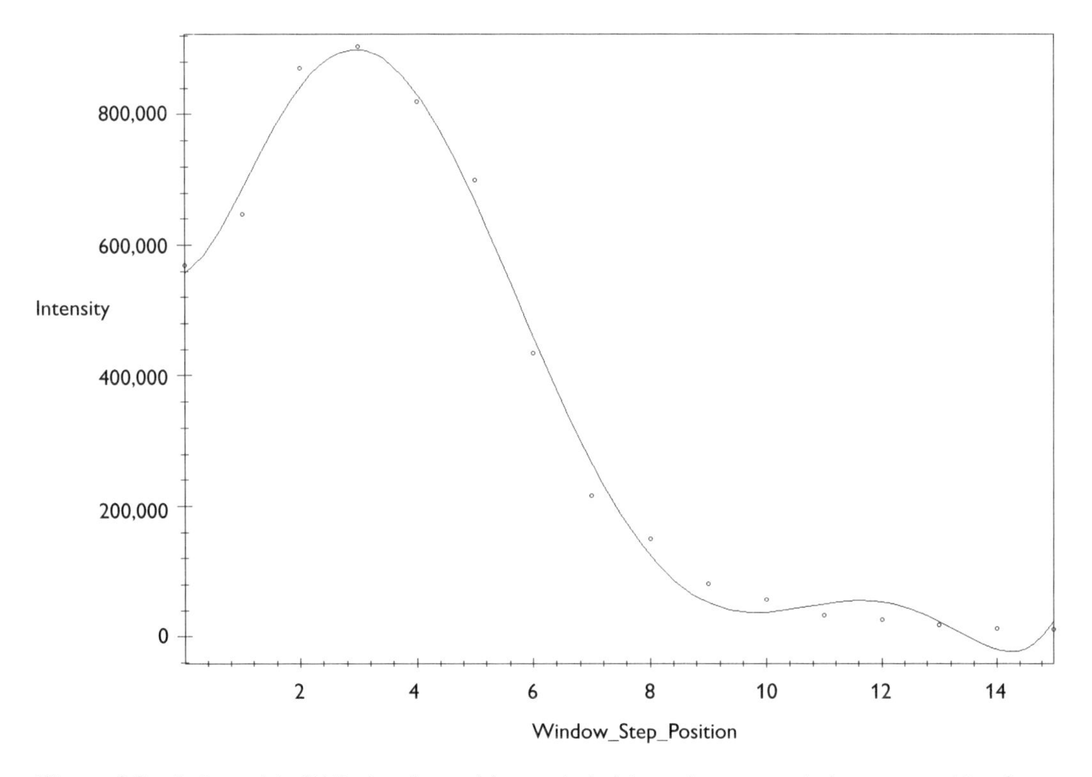

Figure 3.7 Estimated (solid line) and actual (open circles) intensity versus window step position for general polynomial fit (sixth-order fit).

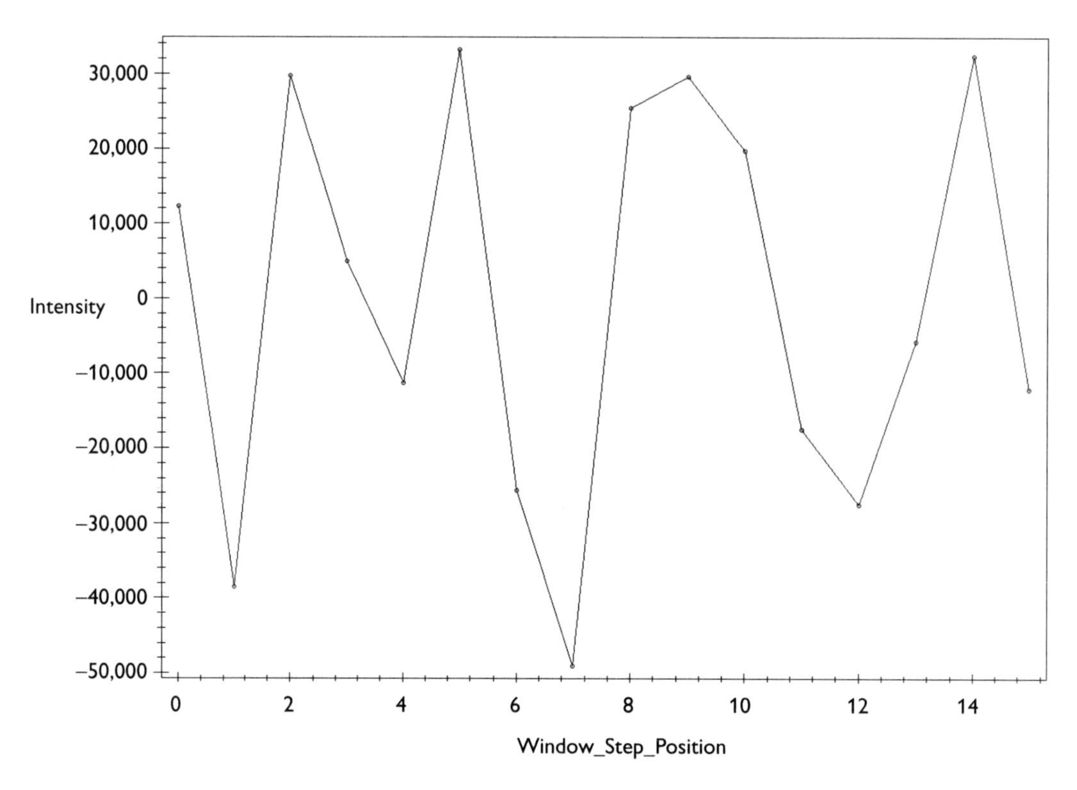

Figure 3.8 Residual error between estimated (solid line) and actual (open circles) intensity versus window step position (sixth-order fit).

```
Estimator := y=Y_Estimated:
Transform_Equation := unapply(rhs(Estimator),x):
Estimated_Data :=
 transform[apply[Transform_Equation]](Xvalues):
Residual_Error := transform[multiapply[(x,y)->x-y]]
 ([Yvalues, Estimated_Data]):
Error_Set := zip((a,b)-[a,b],Xvalues,Residual_Error):
Error_1 := pointplot(Error_Set,style=line,color=black):
Error_2 :=
 pointplot(Error_Set,style=point,symbol=circle,color=black):
display({Error_1,Error_2},axes=boxed,labels=
 [Window_Step_Position,Intensity]);
```

Now we plot residual error versus window step position as a relative percentage error at each step:

```
Percent_Residual_Error := transform[multiapply[(x,y)
-> ((x-y)/x)*100]]([Yvalues,Estimated_Data]):
Percent_Error_Set := zip((a,b)-[a,b],Xvalues,
 Percent_Residual_Error):
Percent_Error_1 := pointplot(Percent_Error_Set,style=line):
Percent_Error_2 :=
 pointplot(Percent_Error_Set,style=point,symbol=circle,
 color=black):
display({ Percent_Error_1,Percent_Error_2},labels=
 [Window_Step_Position,Percent_Er ror],axes=boxed);
```

Figure 3.9 shows a relatively large percent error at the end of the sampling window. As stated earlier, this effect is typical of simple linear regression type filters since there is no extra data beyond the window to average into the final estimation. This causes a sort of burst or accumulation of error to appear at the end of any simple polynomial regression fit. Compare this plot to Figure 3.3 where a second-order fit was accomplished. Notice that the relative percent error was smaller across the window. Clearly the jagged error effect becomes worse with higher order polynomials and data that vary (whether actually or from measurement error) greatly from the previous few data points. Figure 3.3 showed a large jagged relative error at the beginning of the sampling window as opposed to Figure 3.9. This effect happens because there are no "trend" data to start the regression, hence a rather large initial guess can be very wrong at the beginning of the sampling window. This, in effect, is similar to the transient behavior associated with circuits when they are exposed to any sudden change of the input.

The single biggest problem associated with the general polynomial fit is the model's ability to become noisy by responding to nearly every data point. What happens, simply, is that the polynomial regression tries to get every point it can within the confines of the polynomial's order. For instance, if we had a tenth-order polynomial, the estimator curve could have a maximum of nine $(10 - 1 = 9)$ extrema. If we had 9 or fewer data points, this regression would nearly hit every point, thereby neutralizing any smoothing or averaging aspect of the estimator. In short, the estimator filter has too much bandwidth.

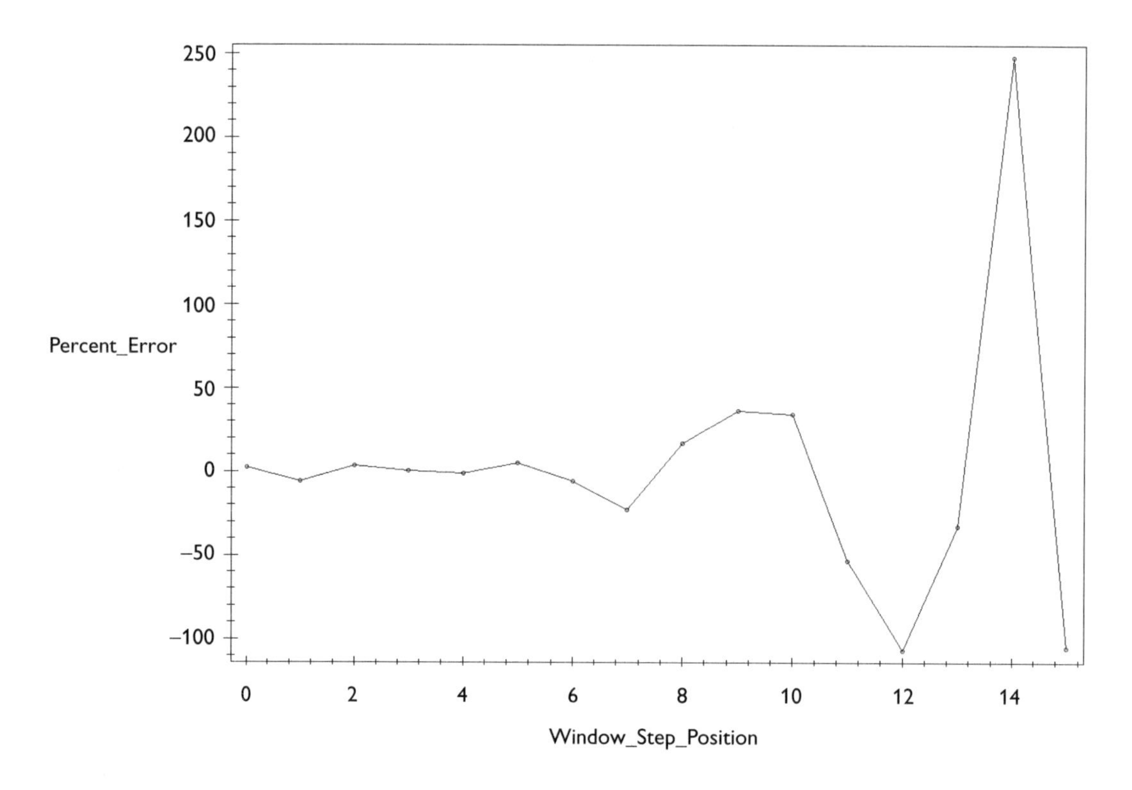

Figure 3.9 Percent relative residual error versus window step position (sixth-order fit).

High-order polynomial regression fit problems

To the pitfall just mentioned that is associated with general polynomial curve fitting, let's take the last set of data and use an artificially high polynomial order, say, 12. Further, let's allow one of the data points to be somewhat of an outlier to show what happens when the bandwidth of the estimator is too high for useful purposes. We will choose a peak data point (Yvalues = 904318) and cut it in half (Yvalues = 452159) to represent a single data point that either exhibited a noise or measurement corruption:

```
with (plots):
with (stats):
Yvalues :=
[569322,647595,871287,452159,820139,700099,434252,216687,
 150058,81671,57118,32746,25717,17639,13063,11527]:
Xvalues := [seq(i,i=0..nops(Yvalues)-1)]:
```

```
eq_fit := fit[least-
square[[x,y],y=a+b*x+c*x^2+d*x^3+e*x^4+f*x^5+g*x^6+h*x^7
+ i*x^8+j*x^9+k*x^10+l*x^11+m*x^12]]([Xvalues,Yvalues]):
a := evalf(coeff(rhs(eq_fit),x,0)):
b := evalf(coeff(rhs(eq_fit),x,1)):
c := evalf(coeff(rhs(eq_fit),x,2)):
d := evalf(coeff(rhs(eq_fit),x,3)):
e := evalf(coeff(rhs(eq_fit),x,4)):
f := evalf(coeff(rhs(eq_fit),x,5)):
g := evalf(coeff(rhs(eq_fit),x,6)):
h := evalf(coeff(rhs(eq_fit),x,7)):
i := evalf(coeff(rhs(eq_fit),x,8)):
j := evalf(coeff(rhs(eq_fit),x,9)):
k := evalf(coeff(rhs(eq_fit),x,10)):
l := evalf(coeff(rhs(eq_fit),x,11)):
m := evalf(coeff(rhs(eq_fit),x,12)):
```

Now that we have the regression coefficients, solve for the multiple peak values (`Y_Max_Values`) and their corresponding window step positions (`X_Max_Values`) associated with the polynomial model:

```
Y_Estimated :=
 a+b*x+c*x^2+d*x^3+e*x^4+f*x^5+g*x^6+h*x^7+i*x^8+j*x^9
 + k*x^10+l*x^11+m*x^12:
Y_Estimated_Derivative := diff(Y_Estimated,x):
X_Peak_Values := [fsolve(Y_Estimated_Derivative=0,x)];
for i from 1 to nops(X_Peak_Values)
 do Y_Estimated_Peak(i) := subs(x=X_Peak_Values[(i)],
 Y_Estimated):
od:
Y_Peak_Values := [seq(Y_Estimated_Peak(i),
i=1..nops(X_Peak_Values))];
```

X_Peak_Values := [−11.70881911 , .3224662904 , 1.578737402 , 3.091536358 , 4.687042269 , 7.743771431 , 8.814838531 , 10.86172372 , 12.26584677 , 13.55914242 , 14.68280892]

Y_Peak_Values := [.2961197602 10^{12}, 107056.6765 , 899412.4643 , 560382.9827 , 773676.2481 , 112923.53 , 143372.0 , −14721. , 82580 , −87660. , 434000]

Again note that some solutions to this twelfth-order problem show strong filter artifact from the regression process. The X_Peak_Values value equal to −.11.70881911 and two Y_Max_Values values (−14721 and −87660) clearly do not represent anything associated with the observed data set. The first Y_Max_Values value is quite high ($.2961197602 \times 10^{12}$), which is a regression artifact when compared to the observed data. Consequently, the reader is again warned against accepting any regression result blindly without verifying the veracity of the output against the real data.

Now that we have the x-y values associated with the polynomial's peaks, let's abstract the polynomial's result for the maximal intensity (Y_Max_Value) and its corresponding window step position (X_Max_Value). These x-y values graphically correspond to the data set peak fairly well as shown when we generate a composite curve fit and data plot (Figure 3.10):

```
Data_Set := zip((a,b)-[a,b],Xvalues,Yvalues):
Data_Plot := pointplot(Data_Set,style=point,symbol= circle,
 color=black):
Estimated_Plot := plot(Y_Estimated,x=0..nops(Yvalues)-1,
 style=line,color=black):
display({Estimated_Plot,Data_Plot},axes=boxed,labels=
 [Window_Step_Position,Intensity]);
```

Note the severe rippling of the curve fit in Figure 3.10 due to the high-order polynomial. Remembering this curve has 12–1 or 11 extrema, 10 are clearly seen in Figure 3.9. Obviously, the eleventh extremum is outside the 16-step window, but we do not know on which side it exists.

Again, plotting the residual error between the actual and estimated data, we can see where the interpolation errors are greatest (Figure 3.11).

Also, when the user uses very high order polynomials, the residual error plot will exhibit a large amount of zigzag behavior.

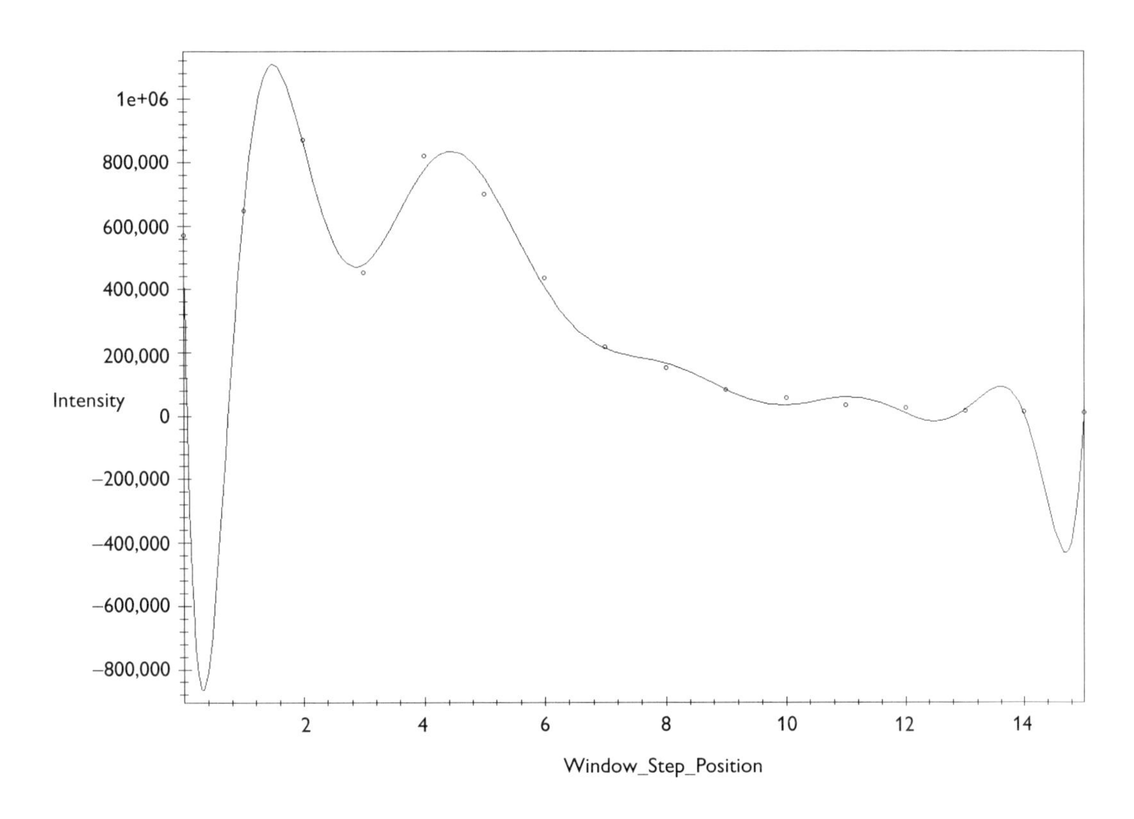

Figure 3.10 Estimated (solid line) and actual (open circles) intensity versus window step position for general polynomial fit (twelfth-order fit).

```
Estimator := y=Y_Estimated:
Transform_Equation := unapply(rhs(Estimator),x):
Estimated_Data :=
 transform[apply[Transform_Equation]](Xvalues):
Residual_Error := transform[multiapply[(x,y)-x-y]]([Yvalues,
 Estimated_Data]):
Error_Set := zip((a,b)-[a,b],Xvalues,Residual_Error):
Error_1 := pointplot(Error_Set,style=line,color=black):
Error_2 :=
 pointplot(Error_Set,style=point,symbol=circle,color=black):
```

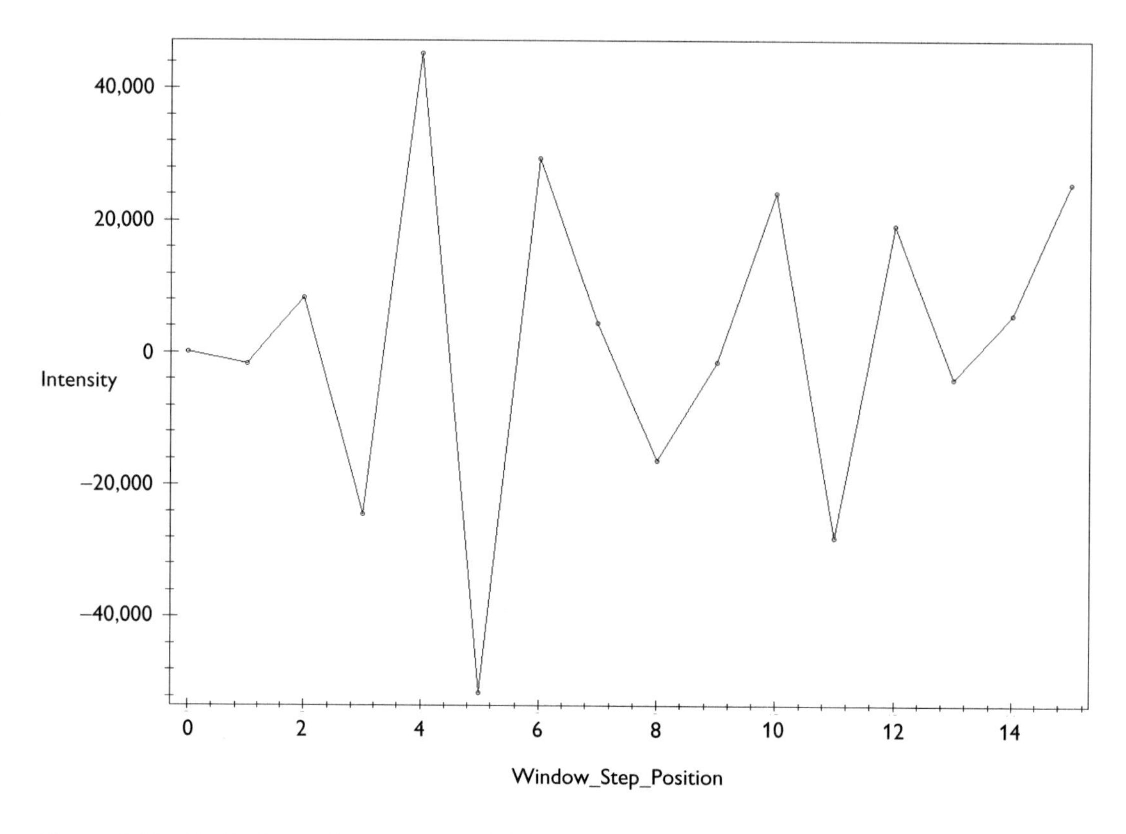

Figure 3.11 Residual error between estimated (solid line) and actual (open circles) intensity versus window step position (twelfth-order fit).

```
display({Error_1,Error_2},labels=[Window_Step_Position,
Intensity], axes=boxed);
```

Again, plotting the residual error versus window step position as a relative percentage error at each step (Figure 3.12):

```
Percent_Residual_Error := transform[multiapply[(x,y)-
 ((x-y)/x)*100]]([Yvalues,Estimated_Data]):
Percent_Error_Set :=
 zip((a,b)-[a,b],Xvalues,Percent_Residual_Error):
Percent_Error_1 := pointplot(Percent_Error_Set,style=line):
Percent_Error_2 :=
 pointplot(Percent_Error_Set,style=point,symbol=circle,
color=black):
```

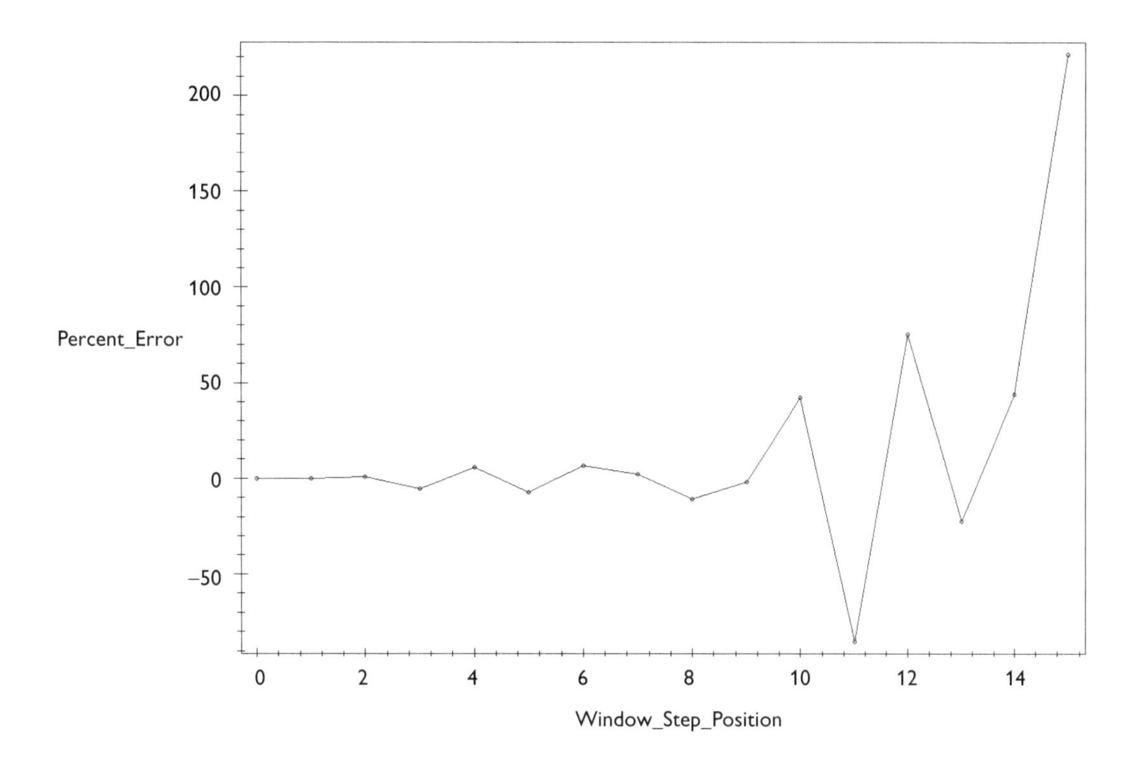

Figure 3.12 Percent relative residual error versus window step position (twelfth-order fit).

```
display( {Percent_Error_1,Percent_Error_2},labels=
[Window_Step_Position,Percent_Error],axes=boxed);
```

Again, note in Figure 3.12 the large jagged percent error toward the end of the window. This indicates that for any window step position an error between estimated and observed data will create a large "reaction" to the high-order polynomial. Again, because this high-order polynomial "filter" has the bandwidth to track the data, it can jump quickly from measure to measure trying to compensate rather than smooth the result.

Quick moral about curve fitting

Finally, the moral of the curve-fitting story is that the reader should never arbitrarily use a higher order estimator than is necessary to smooth the data. Further, before using any regression, linear or otherwise, try to figure what the data are doing in a subjective way before you get into all the mechanics of producing a large Maple session. Sometimes, and it is an art, a little human insight in the beginning is worth hundreds of hours and perhaps tens of thousands of dollars of effort, after the fact.

Conclusion

Maple has allowed us to see the effects of using several different regressors or estimators with a given set of data. What becomes clear is that whatever the user has for data determines which type of curve fitting to use (i.e., simple polynomial, nonlinear, or high-order polynomial). No software package can take the place of human intuition and past experience with data structures when it comes time to use the correct curve-fitting algorithm. However, when one has determined which type or even types have efficacy, Maple can easily get the user up and running both numerically and graphically to produce the required insights and/or algorithm implementation. Contrast what we have done in this chapter with Maple as opposed to more conventional numerical packages on the market. Most C, C++, and nonsymbolic statistical packages can give you curve fitting, but only a symbolic package can give a universal handle on the algorithmic engine in symbolic form that is understandable to the user. This fact is critical to the human comprehension of any underlying process, because, as humans, we are quicker to see a trend via symbolics rather than abstracting information associated with an analysis via a bunch of numbers. Maple allows the user to deal in concepts much more than any standard numerical only analysis package.

References

[1] Wadsworth, Harrison M. (ed.), *Handbook of Statistical Methods for Engineers and Scientists*, New York: McGraw-Hill, 1990.

[2] Fletcher, R., *Practical Methods of Optimization*, 2nd Ed., New York: John Wiley & Sons, 1987.

[3] Draper, N. R., and H. Smith, *Applied Regression Analysis*, New York: John Wiley & Sons, 1981.

Chapter 4

Mathematical models: working with differential equations

An important part of the design process is the modeling and simulation stage using a mathematical prototype. Once the basic framework of a prototype is in place, it is important to be able to interrogate its state at any time during the simulation process, modify the model, and rerun it as necessary. The Maple system enables us to develop complex mathematical models, run them, analyze their output, modify them, and then rerun them easily.

In this chapter we take a brief look at some of the functions that Maple provides for investigating differential equations. We analyze the time response of a tachometer needle using the series methods and develop and investigate a shock absorber model. Maple's ability to deal with both linear and nonlinear systems is addressed. The techniques used in this section are applicable to any dynamic system modeled using differential equations, not just to the control systems that are discussed in the examples.

ODE tools: a tour

The most commonly used tool is the `dsolve` function. In addition, the `DEtools`, `Difforms`, and `Liesymm` packages contain functions that can be used to manipulate both ordinary differential equations (ODEs) and partial differential equations (PDEs). The Liesymm package is not considered further because it is beyond the scope of this book.

The dsolve function

Maple's standard tool for solving ODEs either symbolically or numerically, is the `dsolve` function. In the first example, we find the time solution to a second-order ODE with a triangular wave-forcing function. In this particular example, the triangle wave is approximated with a six-term *Fourier* series:

```
SYS1:=diff(x(t), t, t)+2*diff(x(t), t)+50*x(t) =
sum(100*cos(t*(2*n+1))/(2*n+1)^2, n=0..5);
```

$$\mathrm{SYS1} := \left(\frac{\partial^2}{\partial t^2} x(t)\right) + 2\left(\frac{\partial}{\partial t} x(t)\right) + 50\ x(t) = 100\ \cos(t) + \frac{100}{9}\cos(3t) + 4\ \cos(5t) + \frac{100}{49}\cos(7t) + \frac{100}{81}\cos(9t)$$

When the system is initially at rest, the initial conditions are as follows:

```
ICS1:=D(x)(0)=0, x(0)=0;
```

$$\mathrm{ICS1} := \mathrm{D}(x)(0) = 0, x(0) = 0$$

The `D` or differential operator can be used to compute the derivative of a function. The call `D(f)(x)` is equivalent the call `diff(f(x), x)`:

```
D(x)(t)=diff(x(t), t);
```

$$D(x)(t) = \frac{\partial}{\partial t} x(t)$$

The preceding relationship becomes obvious if we let x equal the tangent and evaluate it:

```
> eval(subs(x=tan, "));
```

$$1 + \tan(t)^2 = 1 + \tan(t)^2$$

Returning to the definition of the initial conditions, we can see that the expression `D(x)(0)` is equivalent to the value of $\frac{d}{dt}x(t)\Big|_{t=0}$.

The time response $x(t)$ is obtained and plotted using `dsolve`, `rhs`, and `plot` as shown. As the function dsolve returns an equation of the form $x(t)$=..., the right-hand side is isolated using `rhs` so that it can be graphed (see Figure 4.1):

```
> plot(rhs(dsolve({SYS1, ICS1}, x(t))), t=0..10,
  labels=['t', 'x(t)']);
```

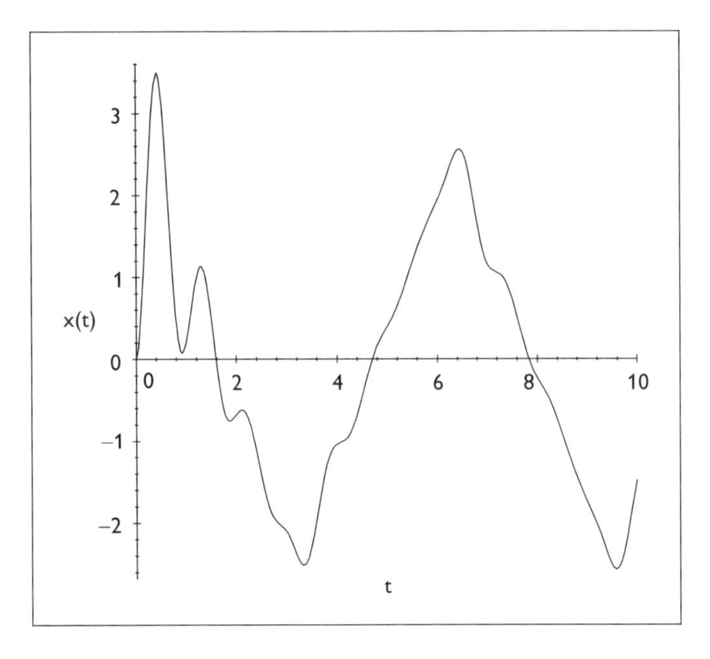

Figure 4.1

Numeric

When a closed-form analytic solution cannot be found, then a numerical solution can be attempted, as the next example shows:

```
> SYS2:=diff(x(t), t)+(2+cos(t^2))*x(t)=2*sin(0.1*t^2),
  x(0)=0;
```

$$SYS2 := \left(\frac{\partial}{\partial t}x(t)\right) + (2 + \cos(t^2))\, x(t) = 2\,\sin(.1t^2),\ x(0) = 0$$

```
> dsolve({SYS2}, x(t));
```

$$x(t) = 2.\int_0^t \sin(.1000000000\, u^2)\, du \int_0^t e^{(2.u)}\, du$$
$$\int_0^t e^{(1.253314137\ \mathrm{FresnelC}(.7978845605\ u))}\, du$$
$$e^{(-2.t\ -\ 1.253314137\ \mathrm{FresnelC}(.7978845605\ t))}$$

By specifying the method as numeric, we turn Maple's ODE solver from a symbolic one into a numerical one:

```
> SOLS:=dsolve({SYS2}, x(t), type=numeric);
  SOLS := proc(rkf45_x) ... end
```

Maple returns a called procedure such that the value of the function can be computed for any specified t. The formal parameter name, in this case `rkf45_x`, indicates which algorithm has been used.

```
> SOLS(10);
```

$$[t = 10,\ x(t) = .1582530550576577]$$

Using this procedure we are able to plot the time response $x(t)$ (see Figure 4.2):

```
> pts:=[seq(subs(SOLS(T/10), [T/10, x(t)]), T=0..100)]:

> plot(pts, labels=['t','x(t)']);
```

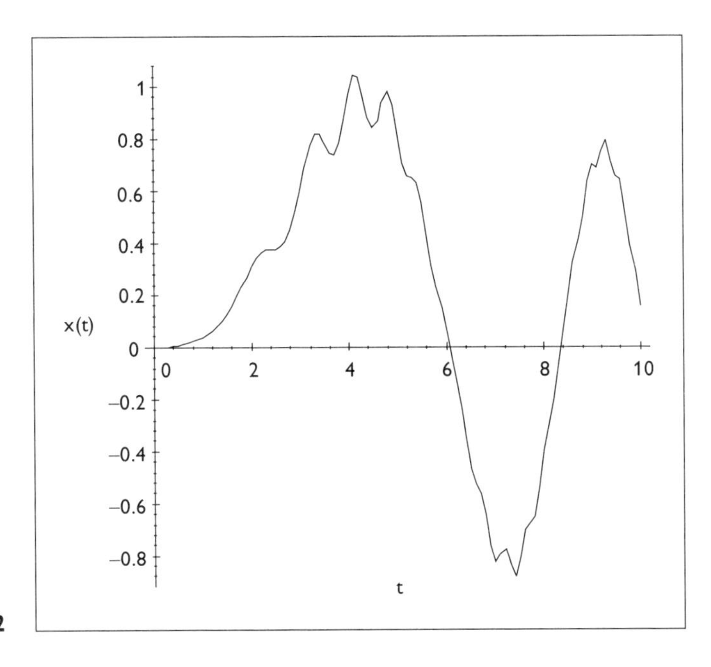

Figure 4.2

The default numerical solver is the Fehlberg fourth-fifth order Runga-Kutta method (see `?dsolve[numeric]`). This algorithm requires four evaluations of the function in order to return an estimate of the dependent variable, which is equivalent to using a Taylor expansion as far as the term in $(x - x_0)^4$. Although the Fehlberg fourth-fifth order Runga-Kutta method is applicable in most cases this is not always the case. For this reason, Maple supports a set of alternative numerical solvers, as shown in Table 4.1. See `?dsolve[algorithm name]` for more information.

The DEtools package

The `DEtools` package continues a suite of tools for investigating PDEs and ODEs graphically.

As before we model a second-order dynamic system with a second-order ODE, which we try to solve analytically using `dsolve`. The initial conditions are y´(0) = 0 and y(0) = 1.

Algorithm name	Method	Comments
dverk78	Seventh-eight order continous Runga-Kutta.	Improved accuracy over fourth-fifth method but with an increased time of computation.
classical	Forward Euler method. Heun formula also known as the trapezoidal or improved Euler method. Improved polygon method or modified Euler method. Adams-Bashford and Adams-Bashford-Moulton methods.	A collection of classical methods using a fixed step size between mesh points. None of the methods employs error correction or estimation to improve accuracy. The default method is Forward Euler. The Adams-Bashford method is a predictor method, whereas the Adams-Bashford-Moulton method is a predictor-corrector.
gear	Gear method.	Uses a variable-size single-step extrapolation method. The two methods are a Burlirsch-Stoer rational extrapolation method (default) and a polynomial extrapolation method.
mgear	Multistep Gear method.	Uses a variable-size multistep algorithm that is applicable to both stiff and nonstiff systems. The step size is determined by evaluating the Jacobian matrix of the system at every step. The Jacobian can be computed either symbolically or by using numerical differencing of the derivatives (default). The Adams predictor-corrector method can also be used to determine the step size.
lsode	The Livermore Stiff ODE solver.	Uses a variable size multistep algorithm to solve stiff systems of ODEs.
taylorseries	Taylor series method.	Used for high-accuracy solutions. Because this method is computationally intensive, computation times can be high.

Table 4.1 Alternative numerical solvers

```
SYS3 := diff(y(t), t$2) + cos(t)*sin(t)*diff(y(t),t) +
  exp(t/10)*y(t) = cos(t)^2;
```

$$SYS3 := \left(\frac{\partial^2}{\partial t^2}\, y(t)\right) + \cos(t)\,\sin(t)\left(\frac{\partial}{\partial t}\, y(t)\right) + e^{(1/10\, t)}\, y(t) = \cos(t)^2$$

```
dsolve({SYS3, D(y)(0)=0, y(0)=1}, y(t));
```

Maple returns null indicating that a solution could not be found. Instead of obtaining the numerical solution to the system and plotting it as we did earlier, we use the general ODE plotter `DEplot` found in the `DEtools` package.

A package in Maple is a collection of functions with a common theme. The functions in a particular package can be accessed in one of two ways: Use their long names (`package_name[function_name](arg1, arg2, .., argn)`) or define their short names in Maple's name space using with and the short function name. Using `with` we load the functions in the `DEtools` package:

```
with(DEtools);
```

[Denormal, DEPlot, DEPlot3d, Dchangevar, PDEchangecoords, PDEplot, autonomous, convertAlg, convertsys, dfieldplot, indicialeq, phaseportrait, reduceOrder, regularsp, translate, untranslate, varparam]

The list that is returned lists the functions that have been loaded from the specified package. These functions can now be called in exactly the same way (using the short name syntax) as the standard Maple functions.

Using `DEplot` we can solve our ODE and plot the result directly. The function's syntax is as follows: an ODE or system of ODEs to be solved, the axes (in this case the abscissa is the independent variable t and the ordinate is the response $y(t)$), the range of the independent variable, and the initial conditions followed by any optional arguments. In this example, we have set the step for the independent variable to be 0.1 (see Figure 4.3):

```
DEplot(SYS3, [t, y], 0..20, {[0, 1, 0]}, stepsize=0.1);
```

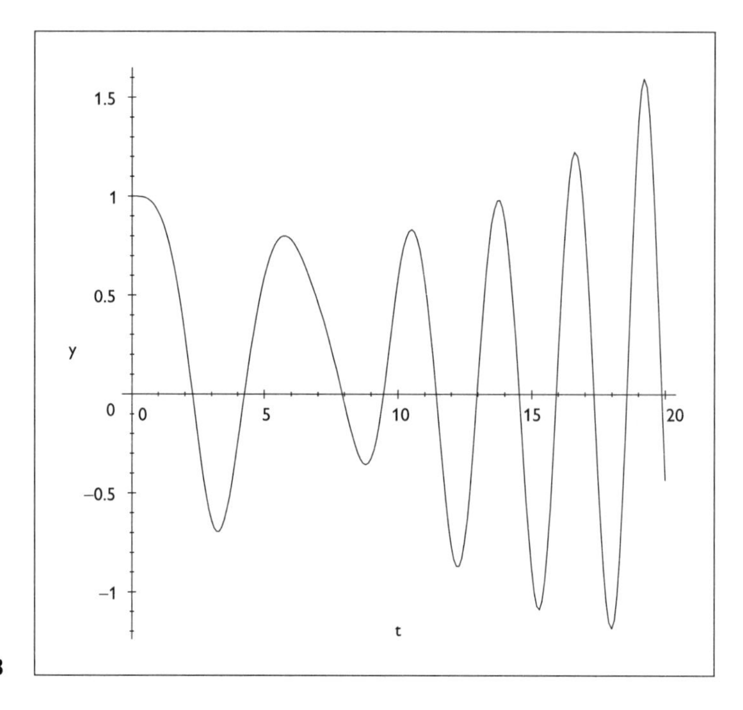

Figure 4.3

We need to exercise some care when using `DEplot` because initial conditions can be specified in one of two ways, either $[t, y(t), y'(t), \ldots]$ or $[y(t) = ic, y'(t) = ic_1, \ldots]$. So, for example, the following are both identical and valid: `[2, -1, 1], [y(2) = -1, D(y)(2) = 1]`.

Phase-Plane Techniques

Systems containing nonlinearities can prove difficult to analyze using conventional methods, and in some cases phase-plane techniques can provide much needed insight into the nonlinear system's behavior. The phase plane provides a bird's-eye view of the possible solutions to the ODE for a given set of initial conditions by showing us the possible trajectories of the system.

In conjunction with the following nonlinear second-order ODE, we use the Maple functions `dfieldplot` and `phaseportrait` to plot a direction field and a trajectory:

```
SYS4:=diff(y(t),t,t)+diff(y(t),t)/(1-y(t))+y(t)=0;
```

$$\mathrm{SYS4} := \left(\frac{\partial^2}{\partial t^2}\, y(t)\right) + \frac{\frac{\partial}{\partial t}\, y(t)}{1 - y(t)} + y(t) = 0$$

The first task is to convert our second-order ODE into two coupled first order ODEs so that we can use both `dfieldplot` and `phaseportrait`. We do this by defining two state variables, $x_1(t)$ and $x_2(t)$:

```
STATE1:=y(t)=x[1](t);
```

$$\text{STATE 1} := y(t) = x_1(t)$$

```
STATE2:=diff(y(t),t)=x[2](t);
```

$$\text{STATE 2} := \frac{\partial}{\partial t} y(t) = x_2(t)$$

By using these variables we can transform the original ODE into two coupled ODEs as shown:

```
'2ND_DE' := subs(STATE2, STATE1, SYS4);
```

$$\text{2ND\_DE} := \left(\frac{\partial}{\partial t} x_2(t)\right) + \frac{x_2(t)}{1 - x_1(t)} + x_1(t) = 0$$

We are now ready to plot the direction field $x_1(t)$ versus $x_2(t)$.

```
dfieldplot({'2ND_DE', '1ST_DE'}, [x[1](t), x[2](t)],
t=10..30, x[1]=-10..10, x[2]=-10..10, color=BLACK);
```

Two things are immediately deducible from the direction field of Figure 4.4: The circular nature of the arrows indicates that limit cycles are possible and that a singularity at $x_1(t)$ equals one.

The next step is to compute the phase portrait of the system with initial conditions of $x_1(0) = 4$ and $x_2(0) = 1$ using `phaseportrait` (see Figure 4.5):

```
phaseportrait(['2ND_DE', '1ST_DE'], [x[1](t), x[2](t)],
t=0..30, {[0,4,1]}, stepsize=0.2, thickness=2, color=BLACK);
```

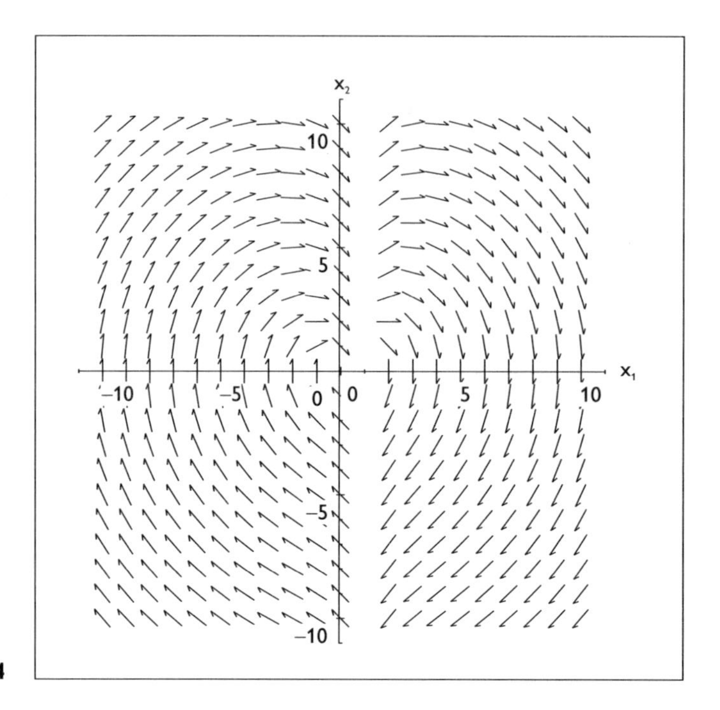

Figure 4.4

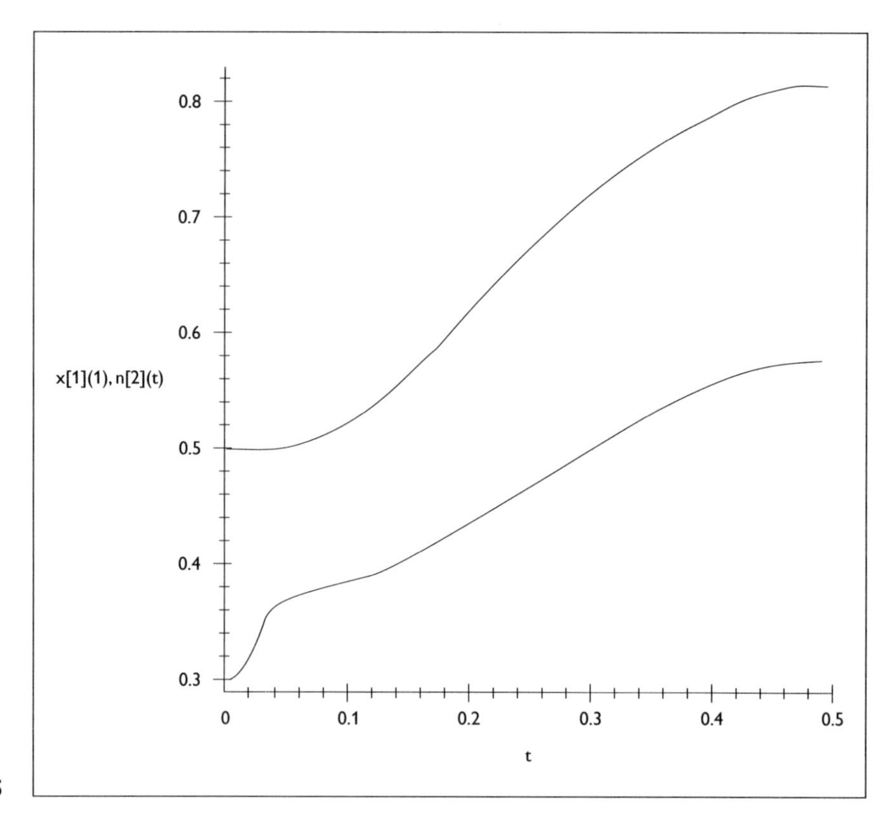

Figure 4.5

Using display, found in the plots package, we can place a trajectory on this direction field (see Figure 4.6).

```
plots[display]({",""}, view=[-6..6, -8..6]);
```

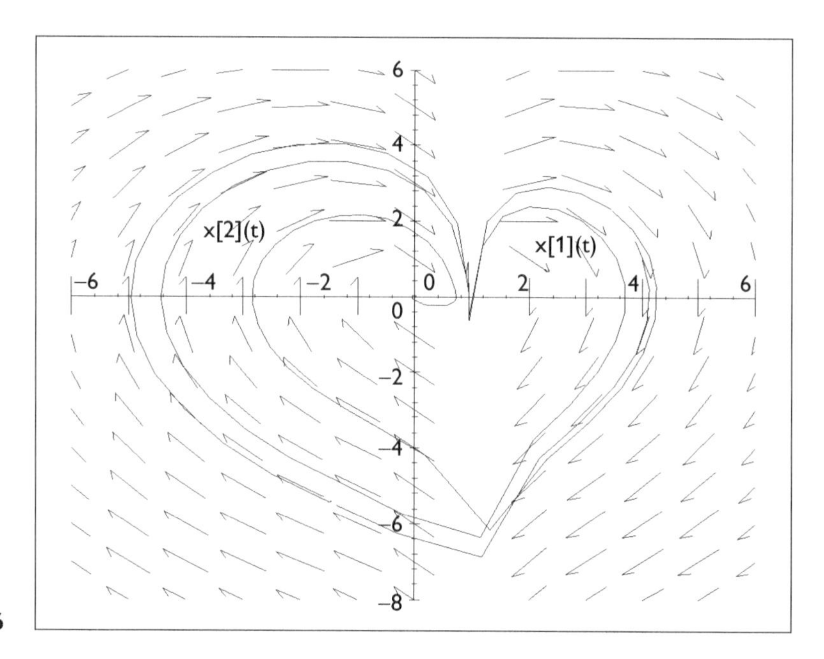

Figure 4.6

We can see with reference to Figure 4.6 that the system, although it does not enter a limit cycle, is stable if very oscillatory. It is also possible to see the effect of the singularity on the system's response.

The difforms package

The difforms package contains a set of functions that enables us to define the variables in a computation, operate on them, and interrogate the resulting functions. This package has its origins more in the mathematician's world than the engineer's in terms of the terminology used. The three functions of most interest to us in this package are d, defform, and &^, which computes the exterior derivative of an expression, defines the basic variables used in a computation, and represents the wedge product, respectively. Using the difforms package a formal framework can be quickly formed that enables us to describe complex dynamic systems using differential equations and the elements that make up differential equations and to then manipulate them. The following code loads three functions from the difforms package:

```
with(difforms, [ defform, d, &^]);
```

$$[\&\hat{}, \text{d}, \text{defform}]$$

First we declare our system of variables:

```
EXPR := defform(A=function, B=scalar, C=scalar, f=scalar):
```

This takes the external derivative

```
d(f*A+&^(B,C));
```

$$(d(f)\ \&\hat{}A) + fd(A) + Cd(B) + Bd(C)$$

Here we take the derivative of a simpler function:

```
d(A(B,C));
```

$$\left(\frac{\partial}{\partial B} A(B, C)\right) d(B) + \left(\frac{\partial}{\partial C} A(B, C)\right) d(C)$$

Here we define known external derivatives:

```
defform(d(B) = X, d(C)=0);
```

Here we view the simplified expression:

```
EXPR;
```

$$\left(\frac{\partial}{\partial B} A(B, C)\right) X$$

Series methods

Although power series methods for solving ODEs are well understood and in many cases are perfectly adequate for finding accurate approximations, they do exhibit some drawbacks: The resultant series must converge and they can be computationally problematic in that they can require the evaluation of high-order derivatives. Maple uses the Frobenius series method, which assumes that a solution of the following form is possible: $y = x^c(a_0 + a_1x + a_2x^2 + a_3x^3 + \ldots + a_rx^r + \ldots)$; see `?dsolve [series]` for more information. In this example, however, we will use the Taylor series, for the sake of clarity, to demonstrate how Maple can be used to manipulate series in order to calculate the time response of a dynamic system described by a differential equation.

This method requires that a Taylor's series can be formed for a function $y = f(t)$ at $t = 0$ by manipulating the higher order derivatives, $f'(t), f''(t), f'''(t)$, ... The general form of the Taylor series can be seen by simply computing the Taylor series for an arbitrary function of t.

```
taylor(f(t),t);
```

$$f(0) + D(f)(0)t + \frac{1}{2}D^{(2)}(f)(0)t^2 + \frac{1}{6}D^{(3)}(f)(0)t^3 + \frac{1}{24}D^{(4)}(f)(0)t^4 + \frac{1}{120}D^{(5)}(f)(0)t^5 + O(t^6)$$

The dynamics of a tachometer needle can be described with the following ODE: $d\theta/dt = 1 + \cos(t/10) - 0.05\theta$ where t is time and θ is the angular displacement of the needle. We solve this ODE using the series methods as follows: Obtain the higher derivatives (the more derivatives, the better the accuracy), reduce them to functions of the first-order derivative, evaluate the derivatives at the point $t = 0$, generate the Taylor's series and substitute the derivatives into the series, and, finally, substitute the initial condition and convert to a polynomial.

With reference to the Taylor series generated earlier we can see that it is fifth order—because the default number of terms calculated by Maple is six, no order number was specified. In the following treatment we only need to calculate the first five derivatives. Using the `D` and `@@` operators we can map the preceding equation into Maple, but before we do we take a quick look at the `@@` operator. The `@@` operator is the infix form of the repeated composition operator, which is used to apply functions and operators repeatedly. So applying the function f four times to the argument arg we get the following:

```
(f@@4)(arg);
```

$$f^{(4)}(\text{arg})$$

By expanding the previous result we can see that the function *f* is first applied to the argument and then the function *f* is applied to the first result followed by a further application of *f* to that result and so on.

Now we can define the ODE describing the systems dynamics in Maple as follows:

```
SYS5 := (D@@(n))(theta)(t) = (D@@(n-1))(1+cos(t/10)-
0.05*theta(t));
```

$$SYS5 := D^{(n)}(\theta)(t) = D^{(n-1)}\left(1 + \cos\left(\frac{1}{10}t\right) - .05\ \theta(t)\right)$$

At first glance this does not resemble the original ODE given earlier. However, a closer look reveals that with n set to one and thinking of `D`(θ)(t) as shorthand for $d\theta/dt$, `SYS5` and the above ODE are the same.

```
subs( n = 1, SYS5 );
```

$$D^{(1)}(\theta)(t) = D^{(0)}\left(1 + \cos\left(\frac{1}{10}t\right) - .05\ \theta(t)\right)$$

Using this general expression we can generate the higher order derivatives with `seq`. However, we do need a trick to be able to do this. We cannot apply the `D` operator, using the repeated composition operator `@@`, to the expression $1+\cos(t/10) - 0.05\theta(t)$. The trick is to represent this as a function of t.

```
TEMP:= t -1 + cos( t/10 ) - 0.05*theta(t);
```

$$TEMP := t \to 1 + \cos\left(\frac{1}{10}t\right) - .05\ \theta(t)$$

Now we are able to generate the higher order derivatives:

```
TERMS := [seq((D@@(n))(theta)(t) = (D@@(n-1))(TEMP)(t),
 n=1..5)];
```

$$\text{TERMS} := \left[D(\theta)(t) = 1 + \cos\left(\frac{1}{10}t\right) - .05\ \theta(t), \right.$$
$$D^{(2)}(\theta)(t) = -\frac{1}{10}\sin\left(\frac{1}{10}t\right) - .05\ D(\theta)(t),$$
$$D^{(3)}(\theta)(t) = -\frac{1}{100}\cos\left(\frac{1}{10}t\right) - .05\ D^{(2)}(\theta)(t),$$
$$D^{(4)}(\theta)(t) = \frac{1}{1000}\sin\left(\frac{1}{10}t\right) - .05\ D^{(3)}(\theta)(t),$$
$$\left. D^{(5)}(\theta)(t) = \frac{1}{10000}\cos\left(\frac{1}{10}t\right) - .05\ D^{(4)}(\theta)(t) \right]$$

It is easy to see from this result that the nth derivative is a function of the $(n-1)^{\text{th}}$ derivative. So the next stage is to step through the list of differential equations, reducing them to functions of the first-order derivative only. We use a do loop for this operation:

```
ANS1 := [TERMS[1]]:
for n from 5 to 2 by -1 do
ANS1:=[op(ANS1),
 subs(op([seq(op(n-N,[TERMS[1..n-1]]), N=1..n-1)]),
TERMS[n])];
od:
ANS1;
```

$$\left[D(\theta)(t) = 1 + \cos\left(\frac{1}{10}t\right) - .05\ \theta(t),\ D^{(5)}(\theta)(t) = \right.$$
$$.00008125000000\ \cos\left(\frac{1}{10}t\right) - .00003750000000\ \sin\left(\frac{1}{10}t\right)$$
$$+ .625\ 10^{-5} - .3125\ 10^{-6}\ \theta(t),\ D^{(4)}(\theta)(t) =$$
$$.0007500000000\ \sin\left(\frac{1}{10}t\right) + .0003750000000\ \cos\left(\frac{1}{10}t\right)$$
$$- .000125 + .625\ 10^{-5}\ \theta(t),\ D^{(3)}(\theta)(t) =$$

$$-.00750000000\ \cos\left(\frac{1}{10}t\right) + .005000000000\ \sin\left(\frac{1}{10}t\right) + .0025$$
$$-\ .000125\ \theta(t),\ D^{(2)}(\theta)(t) = -\frac{1}{10}\sin\left(\frac{1}{10}t\right) - .05$$
$$-\ .05\ \cos\left(\frac{1}{10}t\right) + .0025\ \theta(t)\Bigg]$$

$$\Bigg[D(\theta)(t) = 1 + \cos\left(\frac{1}{10}t\right) - .05\ \theta(t),\ D^{5}(\theta)(t) =$$
$$.00008125000000\ \cos\left(\frac{1}{10}t\right) - .00003750000000\ \sin\left(\frac{1}{10}t\right)$$
$$+\ .625\ 10^{-5} - .3125\ 10^{-6}\ \theta(t),\ D^{(4)}(\theta)(t) =$$
$$.0007500000000\ \sin\left(\frac{1}{10}t\right) + .0003750000000\ \cos\left(\frac{1}{10}t\right)$$
$$-\ .000125 + .625\ 10^{-5}\ \theta(t),\ D^{(3)}(\theta)(t) =$$
$$-.00750000000\ \cos\left(\frac{1}{10}t\right) + .005000000000\ \sin\left(\frac{1}{10}t\right) + .0025$$
$$-\ .000125\ \theta(t),\ D^{2}(\theta)t = -\frac{1}{10}\sin\left(\frac{1}{10}t\right) - .05$$
$$-\ .05\ \cos\left(\frac{1}{10}t\right) + .0025\ \theta(t)\Bigg]$$

Note the way that subs is used for sequential substitution. The order of the substitution equations has been reversed. The next step is to evaluate the derivatives at the point $t = 0$:

```
> AT_ZERO := subs(t=0,ANS1);
```

AT_ZERO :=

$$\Big[D(\theta)(0) = 1 + \cos(0) - .05\ \theta(0),\ D^{(5)}(\theta)(0) =$$

$$.00008125000000\ \cos(0) - .00003750000000\ \sin(0) + .625\ 10^{-5}$$

$$- .3125\ 10^{-6}\ \theta(0),\ D^{(4)}(\theta)(0) = .0007500000000\ \sin(0)$$

$$+ .00037540000000\ \cos(0) - .000125 + .625\,(10 \text{ sup } \{\text{-}5\})\theta(0),$$

$$D^{(3)}(\theta)(0) = -.00750000000\ \cos(0) + .005000000000\ \sin(0)$$

$$+ .0025 - .000125\ \theta(0),$$

$$D^{(2)}(\theta)(0) = -\frac{1}{10}\sin(0) - .05 - .05\ \cos(0) + .0025\ \theta(0)\Big]$$

Once we have generated a Taylor's series in θ(t) we substitute for the derivatives:

```
THE_SERIES:=taylor(theta(t), t);
```

$$\text{THE\_SERIES} := \theta(0) + D(\theta)(0)t + \frac{1}{2}D^{(2)}(\theta)(0)t^2 + \frac{1}{6}D^{(3)}(\theta)(0)t^3 + \frac{1}{24}D^{(4)}(\theta)(0)t^4 + \frac{1}{120}D^{(5)}(\theta)(0)t^5 + O(t^6)$$

```
WITHOUT_DIFFS:= subs(AT_ZERO, THE_SERIES);
```

$$\begin{aligned}\text{WITHOUT\_DIFFS} &:= \theta(0) + (2 - .05\ \theta(0))\ t\\ &+ (-.05000000000 + .001250000000\ \theta(0))\ t^2\\ &+ (-.0008333333335 - .00002083333334\ \theta(0))\ t^3\\ &+ (.00001041666667 + .2604166667\ 10^{-6}\ \theta(0))\ t^4\\ &+ (.7291666666\ 10^{-6} - .2604166667\ 10^{-8}\ \theta(0))\ t^5 + O(t^6)\end{aligned}$$

Here we substitute the initial condition and convert the resulting expression to a polynomial by removing the order term:

```
WITH_IC := subs(theta(0) = 1, WITHOUT_DIFFS);
```

$$WITH\_IC := 1 + 1.95\ t - .04875000000\ t^2 - .0008541666668\ t^3 + .00001067708334\ t^4 + .7265624999\ 10^{-6}\ t^5 + O(t^6)$$

```
SERIES_APPROX := convert(WITH_IC, polynom);
```

$$SERIES\_APPROX := 1 + 1.95t - .04875000000t^2 - .0008541666668t^3 + .00001067708334t^4 + .7265624999\ 10^{-6}t^5$$

We now plot the response in Figure 4.7.

```
plot(SERIES_APPROX, t=0..40, labels=['t','Theta']);
```

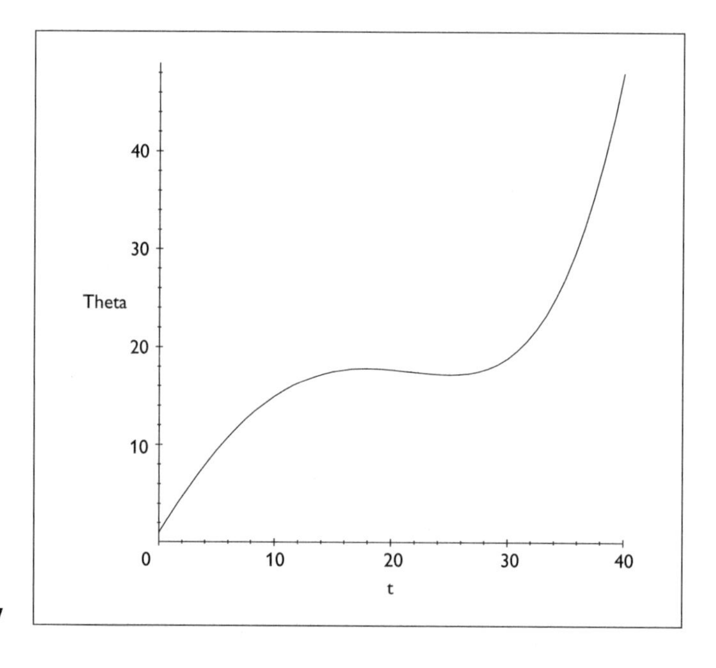

Figure 4.7

As previously mentioned series techniques only provide approximate solutions to ODEs; the accuracy of the solution is determined by the type of series used and the number of terms in the series. Bearing this in mind, the accuracy of the time response of the tachometer needle appears to alter over its range. Within the range $t = 0..20$ the response is as we would expect with the needle's angular position increasing, overshooting, and then, following some oscillations, finally settling at its final position. It is in the range $t = 20..40$ that the needle's response is not as we would expect as the angular position grows in an unbounded fashion. This accuracy problem is common with series methods because the series in question is only guaranteed to be accurate for a limited range about the point of expansion, in out case $t = 0$. It is possible to improve the accuracy of a series method solution by using an alternative series, as Maple does. We increase the number of terms used and change the point about which the function is expanded. This last method means that a piecewise linear approximation to the solution can be constructed in an iterative fashion. In this particular example, however, an analytic solution exists so we can compare that with the approximate solution obtained using the series approach. First we save the plot of the series solution, changing the line style to dashed, so that we can compare it with the analytic solution.

```
PLOT1 := plots[display]({"}, linestyle=2):
```

Using `dsolve` we calculate the exact solution before displaying the two responses on the same graph (see Figure 4.8):

```
EXACT := dsolve( {SYS5, theta(0)=1}, theta(t));
```

$$\mathrm{EXACT} := \theta(t) = 20. + 4.\ \cos(.1000000000t) + 8.\ \sin(.1000000000t) - 23.\ e^{(-.05000000000t)}$$

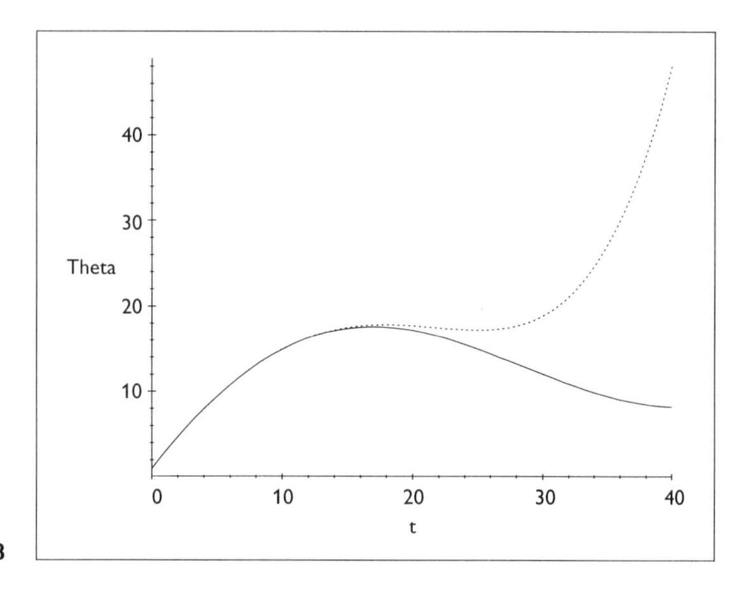

Figure 4.8

This plots the approximate and the exact solutions:

```
plots[display]({plot(rhs(EXACT), t=0..40), PLOT1},
 labels=['t','Theta']);
```

Modeling dynamic systems

Using Maple's ability to represent and manipulate symbolic quantities, we can quickly describe a dynamic system, using differential equations, and analyze it. In many instances, we will be able to obtain the exact closed-form solutions describing its behavior.

A simple shock absorber

In our first example, we consider a mass (m_1), a spring (constant k_1) of length s_1, and a damper (frictional constant b_1) arrangement set up as a simple shock absorber (Figure 4.9). In this particular example, we are interested in determining the behavior over time of the center of the mass (x_1) following a disturbance in the reference x_0.

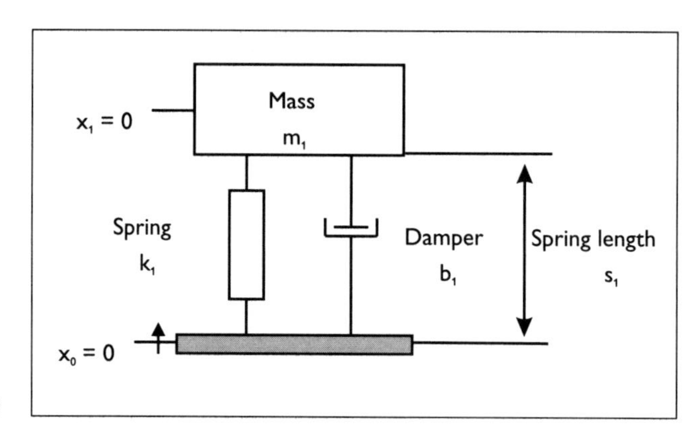

Figure 4.9

By equating the forces acting on the mass m_1 we can determine the equations of motion that describe the system's behavior, in this case, over time.

```
mass[1]:= 0 = m[1]*diff(x[1](t),t,t) + b[1]*diff(x[1](t),t)
 - k[1]*(s[1]-x[1](t)+x[0](t));
```

$$\text{mass}_1 := 0 + m_1\left(\frac{\partial^2}{\partial t^2}x_1(t)\right) + b_1\left(\frac{\partial}{\partial t}x_1(t)\right) - k_1\left(s_1 - x_1(t) + x_0(t)\right)$$

This is the general equation describing the system's motion so we need to define the parameter values and the initial conditions before we proceed. The system parameters are as shown, the system is assumed to be initially at rest, and the forcing function is $\sin^2(2.7t)/2$. This particular forcing function has been chosen to simulate a series of impulses.

```
params:=[m[1]=1, k[1]=1, b[1]=1, s[1]=1,
x[0](t)=0.5*sin(t*2.7)^2];
```

$$\text{params} := \left[m_1 = 1,\ k_1 = 1,\ b_1 = 1,\ s_1 = 1,\ x_0(t) = .5\ \sin(2.7\ t)^2\right]$$

```
ics:=D(x[1])(0)=0, x[1](0)=1;
```

$$\text{ics} := D(x_1)(0) = 0,\ x_1(0) = 1$$

These are substituted into the second-order ODE describing the shock absorber, which is then solved analytically using `dsolve`:

```
{subs(params, mass[1]), ics}, {x[1](t)}
```

$$\left\{D(x_1)(0) = 0,\ x_1(0) = 1,\ 0 = \left(\frac{\partial^2}{\partial t^2}x_1(t)\right) + \left(\frac{\partial}{\partial t}x_1(t)\right) - 1 + x_1(t) - .5\ \sin(2.7\ t)^2\right\},\ \{x_1(t)\}$$

The `convert( ..., rational)` is necessary because `dsolve` is unable to deal with floating-point numbers in the ODE's coefficients.

```
> simplify(dsolve(op(convert(["], rational))));
```

$$x_1(t) = \frac{2200}{5137841}\sin\left(\frac{1}{2}\sqrt{3}t\right)\sin\left(\frac{1}{2}\sqrt{3}t+\frac{27}{5}t\right)$$

$$+\frac{3375}{4110728}\cos\left(\frac{1}{2}\sqrt{3}t\right)\sin\left(\frac{1}{2}\sqrt{3}t+\frac{27}{5}t\right)$$

$$-\frac{3375}{4110728}\cos\left(\frac{1}{2}\sqrt{3}t\right)\sin\left(\frac{1}{2}\sqrt{3}t-\frac{27}{5}t\right)$$

$$+\frac{2200}{5137841}\cos\left(\frac{1}{2}\sqrt{3}t\right)\cos\left(\frac{1}{2}\sqrt{3}t-\frac{27}{5}t\right)$$

$$+\frac{2200}{5137841}\cos\left(\frac{1}{2}\sqrt{3}t\right)\cos\left(\frac{1}{2}\sqrt{3}t+\frac{27}{5}t\right)$$

$$+\frac{2200}{5137841}\sin\left(\frac{1}{2}\sqrt{3}t\right)\sin\left(\frac{1}{2}\sqrt{3}t-\frac{27}{5}t\right)+\frac{5}{4}$$

$$+\frac{9425}{6166092}\cos\left(\frac{1}{2}\sqrt{3}t\right)\sin\left(\frac{1}{2}\sqrt{3}\,t-\frac{27}{5}t\right)\sqrt{3}$$

$$+\frac{64485}{4110728}\cos\left(\frac{1}{2}\sqrt{3}\,t\right)\cos\left(\frac{1}{2}\sqrt{3}t-\frac{27}{5}t\right)\sqrt{3}$$

$$+\frac{9425}{6166092}\cos\left(\frac{1}{2}\sqrt{3}t\right)\sin\left(\frac{1}{2}\sqrt{3}\,t+\frac{27}{5}t\right)\sqrt{3}$$

$$-\frac{64485}{4110728}\cos\left(\frac{1}{2}\sqrt{3}\,t\right)\cos\left(\frac{1}{2}\sqrt{3}t+\frac{27}{5}t\right)\sqrt{3}$$

$$-\frac{9425}{6166092}\sin\left(\frac{1}{2}\sqrt{3}t\right)\cos\left(\frac{1}{2}\sqrt{3}\,t+\frac{27}{5}t\right)\sqrt{3}$$

$$+\frac{64485}{4110728}\sin\left(\frac{1}{2}\sqrt{3}\,t\right)\sin\left(\frac{1}{2}\sqrt{3}t-\frac{27}{5}t\right)\sqrt{3}$$

$$-\frac{9425}{6166092}\sin\left(\frac{1}{2}\sqrt{3}t\right)\cos\left(\frac{1}{2}\sqrt{3}\,t-\frac{27}{5}t\right)\sqrt{3}$$

$$- \frac{64485}{4110728} \sin\left(\frac{1}{2}\sqrt{3}\,t\right) \sin\left(\frac{1}{2}\sqrt{3}\,t + \frac{27}{5}t\right)\sqrt{3}$$

$$- \frac{3375}{4110728} \cos\left(\frac{1}{2}\sqrt{3}\,t\right) \sin\left(\frac{1}{2}\sqrt{3}\,t + \frac{27}{5}t\right)$$

$$+ \frac{3375}{4110728} \cos\left(\frac{1}{2}\sqrt{3}\,t\right) \sin\left(\frac{1}{2}\sqrt{3}\,t - \frac{27}{5}t\right)$$

$$+ \frac{531441}{2055364} e^{(-1/2 t)} \cos\left(\frac{1}{2}\sqrt{3}\right) - \frac{164997}{2055364}\sqrt{3}\, e^{(-1/2 t)} \sin\left(\frac{1}{2}\sqrt{3}\,t\right)$$

As we can see, the solution of even simple-looking ODEs can be complex. Here we display both the time response of the system and the forcing function together:

```
Plots[display](array(1..2,1..1,[[plot(rhs("), t=0..10,
 labels=['t','o/p'])],[ plot(0.5*sin(2.7*t)^2, t=0..10,
 labels=['t','i/p'])]]));
```

We can see in Figure 4.10 that this system would not make a particularly good shock absorber because its response time is too slow. With reference to the time response, we can deduce that the shock absorber is

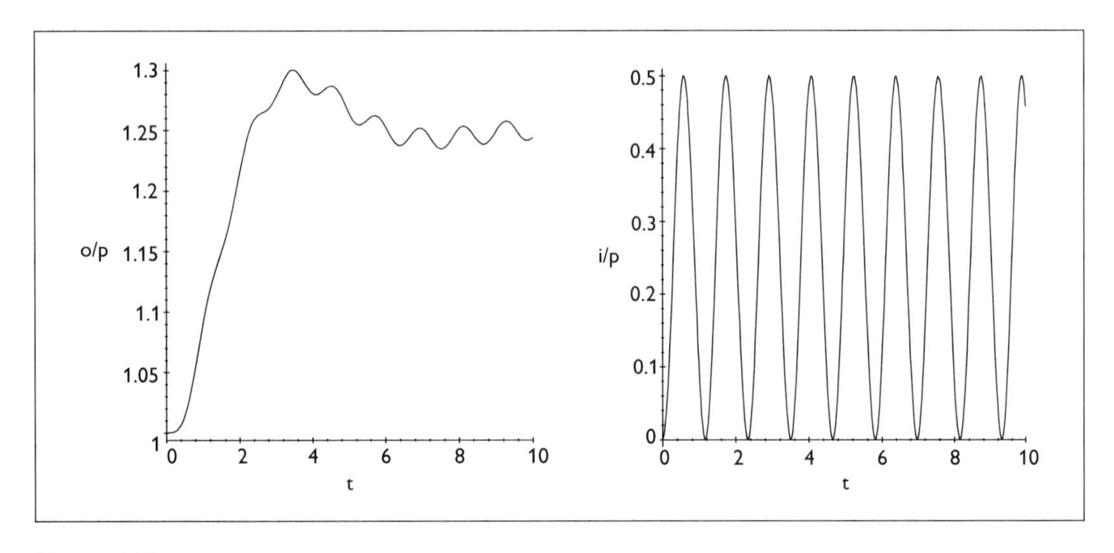

Figure 4.10

bouncing from high-spot to high-stop as the steady-state output is centered on the dc value of the input signal and tracks the oscillations at the input's fundamental frequency.

Using the Heaviside *function*

In many practical systems the forcing function to a system will be discontinuous in nature. For example, by using the Heaviside function we can model step, pulse, delayed, and windowed forcing functions.

In the previous example, we approximated a pulse train forcing function with a $\sin^2 t$ function. Here, instead we can use Heaviside functions as shown:

```
f(t):=sum(Heaviside (t-n)*(-1)^(n+1), n=1..10);
```

$$f(t) := \text{Heaviside}(t - 1) - \text{Heaviside}(t - 2) + \text{Heaviside}(t - 3) - \text{Heaviside}(t - 4) + \text{Heaviside}(t - 5) - \text{Heaviside}(t - 6) + \text{Heaviside}(t - 7) - \text{Heaviside}(t - 8) + \text{Heaviside}(t - 9) - \text{Heaviside}(t - 10)$$

Here we plot the forcing function *f(t)*. We save this plot for future use by assigning it to the variable `p1` (see Figure 4.11):

```
plot(f(t), t=0..20, labels=['t','f(t)']);
```

```
p1:=":
```

The parameter definition is modified to include the forcing function *f(t):*

```
params:=[m[1]=1, g=10, k[1]=1, b[1]=1, x[0](t)=f(t),
 s[1]=1]:
```

We can now substitute the new parameter list into the ODE describing the mass's motion developed above and then solve it.

```
{subs(params, mass[1]), ics}, {x[1](t)};
dsolve(");
p2:=plot(rhs("), t=0..20):
```

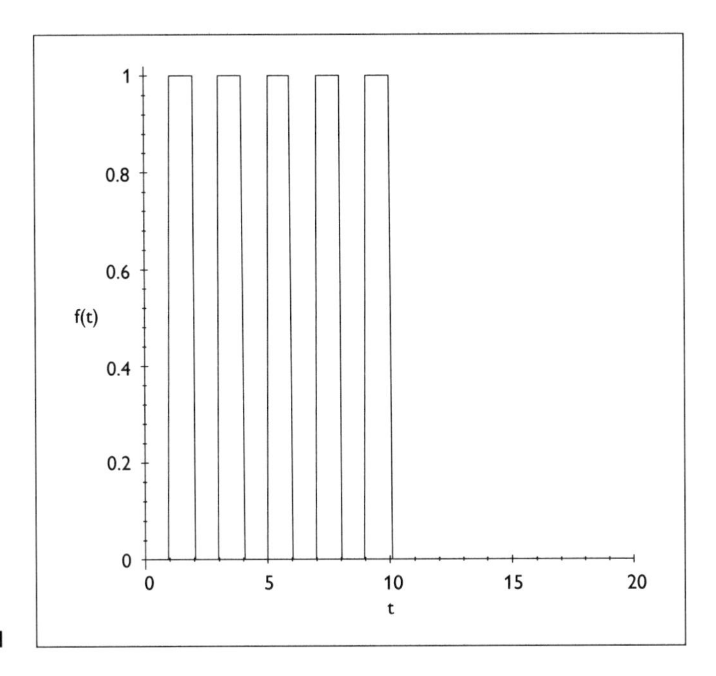

Figure 4.11

The forcing function and system's time response are displayed together in Figure 4.12 using the `display` function found in the `plots` package.

```
plots[display]({p2,p1}, labels=['t','o/p']);
```

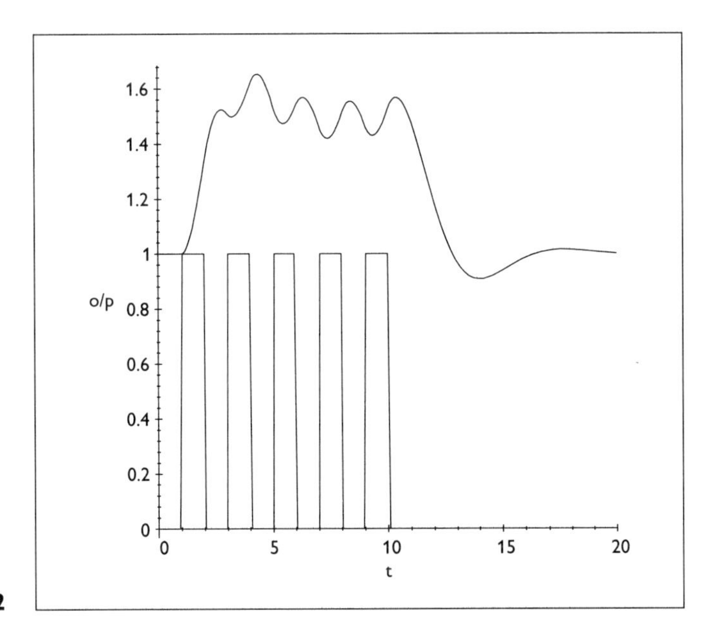

Figure 4.12

A twin mass shock absorber

In the next example, we model a car tire-spring-damper assembly with two masses (m_1 and m_2), two springs (constants k_1 and k_2) of length s_1 and s_2, respectively, and two dampers (frictional constants b_1 and b_2) arranged as shown in Figure 4.13. The bottom mass-spring-damper arrangement models the tire whereas the top mass-spring-damper configuration models the car's shock absorber assembly. In this particular example, we are interested in determining the behavior over time of both of the center of the masses (x_1 and x_2) following a disturbance in the reference x_0.

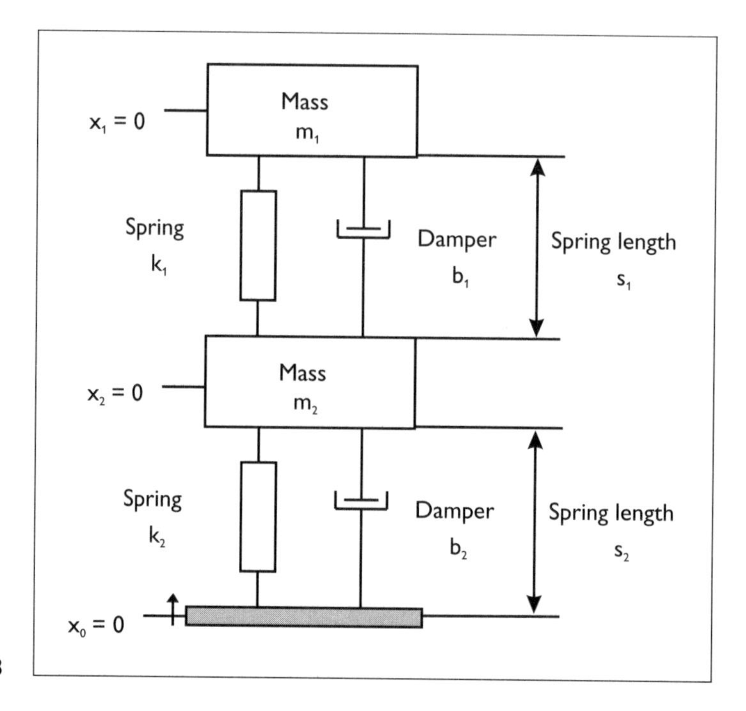

Figure 4.13

By equating the forces acting on both of the masses, we can determine the equations of motion that describe the system's behavior over time. In this example, we are faced with two coupled ODEs that must be solved simultaneously.

```
mass[1]:=m[1]*diff(x[1](t),t,t)-k[1]*(s[1]-(x[1](t)
x[2](t)))+b[1]*diff(x[1](t)-x[2](t),t)=0;
```

$$\text{mass}_1 := m_1 \left(\frac{\partial^2}{\partial t^2} x_1(t) \right) - \text{k}_1 \left(s_1 - x_1(t) + x_2(t) \right) + \text{b}_1 \left(\left(\frac{\partial}{\partial t} x_1(t) \right) - \left(\frac{\partial}{\partial t} x_2(t) \right) \right) = 0$$

```
mass[2]:=m[2]*diff(x[2](t),t,t)-k[1]*(-s[1]+x[1](t)
 x[2](t))-b[1]*diff(x[1](t)-x[2](t),t) -k[2]*(s[2]
 x[2](t)+x[0](t))+b[2]*diff(x[2](t)-x[0](t),t)=0;
```

$$\text{mass}_2 := m_2 \left(\frac{\partial^2}{\partial t^2} x_2(t) \right) - \text{k}_1 \left(-s_2 + x_2(t) - x_2(t) \right) - \text{b}_1 \left(\left(\frac{\partial}{\partial t} x_1(t) \right) - \left(\frac{\partial}{\partial t} x_2(t) \right) \right) - \text{k}_2 \left(s_2 - x_2(t) + x_0(t) \right) + \text{b}_2 \left(\left(\frac{\partial}{\partial t} x_2(t) \right) - \left(\frac{\partial}{\partial t} x_0(t) \right) \right) = 0$$

As in the previous example, these are the general equations so we will need to assign values to the system parameters for mass m_i, damping b_i, and the spring constants k_i. The forcing function $x_0(t)$ is also defined.

```
params:=[m[1]=200, m[2]=10, k[1]=25000, k[2]=15000,
 b[1]=50,b[2]=1000,x[0](t)=0.2, s[1]=0.2, s[2]=0.3];
```

$$\text{params} := \left[m_1 = 200,\ m_2 = 10,\ \text{k}_1 = 25000,\ \text{k}_2 = 15000,\ \text{b}_1 = 50\ \text{b}_2 = 1000,\ x_0(t) = .2,\ s_2 = .3 \right]$$

The final set of parameters that needs to be defined are the initial conditions:

```
ics:=x[1](0)=0.5, x[2](0)=0.3, D(x[1])(0)=0, D(x[2])(0)=0;
```

$$\text{ics} := x_1(0) = .5,\ x_2(0) = .3,\ D(x_1)(0) = 0,\ D(x_2)(0) = 0$$

The ODEs and the parameter values are brought together using the substitution function `subs` to form the equations that we will be manipulating further:

```
eqns:=subs(params,{mass[1], mass[2]});
```

$$eqns := \left\{200\left(\frac{\partial^2}{\partial t^2}\, x_1(t)\right) - 5000.0 + 25000\, x_1(t) - 25000\, x_2(t) + 50\left(\frac{\partial}{\partial t}\, x_2(t)\right) - 50\left(\frac{\partial}{\partial t}\, x_2(t)\right) = 0,\ 10\left(\frac{\partial^2}{\partial t^2}\, x_1(t)\right) - 2500.0 - 25000\, x_1(t) + 40000\, x_2(t) - 50\left(\frac{\partial}{\partial t} x_1(t)\right) + 1050\left(\frac{\partial}{\partial t} x_2(t)\right) - 1000\left(\frac{\partial}{\partial t}\, .2\right) = 0\right\}$$

In this particular case we will use Laplace transform methods to solve the variables $x_{i(t)}$. The Laplace transform functions are found in the integral transforms package `inttrans`, which we load using `with`. If you are concerned with system resources (i.e., unnecessarily loading functions into Maple work space that will not be used) you can load only the Laplace transform functions by using `with (inttrans, [laplace, invlaplace]):`

```
with(inttrans):
```

We can now take the Laplace transforms of the system's differential equations:

```
laplace(eqns, t, s);
```

$$\left\{200.\left(\text{laplace}\,(x_1(t), t, s)\, s - 1.\, x_1(0)\right)s - 200.\, D(x_1)(0) - 5000.\,\frac{1}{s} + 25000.\,\text{laplace}\left(x_1(t), t, s\right) - 25000.\,\text{laplace}\left(x_2\,(t), t, s\right) + 50.\,\text{laplace}\left(x_1(t), t, s\right)s - 50.\, x_1(0) - 50.\,\text{laplace}\left(x_2(t), t, s\right)s + 50.\, x_2(0) = 0,\ 10.\left(\text{laplace}\left(x_2(t), t, s\right)s - 1.\, x_2(0)\right)s\right.$$

$$- 10.\ D(x_2)(0) - 2500.\ \frac{1}{s} - 25000.\ \text{laplace}\left(x_1(t), t, s\right)$$
$$+ 40000.\ \text{laplace}\left(x_2(t), t, s\right)s - 50.\ \text{laplace}\left(x_1(t), t, s\right)s$$
$$+ 50.\ x_1(0) + 1050.\ \text{laplace}\left(x_2(t), t, s\right)s - 1050.\ x_2(0) = 0\Big\}$$

We can make these expressions clearer by means of Maple's alias facility. The two unknowns that we are interested in are the Laplace transformed quantities laplace$(x_1(t))$ and laplace$(x_2(t))$, which we will alias to `X1` and `X2`, respectively:

```
alias(X1=laplace(x[1](t), t, s), X2=laplace(x[2](t), t, s));
```

$$I,\ \text{diff1},\ \int 1,\ X1,\ X2$$

Maple returns a sequence of the aliases currently known to the system. Now we substitute for the initial conditions:

```
subs(ics, "");
```

$$\Big\{200.\ (X1\ s - .5)\ s - 5000.\ \frac{1}{s} - 25000.\ X1 + 25000.\ X2 + 50.\ X1\ s$$
$$- 10.0 - 50.\ X2\ s = 0,\ 10.\ (X2\ s - .3)\ s - 2500.\ \frac{1}{s} - 25000.\ X1$$
$$+ 40000.\ X2 - 50.\ X1\ s - 290.0 + 1050.\ X2\ s = 0\Big\}$$

We now can solve for the two unknowns $X1$ and $X2$

```
solve(", {X1, X2});
```

$$\Big\{X1 = .5000000000\,\frac{16600.\ s^2 + .1050000\ 10^7 + 52100.\ s + 421.\ s^3 + 4.\ s^4}{s\left(750000. + 51500.\ s + 16600.\ s^2 + 4.\ s^4 + 421.\ s^3\right)},$$
$$X2 = .3000000000\,\frac{4.\ s^4 + 421.\ s^3 + 20600.\ s^2 + 52500.\ s + .1250000\ 10^7}{s\left(750000. + 51500.\ s^2 + 16600.\ s^2 + 4.\ s^4 + 421.\ s^3\right)}\Big\}$$

Using map in conjunction with invlaplace, we take the inverse transform of both solutions in a single step and obtain the time responses $x_i(t)$:

```
map(invlaplace, ", s, t);
```

$$\{x1(t) = .7000000000 - .003867611722\, e^{(-51.60982474\, t)} \cos(35.05255107\, t) + .001213402066\, e^{(-51.60982474\, t)} \sin(35.05255107\, t) - .1961323883\, e^{(-1.015175265\, t)} \cos(6.866003481\, t) - .06426567528\, e^{(-1.015175265\, t)} \sin(6.866003481\, t),\ x_2(t) = .5000000000 - .08350768725\, e^{(-51.60982474\, t)} \cos(35.05255107\, t) - .1143131946\, e^{(-51.60982474\, t)} \sin(35.05255107\, t) - .1164923127\, e^{(-1.015175265\, t)} \cos(6.866003481\, t) - .06133234948\, e^{(-1.015175265\, t)} \sin(6.866003481\, t)\}$$

Maple returns a set of solutions which we map onto a list of solutions a shown.

```
sols  :=subs( ", [x[1](t), x[2] (t)]);
```

$$sols := [\, .7000000000 - .003867611722\, e^{(-51.60982474\, t)} \cos(35.05255107\, t) + .001213402066\, e^{(-51.60982474\, t)} \sin(35.05255107\, t) - .1961323883\, e^{(-1.015175265\, t)} \cos(6.866003481\, t) - .06426567528\, e^{(-1.015175265\, t)} \sin(6.866003481\, t),\ .5000000000 - .08350768725\, e^{(-51.60982474\, t)} \cos(35.05255107\, t) - .1143131946\, e^{(-51.60982474\, t)} \sin(35.05255107\, t) - .1164923127\, e^{(-1.015175265\, t)} \cos(6.866003481\, t) - .06133234948\, e^{(-1.015175265\, t)} \sin(6.866003481\, t)]$$

Now we can plot the response of the system for an input step of 0.2 units in amplitude (Figure 4.14):

```
plot({sols[1], sols[2]},t=0..5,numpoints=100,
 labels=['t','x[1](t), n[2](t)']);
```

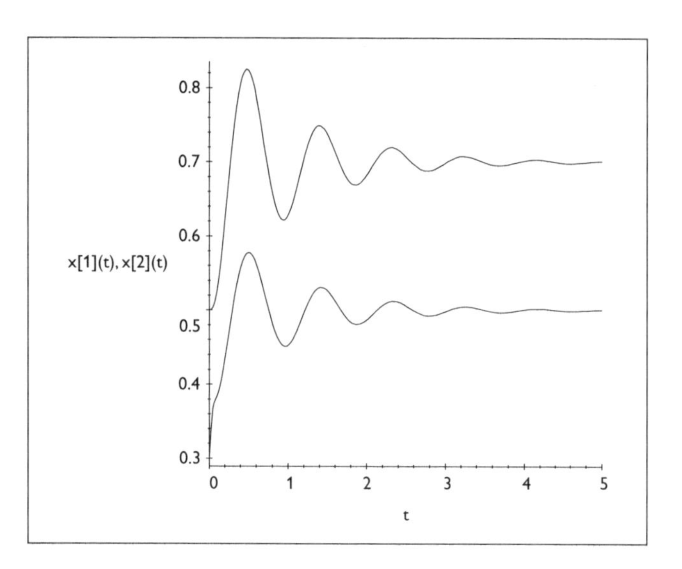

Figure 4.14

At first glance the step response of this system does not seem unusual with a settling time of approximately 3 sec, an overshoot of approximately 70%, and a natural damped natural frequency of about 1.2 Hz. The responses are closely coupled with only a slight delay between the two masses. However, on closer inspection, the initial rises of each curve are not the same. Using Maple we can zoom in on this portion of the response (see Figure 4.15):

```
plot({sols[1], sols[2]},t=0..0.5, numpoints=100,
 labels=['t','x[1](t), n[2](t)']);
```

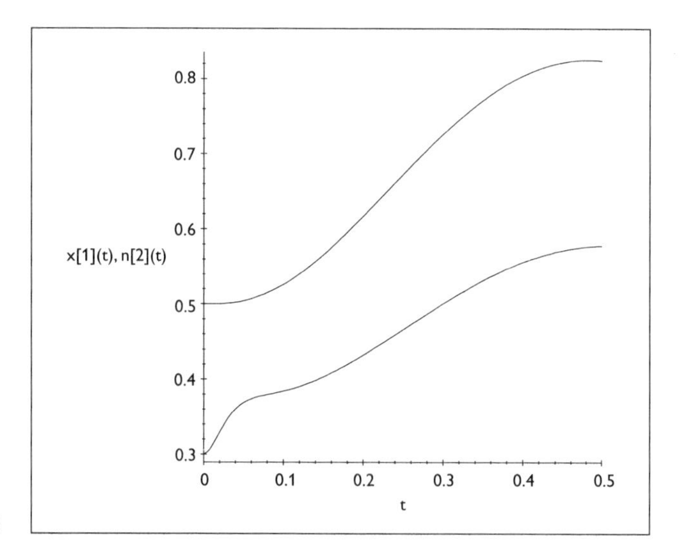

Figure 4.15

Using piecewise

The strange "kink" in the response of mass m_2 is indicative of the presence of high-frequency components. The high-frequency components in the output could lead to unsuitable operation for a shock absorber assembly when it is subjected to a sequence of impulses such as those that could be encountered when a car passes over a railroad crossing. It is this type of input and how to model it that we consider in this section.

As before we define the parameters for the system but this time we set the forcing function to $f(t)$ as defined below:

```
params:=[m[1]=200, m[2]=10, k[1]=25000, k[2]=15000,
b[1]=50, b[2]=1000, x[0](t)=f(t), s[1]=0.2, s[2]=0.3];
```

$$\text{params} := \left[m_1 = 200, m_2 = 10, \mathrm{k}_1 = 25{,}000, \mathrm{k}_2 = 15{,}000, \mathrm{b}_1 = 50 \right.$$
$$\left. \mathrm{b}_2 = 1000, x_0(t) = f(t), s_1 = .2, s_2 = .3 \right]$$

Using the `piecewise` function we model a sequence of impulses that will become the system's forcing function:

```
f(t):=piecewise(t<=1,0.1,t>1 and t<=1.5,0, t>1.5 and
 t<=1.75, 0.2, t>1.75 and t<=2.2, 0, t>2.4 and t<=2.7,
 -0.2, t>2.7, 0):
```

Here we substitute for the system parameters and the forcing function prior to solving for $x_i(t)$ numerically:

```
sols:=dsolve(subs(params,{mass[1], mass[2], ics}),
 {x[1](t), x[2](t)}, numeric);
```

```
sols := proc(rkf45_x) ... end
```

This time we use `odeplot` to plot the time responses $x_i(t)$ over the range $t = 0..5$ s. The system response along with the forcing function are displayed in Figure 4.16.

```
plots[odeplot](sols, [[t,x[1](t)],[t,x[2](t)]],0..5,
 numpoints=100);
```

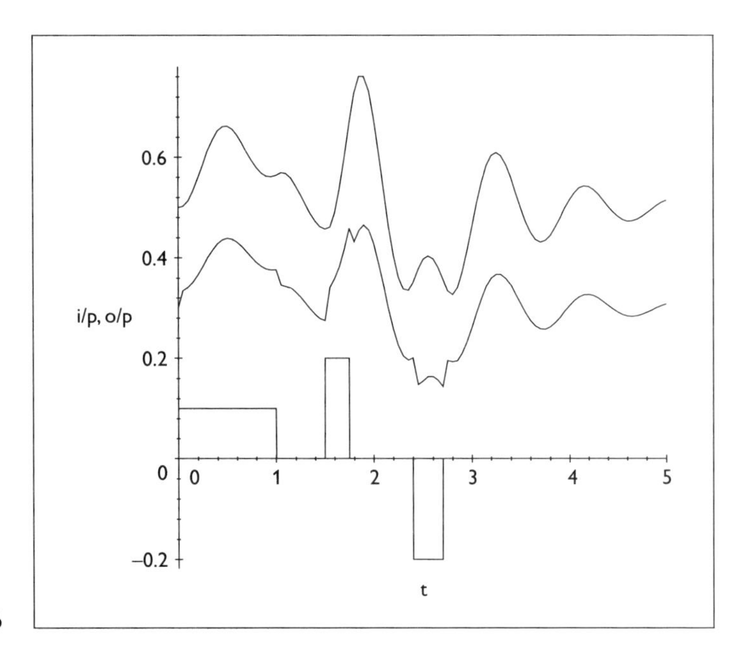

Figure 4.16

As predicted the response of the top mass is smooth despite the impulsive nature of the forcing function, whereas the response of the lower mass exhibits a number of discontinuities corresponding to the rising and falling edges of the forcing function. In a real system, where the lower mass-spring-damper assembly would be tire, such behavior could lead to a loss of traction and hence control of the vehicle.

A nonlinear system

Thus far we have used Maple to investigate linear ODEs but that tools and techniques used are equally applicable to nonlinear ODEs. In some cases it may be possible to obtain an analytical solution to a nonlinear ODE, in which case `dsolve` can still be used. For example,

```
dsolve(diff(y(t),t,t)=y(t)^2, y(t));
```

$$t = \int_0^{y(t)} -3\,\frac{1}{\sqrt{6y1^3 + 9\_C1}}\,dy1 - \_C2,$$

$$t = \int_0^{y(t)} 3\,\frac{1}{\sqrt{6y2^3 + 9\_C1}}\,dy2 - \_C2$$

The number of nonlinear ODEs that can be solved analytically is small, which means that more often than not we are forced to solve nonlinear ODEs numerically. As we have already seen, Maple has a powerful array of numerical solvers that we can apply to nonlinear ODEs. Returning to the linear coupled system of ODEs considered earlier, we can consider a nonlinear system by introducing a nonlinear spring into the top mass-damper-spring assembly of the preceding example. We redefine the force balance equation for the top mass as follows:

```
mass[1]:=m[1]*diff(x[1](t),t,t)-k[1]*(s[1]-(x[1](t)-
x[2](t))^2)+b[1]*diff(x[1](t)-x[2](t),t)=0;
```

$$\text{mass}_1 := m_1\left(\frac{\partial^2}{\partial t^2}\,x_1(t)\right) - \mathrm{k}_1\left(s_1 - \left(x_1(t) - x_2(t)\right)^2\right)$$

$$+\ \mathrm{b}_1\left(\left(\frac{\partial^1}{\partial t^1}\,x_1(t)\right) - \left(\frac{\partial^1}{\partial t^1}\,x_2(t)\right)\right) = 0$$

We use the following system parameters and the original initial conditions to compute the system step response:

```
ics:=x[1](0)=0.5, x[2](0)=0.3, D(x[1])(0)=0, D(x[2])(0)=0:

params:=[m[1]=200, m[2]=10, k[1]=25000, k[2]=15000,
 b[1]=50,b[2]=1000, x[0](t)=0.2, s[1]=0.2, s[2]=0.3]:

sols:=dsolve(simplify(subs(params,{mass[1], mass[2],
 ics})), {x[1](t), x[2](t)}, numeric);
```

sols := proc(rkf45_x) ... end

Note the use of `simplify` to tidy up the differential equations prior to them being passed to `dsolve`.

```
plots[odeplot](sols, [[t,x[1](t)],[t,x[2](t)]],0..5,
 numpoints=100, labels=['t','x1(t), x2(t)']);
```

The response obtained (Figure 4.17) for the nonlinear system is obviously different from that obtained for the linear system in that the displacement of both masses is much larger and the settling time in increased. The kink in the lower traces is not, however, as pronounced in the nonlinear system as in the linear one.

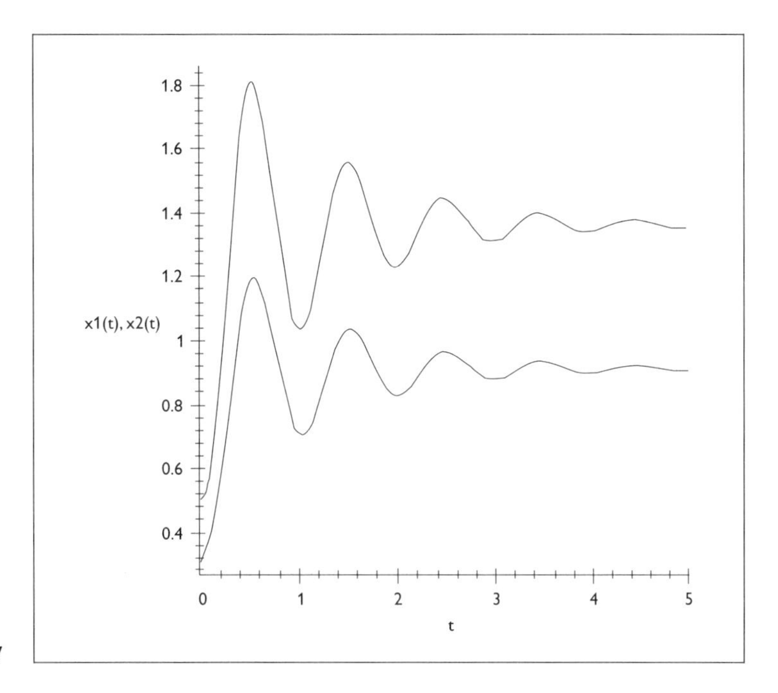

Figure 4.17

The ability to deal with nonlinear components in our models enables us to investigate how systems will react to marked changes in the system's parameters. For example, what happens to the system if a spring or damper fails? In this our final example in this section, we investigate how the system's response changes with regard to changes in the damping coefficient of the bottom mass-spring-damper assembly. Such a change in the coefficient could be associated with changes in temperature or a punctured tire.

First we define how the damper's coefficient of friction will change over time:

```
B:=T->piecewise(T>12, 4000, T>8, 1000, T>4, 500,T>0, 0):
```

The profile of the damping coefficient can be plotted as shown in Figure 4.18.

```
plot(B(t),t=0..20, labels=['t','B(t)']);
```

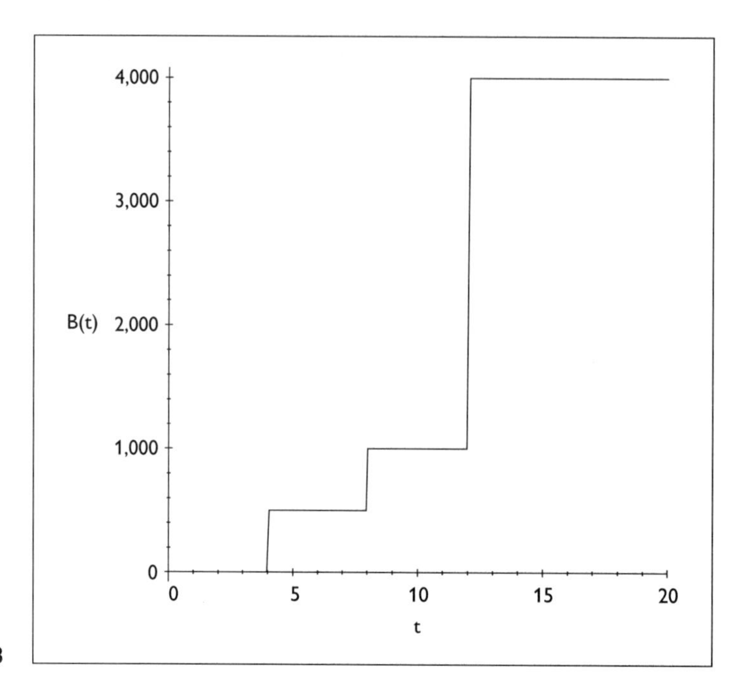

Figure 4.18

Next we define the square-wave forcing function of 4-Hz frequency and peak-to-peak amplitude of 0.4 units:

```
f:=T->piecewise(T>16, 0.2, T>12, -0.2, T>8, 0.2, T>4, -0.2,
 T>0, 0.2):
```

Here we plot the forcing function, which is shown in Figure 4.19:

```
plot(f(t),t=0..20, labels=['t','f(t)']);
```

The parameter list is altered so that the coefficients b_2 and $x_0(t)$ are the functions of time $B(t)$ and $f(t)$, respectively:

```
params:=[m[1]=200, m[2]=10, k[1]=25000, k[2]=15000,
 b[1]=50, b[2]=B(t),x[0](t)=f(t), s[1]=0.2, s[2]=0.3];
```

$$\text{params} := \Big[m_1 = 200, m_2 = 10, \mathrm{k}_1 = 25000, \mathrm{k}_2 = 15000, \mathrm{b}_1 = 50, \mathrm{b}_2 = \text{piecewise}(0, 12 < t, 4000, 8 < t, 1000, 4 < t, 500, 0 < t, 0, \text{undefined}), x_0(t) = \text{piecewise}(0, 16 < t, .2, 12 < t, -.2, 8 < t, .2\ 4 < t, -.2, 0 < t, .2, \text{undefined}), s_1 = .2, s_2 = .3 \Big]$$

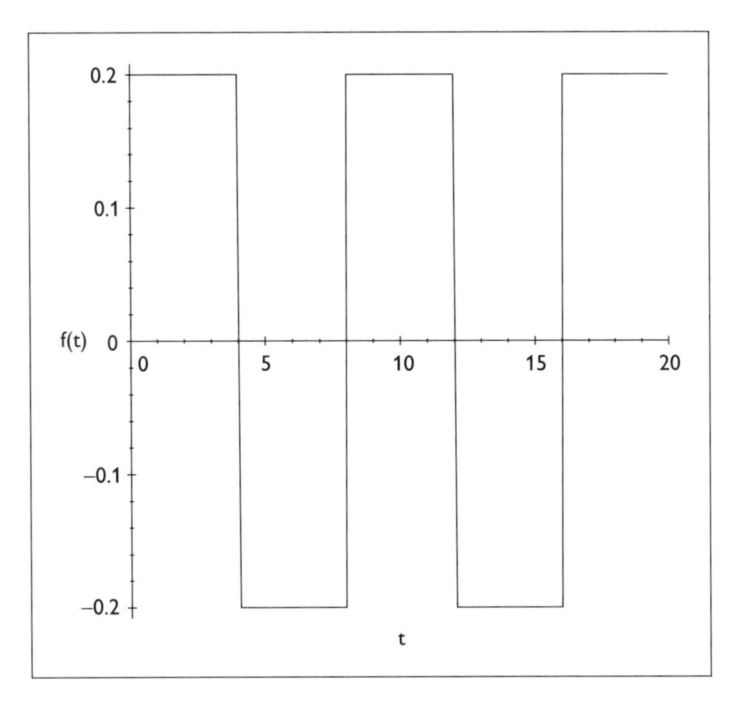

Figure 4.19

We can now solve this numerically

```
sols:=dsolve(subs(params,{mass[1], mass[2], ics}),
 {x[1](t), x[2](t)}, numeric);
```

```
sols := proc(rkf45_x) ... end
```

and plot the result (Figure 4.20) as before using `odeplot`:

```
plots[odeplot](sols,[[t,x[1](t)],[t,x[2](t)]],0..20,
 numpoints=100, labels=['t','x1(t), x2(t)']);
```

This previous plot demonstrates that this particular system is closely coupled. As the tire becomes more and more stiff, the oscillations of both the masses reduces until, in the range of $t = 16..20$, the majority of the damping is supplied by the top damper. The data displayed in Figure 4.20 is a little cluttered. By altering the representation to a three-dimensional surface, the oscillations and their effect on the system's step response as the damping coefficient varies can be clearly seen.

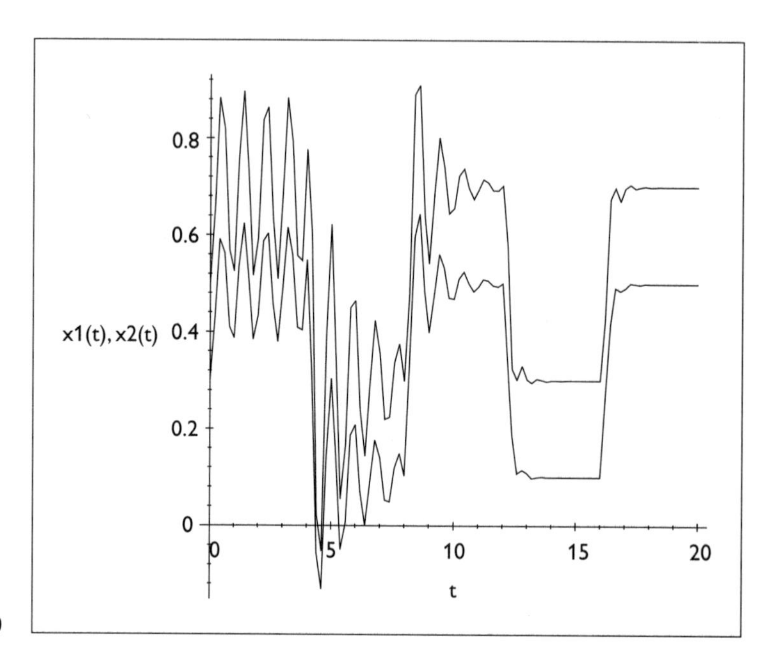

Figure 4.20

First we redefine the function *B(t)* so that it has four bands and the parameter list so that the forcing function is a step of amplitude 0.2 units:

```
B:=T->piecewise(T>3, 4000, T>2, 1000, T> 1, 500, T>0, 0):
params:=[m[1]=200, m[2]=10, k[1]=25000, k[2]=15000,
 b[1]=50,b[2]=B,x[0](t)=0.2, s[1]=0.2, s[2]=0.3]:
```

Next we define a function that returns a Maple procedure. The procedure is the numeric solution, for a given damping coefficient, to the nonlinear ODE:

```
sols:=(y)-dsolve(subs(params, B=B(y),{mass[1], mass[2],
 ics}), {x[1](t),x[2](t)}, numeric);
```

$$\text{sols} := y \rightarrow \text{dsolve}\left(\text{subs}\left(\text{params}, B = B(y), \{\text{ics}, \text{mass}_1, \text{mass}_2\}\right), \{x_1(t), x_2(t)\}, \text{numeric}\right)$$

This is called in the following manner:

```
sols(3);
```

```
proc(rkf45_x) ... end
```

By using two nested `for` loops we construct the surface. The outer loop computes the step response for each band in the damping function, and the inner loop copies the response to thicken the band. Finally, the computed data are displayed using the `surfdata` function found in the `plots` package (Figure 4.21).

```
ANS:=NULL:
for n in [0,1,2,3] do
 TEMP:=plots[odeplot](sols(n), [t, x[1](t)], 0..5,
  umpoints=100);
  'EMP:=op(1,op(1,[op(TEMP)]));
 ANS:=ANS,map((x,y)-[y, op(x)],TEMP,2*n);
 for nn to 2 do
  ANS:=ANS,map((x,y)-[y, op(x)],TEMP,2*n+nn);
 od;
od:
plots[surfdata]([ANS], color=BLACK, orientation=[25, 60],
axes=FRAME,labels=['b1','t','x1(t)']);
```

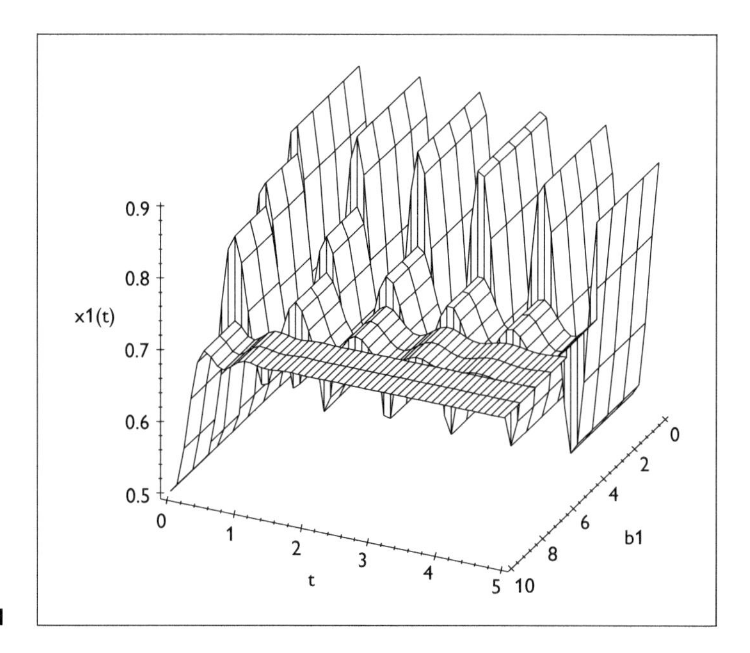

Figure 4.21

Chapter 5

Continuous control application theory

Linear control system analysis

THE MOST COMMON use for Maple's linear algebra capabilities are those applications where matrix algebra is advantageous when dealing with systems that have many constituent equations describing the underlying dynamics. These systems arise in particular when dealing with circuits (mesh and nodal equations) and general control system analysis (stability and sensitivity analysis with many variables).

Many extremely valuable references are available that discuss classical control techniques [1–4]. Most of these texts introduce the use of matrices and linear algebra since most control system applications, whether done in the frequency or the time domain, are described and solved using these mathematical methods.

We deal here with two separate approaches for solving a standard linear controller, namely, the frequency-domain method and the time-domain method. The best way to perform this comparison is to analyze a control problem and compare the desired results, both graphically and analytically. This comparison will demonstrate that linear algebra is a very powerful method for obtaining needed solutions. However, and more importantly, some basic matrix methods coupled with Maple give the user an ability to *play* with internal system dynamics symbolically, hence greatly helping the user understand the underlying general dynamics associated with any variable of any given controller.

We start our discussion with the general feedback model as shown in Figure 5.1. No matter how complicated the controller becomes, most detailed analyses break the bigger system into smaller subsystems as shown in Figure 5.1. Consequently, the importance of knowing how to use Maple to set up, solve, and manipulate variables within the system under investigation is what this chapter is about.

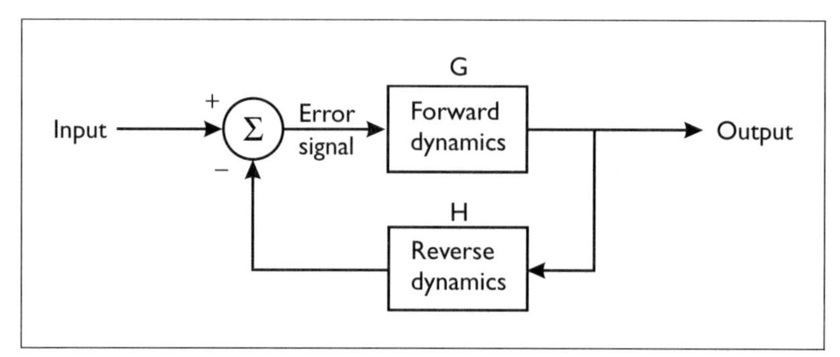

Figure 5.1 Standard feedback controller model.

Figure 5.1 shows the basic feedback controller with designated input, output, feed-forward (forward dynamics), and feedback (reverse dynamics) paths. These are described as follows:

1. *Forward dynamics.* Sometimes called the *plant.* This block represents the dynamics associated with the power or overall system actuator. It is also called the effector or motivator branch.
2. *Reverse dynamics.* Sometimes called the *compensator.* This block represents the dynamics associated with the sensor of the overall system actuator. It is also called the affector or sensor branch.
3. *Summation* ($\sum$). This block creates the difference between the forward and reverse system components. This difference signal is known as the control system error signal.

We start our analysis of the frequency-domain approach with a simple example and compare the results with the time-domain analysis given later.

Frequency-domain approach

Generally, the block components shown in Figure 5.1 are transformed via Laplace transforms for analysis [1–4]. This transform technique allows the user to describe the system dynamics in algebraic terms rather than integrodifferential time-invariant equations. Thus, the Laplace transform of a time function, *f(t)*, becomes *F(s)* as defined by

$$\text{Laplace}[f(t)] = F(s)$$

Mathematically, this is achieved by the following transform:

$$\text{Laplace}[f(t)] = \int_0^\infty f(t)e^{-st}\, dt \equiv F(s)$$

where $s = j\omega$ and j represents the imaginary value of $\sqrt{-1}$.

As a consequence, the system blocks are considered linear, although this graphic representation can certainly be utilized with nonlinear dynamics. However, Laplace transform techniques cannot be used in these cases [1–3].

Let's borrow some system dynamics from an air stabilizer design used in aerodynamics [2] (see Table 5.1). The overall closed-loop or system transfer function, assuming no loading between blocks, is defined as

Table 5.1 Stabilizer controller transfer functions

Control branch	**Laplace transform**
Forward dynamics	$G(s) = \dfrac{2(s+2)^2}{2s^2 + 3s + 2}$
Reverse dynamics	$H(s) = \dfrac{1}{2s}$

The *no loading* between system blocks means that no individual block's output is affected by the input of the following system block.

$$\text{Overall transfer function} = \frac{G(s)}{1 + G(s)H(s)}$$

Incorporating this expression via Maple yields,

```
G := 2*(s+2)^2/(2*s^2+3*s+2):
H := 1/(2*s):
System_Xfer_Function := simplify(G/(1+G*H));
```

$$\text{System\_Xfer\_Function} := \frac{s(s+2)^2}{s^3 + 2\,s^2 + 3\,s + 2}$$

Now, let's give the system an input step function:

$$\text{Input} = \frac{1}{s}$$

The output response is determined by multiplying the system with the input function, or

$$\text{Output} = (\text{System transfer function}) \times (\text{Input})$$

Hence,

```
Input := 1/s:
System_Response := simplify(System_Xfer_Function*Input);
```

$$\text{System\_Response} := \frac{(s+2)^2}{s^3 + 2\,s^2 + 3\,s + 2}$$

Now we invoke the Laplace transform within the `inttrans` library, because we will be converting between the frequency and time domains:

```
with(inttrans):
```

Computing the inverse Laplace transform of the `System_Response`, we obtain

```
System_Response_Time_Domain := evalf(invlaplace(System_
Response,s,t));
```

$$\text{System\_Response\_Time\_Domain} := .5000000000\ e^{(-1.\ t)} + 2.078804601\ e^{(-.5000000000\ t)} \sin(1.322875656\ t) + .5000000000\ e^{(-.5000000000\ t)} \cos(1.322875656\ t)$$

Plotting the output response (Figure 5.2), we see the time-domain system's response to a unit step function as

```
with(plots):
plot (System_Response_Time_Domain,t=0..10,axes=
normal,color=black, labels=[time,response]);
```

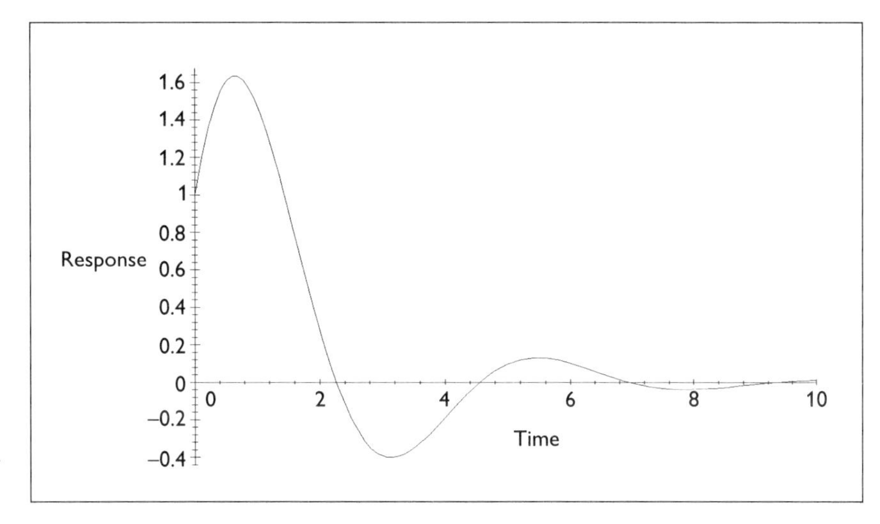

Figure 5.2 Time-domain step response for controller example.

At this point, engineers are interested in the roots of the closed-loop Laplace transfer function expression. The roots of the Laplacian polynomial function, `System_Response`, give the stability and response characteristics of the individual components that comprise the overall closed-loop transfer function. The following Maple structure will factor the polynomial for the characteristic roots:

```
Factored_Denominator := factor(denom(System_Response));
```

$$\text{Factored\_Denominator} := (s + 1)\ (s^2 + s + 2)$$

Maple was able to factor the third-order polynomial, but could not factor the composite quadratic term because that term contains complex roots. However, we can have Maple directly solve the denominator expression for the three roots with the solve command:

```
System_Roots := [evalf(solve(Factored_Denominator=0,s))];
```

$$System\_Roots := [-1., -.5000000000 + 1.322875656\, I, -.5000000000 - 1.322875656\, I]$$

Extracting and separating the roots from the `System_Roots` expression:

```
Root_1 := op(1, System_Roots);
Root_2 := op(2, System_Roots);
Root_3 := op(3, System_Roots);
```

$$Root\_1 := -1.$$
$$Root\_2 := -.5000000000 + 1.322875656\, I$$
$$Root\_3 := -.5000000000 - 1.322875656\, I$$

Now that we have the roots of the characteristic equation of the system, we can see that none of the roots is in the right half plane (i.e., none of the real aspects of the complex roots exhibits a positive number), hence the system is stable. Stability is defined when a bounded input produces a bounded output. If a system exhibits positive real roots, then the output grows exponentially with time for any input [1–3]. Further, [1–3] indicate the following about the three roots:

Root_1 → Pure exponential decay term
Root_2, Root_3 → Exponentially decaying sinusoidal term

In fact, we could have written the general time-domain response directly from these root expressions. Hence, the general form would have been

$$\text{Output}\ (t) = K_1 e^{([\text{Root\_1}])t} + e^{(\text{Re}[\text{Root\_2, 3}])}\left(K_2 \sin(|\text{Im}[\text{Root\_2, 3}]|\,t) + K_3 \cos(|\text{Im}[\text{Root\_2, 3}]|\,t)\right)$$

where

Re[Root_2,3] $\rightarrow$ Real part of either `Root_2` or `Root_3`
Im[Root_2,3] $\rightarrow$ Imaginary part of either `Root_2` or `Root_3`
K_1, K_2, K_3 $\rightarrow$ Coefficients dependent on intial conditions and characteristic roots

Partial fraction expansion

A very useful way to derive the Laplace transforms of complex rational polynomials is to use a method known as partial fraction expansion [1–4]. This method allows the user to deal with smaller rational polynomials that transform easier via the Laplace transform definition. The basic idea here is that an Nth-order rational polynomial is equivalent to N first-order rational polynomials.

Certain polynomials can have their partial fraction expression obtained via Maple; however, for the simple approach, the roots must not be complex (i.e., they must be real numbers). If the roots are complex or purely imaginary, then you should use the `complex` option within the `convert` command argument when using the `parfrac` option. We demonstrate this and another approach for handling complex roots in the next subsection.

Real and distinct roots

Consider the following Laplacian transfer function:

```
Transfer_Function := (s+5)/(s^3+6*s^2+11*s+6);
```

$$\text{Transfer\_Function} := \frac{s + 5}{s^3 + 6\,s^2 + 11\,s + 6}$$

using the `convert` command with the `parfrac` option:

```
Partial_Fraction_Form := convert(Transfer_Function,
 parfrac,s);
```

$$\text{Partial\_Fraction\_Form} := \frac{1}{s+3} - 3\,\frac{1}{s+2} + 2\,\frac{1}{s+1}$$

Now, we can take the Laplace transform of each transfer function contributor, multiply each term by the input driving function, and add them to obtain the complete output time-domain response (linearity property). To illustrate this, let's assign new Maple variables to each contributor:

```
First_Term := op(1,Partial_Fraction_Form);
Second_Term := op(2,Partial_Fraction_Form);
Third_Term := op(3,Partial_Fraction_Form);
```

$$\text{First\_Term} := \frac{1}{s+3}$$

$$\text{Second\_Term} := -3\,\frac{1}{s+2}$$

$$\text{Third\_Term} := 2\,\frac{1}{s+1}$$

Now assume the input is a unit step:

```
Unit_Step_Input := 1/s:
```

Then multiply each contributor with the input transform:

```
First_Term_Laplace := First_Term*Unit_Step_Input;
Second_Term_Laplace := Second_Term*Unit_Step_Input;
Third_Term_Laplace := Third_Term*Unit_Step_Input;
```

$$\text{First\_Term\_Laplace} := \frac{1}{(s+3)\,s}$$

$$\text{Second\_Term\_Laplace} := -3\,\frac{1}{(s+2)\,s}$$

$$\text{Third\_Term\_Laplace} := 2\,\frac{1}{(s+1)\,s}$$

take the inverse transform of each contributor:

```
with(inttrans):
First_Term_Time := invlaplace(First_Term_Laplace,s,t);
Second_Term_Time := invlaplace(Second_Term_Laplace,s,t);
Third_Term_Time := invlaplace(Third_Term_Laplace,s,t);
```

$$\text{First\_Term\_Time} := -\frac{1}{3}\,e^{(-3\,t)} + \frac{1}{3}$$

$$\text{Second\_Term\_Time} := \frac{3}{2}\,e^{(-2\,t)} - \frac{3}{2}$$

$$\text{Third\_Term\_Time} := -2\,e^{(-t)} + 2$$

and add all the output time-domain responses:

```
System_Response_Time := First_Term_Time+Second_Term_Time
 +Third_Term_Time;
```

$$\text{System\_Response\_Time} := -\frac{1}{3}\,e^{(-3\,t)} + \frac{5}{6} + \frac{3}{2}\,e^{(-2\,t)} - 2\,e^{(-t)}$$

Compare this to the direct approach for obtaining the output response to a unit step input function:

```
System_Response_Time_Direct := invlaplace(Transfer_
 Function*Unit_Step_Input,s,t);
```

$$\text{System\_Response\_Time\_Direct} := -\frac{1}{3}\,e^{(-3\,t)} + \frac{5}{6} + \frac{3}{2}\,e^{(-2\,t)} - 2\,e^{(-t)}$$

This last expression is identical to the previously obtained one except that the ability to see the individual effects of the roots is not directly observed. Obviously this method is especially useful when the reader needs to perform individual root analyses as they apply to the output's response.

Real and nondistinct roots

What about Maple's ability to abstract the roots of a polynomial if the roots are real, but nondistinct (repetitive)? Let's create a Laplacian output function that exhibits two repetitive roots at $s = -2$, a distinct root at $s = -3$, and a zero at $s = -1$, namely,

$$\text{System\_Response} = \frac{s + 1}{(s + 2)^2 \ (s + 3)}$$

Enter the function into Maple,

```
System_Response := (s+1)/((s+2)^2*(s+3));
```

$$\text{System\_Response} = \frac{s + 1}{(s + 2)^2 \ (s + 3)}$$

perform the partial fraction expansion,

```
Partial_Fraction_Form := convert(System_Response,parfrac,s);
```

$$\text{Partial\_Fraction\_Form} := -\frac{1}{(s + 2)^2} + 2 \frac{1}{s + 2} - 2 \frac{1}{s + 3}$$

which is correct if you were to do a hand calculation, and then compare it to the coefficient solutions (A, B, C) of the following expression, which is generated by the residue theorem [1–3]:

$$\text{System\_Response} = \frac{s + 1}{(s + 2)^2(s + 3)} = \frac{A}{s + 2} + \frac{B}{(s + 2)^2} + \frac{C}{s + 3}$$

When compared to the `Partial_Fraction_Form`, this would indicate the following values for the coefficients:

$$A = 2$$
$$B = -1$$
$$C = -2$$

To prove this fact, let's perform the analysis in Maple. We start by defining the left- and right-hand sides of the last `System_Response` equation:

$$f1 \rightarrow \text{Left-hand side}$$
$$f2 \rightarrow \text{Right-hand side}$$

Hence,

```
f1 := (s+1)/((s+2)^2*(s+3));
f2 := A/(s+2)+B/(s+2)^2+C/(s+3);
```

$$f1 := \frac{s+1}{(s+2)^2\,(s+3)}$$
$$f2 := \frac{A}{s+2} + \frac{B}{(s+2)^2} + \frac{C}{s+3}$$

Then use Maple's do loop capability to generate the different simultaneous equations (i.e., three different solutions for three different values) required to solve for the three unknown coefficients:

```
for i from 0 to 2 do
F1(i) := subs(s=i,f1):
F2(i) := subs(s=i,f2):
Equation(i) := F1(i)-F2(i):
od:
```

Now with the three simultaneous equations, `Equation(i)`, generated and stored in Maple's memory, we regenerate the subscripted equations, `Equations`, into a list with the `seq` command. Then we abstract each operand within the `Equations` list with the `op` command, and use Maple's `solve` command to generate another list comprised of the coefficient solutions, `Solutions`:

```
Equations := [seq(Equation(i),i=0..2)]:
Eq_1 := op(1,Equations):
Eq_2 := op(2,Equations):
Eq_3 := op(3,Equations):
Solutions := solve({Eq_1=0,Eq_2=0,Eq_3=0},{A,B,C}):
```

Finally, abstracting the results and reassociating those values back with the original unknown variables, we obtain the numerical results of the unknown partial fraction coefficients (A, B, C):

```
Results := subs(Solutions,[A,B,C]):
A := op(1,Results);
B := op(2,Results);
C := op(3,Results);
```

$$A := 2$$
$$B := -1$$
$$C := -2$$

These values agree with the values obtained using Maple's `convert[parfrac]` command. Hence, the reader can see that Maple performs a significant amount of internal computation to obtain the partial fraction values of a given function with the `convert[parfrac]` command.

Continuing, let's isolate the individual Laplacian terms by assigning them variable names:

```
First_Term_Laplace := A/(s+2);
Second_Term_Laplace := B/(s+2)^2;
Third_Term_Laplace := C/(s+3);
```

$$\text{First\_Term\_Laplace} := 2\,\frac{1}{s+2}$$
$$\text{Second\_Term\_Laplace} := -\frac{1}{(s+2)^2}$$
$$\text{Third\_Term\_Laplace} := -2\,\frac{1}{s+3}$$

Now, take the inverse Laplace transform of each partial fraction term:

```
First_Term_Time := invlaplace(First_Term_Laplace,s,t);
Second_Term_Time := invlaplace(Second_Term_Laplace,s,t);
Third_Term_Time := invlaplace(Third_Term_Laplace,s,t);
```

$$\text{First\_Term\_Time} := 2\,e^{(-2\,t)}$$
$$\text{Second\_Term\_Time} := -t\,e^{(-2\,t)}$$
$$\text{Third\_Term\_Time} := -2\,e^{(-3\,t)}$$

The final output time-domain expression is the sum of all partial fraction terms, hence the reader can see that the sum of these time-domain expressions is identical to the direct inverse transform of the `Partial_Fraction_Form` obtained with Maple:

```
System_Response_Time :=
 invlaplace(Partial_Fraction_Form,s,t);
```

$$\text{System\_Response\_Time} := -t\, e^{(-2\, t)} + 2\, e^{(-2\, t)} - 2\, e^{(-3\, t)}$$

or the inverse of the `System_Response`:

```
System_Response_Time_Direct :=
 invlaplace(System_Response,s,t);
```

$$\text{System\_Response\_Time\_Direct} := -t\, e^{(-2\, t)} + 2\, e^{(-2\, t)} - 2\, e^{(-3\, t)}$$

or reiterating our hand-calculated version (now that Maple *remembers* the computed partial fraction coefficients):

```
System_Response_Time_HC := invlaplace(f2,s,t);
```

$$\text{System\_Response\_Time\_HC} := -t\, e^{(-2\, t)} + 2\, e^{(-2\, t)} - 2\, e^{(-3\, t)}$$

which is identical to the two previous time-domain inverse transforms.

The double root at $s = -2$ caused the $te^{(-2t)}$ and $e^{(-2t)}$ form in the output, whereas the real and distinct root at $s = -3$ directly created the $e^{(-3t)}$ term. The reason we know this is because each of the time-domain terms, `First_Term_Time`, etc., corresponds to each of the frequency-domain described roots, `First_Term`, etc., which is directly observed from the Maple commands previously shown, i.e.,

```
First_Term_Laplace := A/(s+2);
Second_Term_Laplace := B/(s+2)^2;
Third_Term_Laplace := C/(s+3);
```

$$First\_Term\_Laplace := 2 \frac{1}{s+2}$$

$$Second\_Term\_Laplace := -\frac{1}{(s+2)^2}$$

$$Third\_Term\_Laplace := -2 \frac{1}{s+3}$$

yielded

```
First_Term_Time := invlaplace(First_Term_Laplace,s,t);
Second_Term_Time := invlaplace(Second_Term_Laplace,s,t);
Third_Term_Time := invlaplace(Third_Term_Laplace,s,t);
```

$$First\_Term\_Time := 2\ e^{(-2\ t)}$$

$$Second\_Term\_Time := -t\ e^{(-2\ t)}$$

$$Third\_Term\_Time := -2\ e^{(-3\ t)}$$

Hence,

Frequency domain ⟶	**Time domain**
$First\_Term\_Laplace := 2 \frac{1}{s+2}$	$First\_Term\_Time := 2\ e^{(-2\ t)}$
$Second\_Term\_Laplace := -\frac{1}{(s+2)^2}$	$Second\_Term\_Time := -t\ e^{(-2\ t)}$
$Third\_Term\_Laplace := -2 \frac{1}{s+3}$	$Third\_Term\_Time := -2\ e^{(-3\ t)}$

Remember, when Maple solves an equation or provides a transform of some function, the order of the result may not (and usually does not) correspond to the order in which the operands were entered into a Maple session for computation. Therefore, be careful of your Maple associations.

Complex roots

Maple can create a partial fraction expression directly from expressions containing complex roots. However, the user must specify the complex option within the `convert[parfrac]` command.

Consider the transfer function from the system that was stated earlier:

$$\text{System\_Xfer\_Function} = \frac{s(s+2)^2}{s^3 + 2s^2 + 3s + 2}$$

Since we already know that Maple can directly solve for the roots, let's obtain the system's output response to a unit step function via partial fraction expansion. Hence, if

$$\text{System\_Response} = \frac{s(s+2)^2}{s^3 + 2s^2 + 3s + 2}\left(\frac{1}{s}\right) = \frac{(s+2)^2}{s^3 + 2s^2 + 3s + 2}$$

Let's attempt to perform the partial fraction expansion directly without the `complex` option:

```
System_Response := ((s+2)^2)/(s^3+2*s^2+3*s+2):
System_Response_Partial_Fraction := convert(System_
 Response,parfrac,s);
```

$$\text{System\_Response\_Partial\_Fraction} := \frac{1}{2}\frac{1}{s+1} + \frac{1}{2}\frac{6+s}{s^2+s+2}$$

and now with the complex option:

```
System_Response_Partial_Fraction_C := convert(System_
 Response,parfrac,s,complex);
```

$$\text{System\_Response\_Partial\_Fraction\_C} := .49998\frac{1}{s+1}$$

$$+ \frac{.25001 + 1.0394\,I}{s + .50000 + 1.3229\,I} + \frac{-.25001 + 1.0394\,I}{-1.\,s - .50000 + 1.3229\,I}$$

Depending on what the user desires, Maple has yielded either the quadratic or fully factored complex partial fraction form. Maple can easily convert either expression into the correct time-domain expression, hence, the quadratic form:

```
System_Response_Time := invlaplace(System_Response_
Partial_Fraction,s,t);
```

$$\text{System\_Response\_Time} := \frac{1}{2}\,e^{(-t)} + \frac{1}{2}\,e^{(-\frac{1}{2}\,t)}\cos\left(\frac{1}{2}\sqrt{7}\,t\right) + \frac{11}{14}\,e^{(-\frac{1}{2}\,t)}\sqrt{7}\,\sin\left(\frac{1}{2}\sqrt{7}\,t\right)$$

or the complex form:

```
System_Response_Time_C := invlaplace(System_Response_
Partial_Fraction,s,t);
```

$$\text{System\_Response\_Time\_C} := .49998\,e^{(-1.\,t)} + (.25001 + 1.0394\,I)\,e^{((-.50000\,-\,1.3229\,I)\,t)} + (.25001 - 1.0394\,I)\,e^{((-.50000\,+\,1.3229\,I)\,t)}$$

We need to eliminate the explicit imaginary terms from the `System_Response_Time_C` expression. To do so, we use the `evalc` command:

```
System_Response_Time_C1 := evalc(invlaplace(System_
Response_Partial_Fraction,s,t));
```

$$\text{System\_Response\_Time\_C1} := .49998\,e^{(-1.\,t)} + .50002\,e^{(-.50000\,t)}\cos(1.3229\,t) + 2.0788\,e^{(-.50000\,t)}\sin(1.3229\,t)$$

or we could have converted the inverse to a trig with the `simplify` and `convert[trig] command[option]`:

```
System_Response_Time_C2 := simplify(convert(invlaplace
 (System_Response_Partial_Fraction,s,t),trig));
```

$$\begin{aligned}\text{System\_Response\_Time\_C2} &:= .49998\ \cosh(t)\\ &- .49998\ \sinh(t) + .50002\ \cosh(.50000\ t)\ \cos(1.3229\ t)\\ &- .50002\ \sinh(.50000\ t)\ \cos(1.3229\ t)\\ &+ 2.0788\ \cosh(.50000\ t)\ \sin(1.3229\ t)\\ &- 2.0788\ \sinh(.50000\ t)\ \sin(1.3229\ t)\end{aligned}$$

Clearly, the `System_Response_Time_C2` expression is not as familiar to work with as either the `System_Response_Time_C1` or `System_Response_Time` expressions. However, these expressions are approximately the same. The reason for the word *approximate* is due to the floating-point operations that Maple performs to obtain the complex computation.

An alternate and manual method for obtaining the time-domain response of `System_Response` is to abstract the root denominators of the partial fraction expansion, `System_Response_Partial_Fraction`, by initially factoring the denominator of the `System_Response` and obtaining the necessary terms to perform a general coefficient partial fraction expansion form:

```
System_Response := ((s+2)^2)/(s^3+2*s^2+3*s+2):
System_Response_Partial_Fraction := convert(System_Response,
 parfrac,s);
```

$$\text{System\_Response\_Partial\_Fraction} := \frac{1}{2}\,\frac{1}{s+1} + \frac{1}{2}\,\frac{6+s}{s^2+s+2}$$

As expected, the quadratic is left in the expansion. Now let's separate the denominator terms to deal with them more easily (we also divide by the common scalar so as to isolate the polynomials in s):

```
First_Root := denom(op(1,System_Response_Partial_
 Fraction))/2;
Quadratic_Root := denom(op(2,System_Response_Partial_
 Fraction))/2;
```

$$\text{First\_Root} := s + 1$$

$$\text{Quadratic\_Root} := s^2 + s + 2$$

As stated previously, we can obtain a valid partial fraction expansion of the System_Response when stated as follows:

$$\text{System\_Response} = \frac{(s + 2)^2}{s^3 + 2s^2 + 3s + 2} = \frac{A}{s + \text{Root\_1}} + \frac{B}{s + \text{Root\_2}} + \frac{C}{s + \text{Root\_3}}$$

where A, B, C, Root_1, Root_2, Root_3 can be real or complex quantities. The general Laplacian form obtained from this approach is

$$\frac{X}{s + \text{Root\_Z}} \xrightarrow{\text{Inverse Laplace}} Xe^{\text{Root\_Z}}$$

Continuing with our problem, we now solve for the roots of both first-order (FRoot_Output) and quadratic (QRoots_Output) terms:

```
FRoot_Output := [solve(First_Root=0,s)];
QRoots_Output := [solve(Quadratic_Root=0,s)];
```

$$\text{FRoot\_Output} := [-1]$$

$$\text{QRoots\_Output} := \left[-\frac{1}{2} + \frac{1}{2} I \sqrt{7}, -\frac{1}{2} - \frac{1}{2} I \sqrt{7}\right]$$

We abstract and assign the roots to some interim Maple variables,

```
FRoot := op(1,FRoot_Output);
QRoot_1 := op(1,QRoots_Output);
QRoot_2 := op(2,QRoots_Output);
```

$$\text{FRoot1} := -1$$

$$\text{Qroots\_1} := -\frac{1}{2} + \frac{1}{2} I \sqrt{7}$$

$$\text{Qroot\_2} := -\frac{1}{2} - \frac{1}{2} I \sqrt{7}$$

generate the quadratic root's partial fraction expansion along with the first-order root operands,

```
PFraction_1 := A/(s-FRoot);
PFraction_2 := B/(s-QRoot_1);
PFraction_3 := C/(s-QRoot_2);
```

$$Pfraction\_1 := \frac{A}{s+1}$$

$$Pfraction\_2 := \frac{B}{s+\frac{1}{2}-\frac{1}{2}I\sqrt{7}}$$

$$Pfraction\_3 := \frac{C}{s+\frac{1}{2}+\frac{1}{2}I\sqrt{7}}$$

and add the individual partial fraction terms to obtain the total Laplace output response for the input unit step function:

```
Partial_Fraction_Output := PFraction_1+PFraction_2+
 PFraction_3;
```

$$Partial\_Fraction\_Output :=$$
$$\frac{A}{s+1}+\frac{B}{s+\frac{1}{2}-\frac{1}{2}I\sqrt{7}}+\frac{C}{s+\frac{1}{2}+\frac{1}{2}I\sqrt{7}}$$

Solving for the A,B,C coefficients a little differently than before, we generate a set of equations with arbitrarily assigned values of the Laplace variable, s:

```
PF_Output_1 := subs(s=1,System_Response)-subs(s=1,Partial_
 Fraction_Output);
PF_Output_2 := subs(s=2,System_Response)-subs(s=2,Partial_
 Fraction_Output);
PF_Output_3 := subs(s=3,System_Response)-subs(s=3,Partial_
 Fraction_Output);
```

$$PF\_Output\_1 := \frac{9}{8} - \frac{1}{2}A - \frac{B}{\frac{3}{2} - \frac{1}{2}I\sqrt{7}} - \frac{C}{\frac{3}{2} + \frac{1}{2}I\sqrt{7}}$$

$$PF\_Output\_2 := \frac{2}{3} - \frac{1}{3}A - \frac{B}{\frac{5}{2} - \frac{1}{2}I\sqrt{7}} - \frac{C}{\frac{5}{2} + \frac{1}{2}I\sqrt{7}}$$

$$PF\_Output\_3 := \frac{25}{56} - \frac{1}{4}A - \frac{B}{\frac{7}{2} - \frac{1}{2}I\sqrt{7}} - \frac{C}{\frac{7}{2} + \frac{1}{2}I\sqrt{7}}$$

and solve the simultaneous equations:

```
Solutions   :=solve({PF_Output_1=0,PF_Output_2=0,
PF_Output_3=0},{A,B,C}) ;
```

$$Solutions := \left\{A = \frac{1}{2}, C = \frac{1}{4} + \frac{11}{28}I\sqrt{7}, B = \frac{1}{4} - \frac{11}{28}I\sqrt{7}\right\}$$

We abstract the results and assign the values to the appropriate coefficient:

```
Results := subs(Solutions,[A,B,C]):
A := op(1,Results);
B := op(2,Results);
C := op(3,Results);
```

$$A := \frac{1}{2}$$

$$B := \frac{1}{4} - \frac{11}{28}I\sqrt{7}$$

$$C := \frac{1}{4} + \frac{11}{28}I\sqrt{7}$$

and restate the original output partial fraction function to obtain Maple's response:

```
Partial_Fraction_Output;
```

$$\frac{1}{2}\frac{1}{s+1}+\frac{\frac{1}{4}-\frac{11}{28}I\sqrt{7}}{s+\frac{1}{2}-\frac{1}{2}I\sqrt{7}}+\frac{\frac{1}{4}+\frac{11}{28}I\sqrt{7}}{s+\frac{1}{2}+\frac{1}{2}I\sqrt{7}}$$

Since we have complex terms in the coefficients, let's simply use Maple's `evalc` command when taking the inverse Laplace transform to interpret these automatically into contributing output response phase terms:

```
System_Response_Time :=
 evalc(invlaplace(Partial_Fraction_ Output,s,t));
```

$$\text{System\_Response\_Time} := \frac{1}{2}e^{(-t)}+\frac{1}{2}e^{(-1/2\,t)}\cos\left(\frac{1}{2}t\sqrt{7}\right)+\frac{11}{14}\sqrt{7}\,e^{(-1/2\,t)}\sin\left(\frac{1}{2}t\sqrt{7}\right)$$

Finally, plotting the unit step response,

```
with(plots):
plot(System_Response_Time,t=0..10,color=black,axes=normal,
 labels=[Time,Response]);
```

As Figure 5.3 shows, the result is identical to the previously plotted Figure 5.2.

Let's compare this final result with Maple's direct inverse Laplace transform of the `System_Response`:

```
Direct_Result := invlaplace(System_Response,s,t);
```

$$\text{Direct\_Result} :=$$
$$\frac{1}{2}e^{(-t)}+\frac{1}{2}e^{(-1/2\,t)}\cos\left(\frac{1}{2}t\sqrt{7}\right)+\frac{11}{14}\sqrt{7}\,e^{(-1/2\,t)}\sin\left(\frac{1}{2}t\sqrt{7}\right)$$

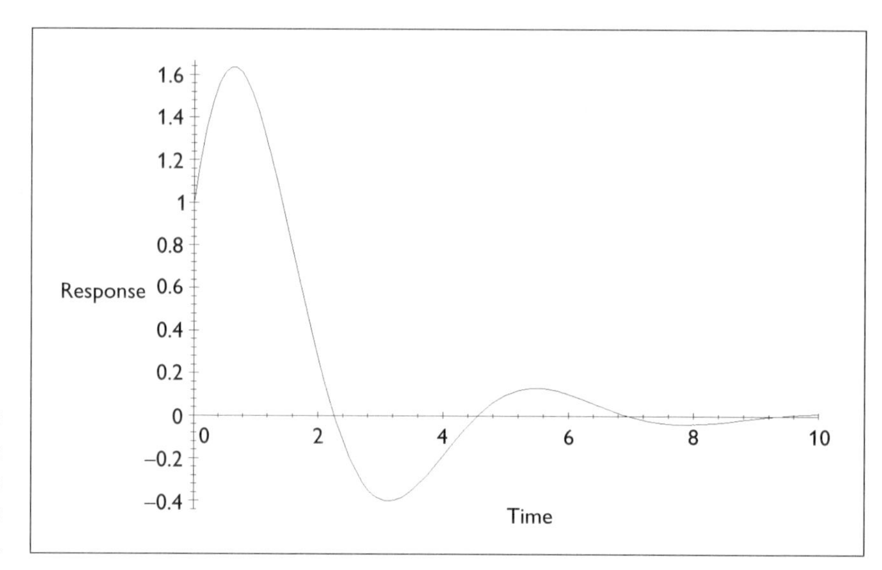

Figure 5.3 Partial fraction expansion solution to unit step response.

The reader will see that the `Direct_Result` is identical to the partial fraction result, `System_Response_Time`.

In summary, the use of partial fraction expansion is useful when analyzing the effect of individual terms or roots associated with any linear system function. Originally, partial fractions were used when symbolic mathematics packages were not available to students and professionals and obtaining time-domain solutions from large Laplace transforms was either impossible or cumbersome. Partial fraction expansion breaks the problem down into smaller transforms. In this way, the student or professional can obtain the inverse by hand and/or look-up tables.

Time-domain approach

The state space approach to analyzing control systems has become extremely popular due to the advent of computers. In fact, this method has almost become the *de facto* standard given the prevalence of the personal computer.

State space techniques use matrix representation of system parameters to ascertain transient, steady-state, and general dynamic responses of very high order linear and some nonlinear systems [1,3,5,6].

Time-invariant versus time-variant systems

For the remainder of this chapter, we examine constant coefficient matrices (i.e., linear time-invariant systems). Time-dependent coefficient or linear time-variant systems can require the users to perform a similarity or other type of transform [3–6] of the state space vector equation, which the authors do not want to involve the reader with at this time.

The authors do not want to confuse the reader by delving into some odd systems that require this or other transformation techniques while studying the main chapter. We only want to expose the reader to the basic approach of state space using Maple from which all other techniques are usually derived or based. If the reader is more interested in this and other matrix transformations, they are referred to the cited references. A basic similarity transformation is depicted in Appendix A at the end of the book.

Analysis of a time-invariant system: fundamentals

Our first approach is to characterize the system by the use of state variables, which are generated by choosing specific *states* or nodes (x_i) of interest in the system under consideration. One starts by creating a set of state equations and putting them into a *normal* equation form, which simply means the set of simultaneous equations representing a linear transformation of coupled dependent variables, hence,

$$\begin{aligned}
\dot{x}_1 &= a_{11}x_1 + a_{12}x_2 + a_{13}x_3 + \ldots \\
&\quad + a_{1N}x_N + b_{11}u_1 + \ldots + b_{1M}u_M \\
\dot{x}_2 &= a_{21}x_1 + a_{22}x_2 + a_{23}x_3 \ldots \\
&\quad + a_{2N}x_N + b_{21}u_1 + \ldots + b_{2M}u_M \\
&\ \ \vdots \\
\dot{x}_N &= a_{N1}x_1 + a_{N2}x_2 + a_{N3}x_3 + \ldots \\
&\quad + a_{NN}x_N + b_{N1}u_1 + \ldots + b_{NM}u_M
\end{aligned}$$

which can be implemented into matrix formulation as

$$\frac{d}{dt}\begin{bmatrix} x_1 \\ x_2 \\ x_3 \\ \cdot \\ \cdot \\ \cdot \\ x_N \end{bmatrix} = \begin{bmatrix} a_{11} & a_{12} & a_{13} & \cdot & \cdot & \cdot & a_{1N} \\ a_{21} & a_{22} & a_{23} & \cdot & \cdot & \cdot & a_{2N} \\ a_{31} & a_{32} & a_{33} & \cdot & \cdot & \cdot & a_{3N} \\ \cdot & & & \cdot & & & \cdot \\ \cdot & & & & \cdot & & \cdot \\ \cdot & & & & & \cdot & \cdot \\ a_{N1} & a_{N2} & a_{N3} & \cdot & \cdot & \cdot & a_{NN} \end{bmatrix} \begin{bmatrix} x_1 \\ x_2 \\ x_3 \\ \cdot \\ \cdot \\ \cdot \\ x_N \end{bmatrix} + \begin{bmatrix} b_{11} & b_{12} & \cdot & \cdot & b_{1M} \\ b_{21} & b_{22} & & & b_{2M} \\ b_{31} & b_{32} & & & b_{3M} \\ \cdot & & \cdot & & \cdot \\ \cdot & & & \cdot & \cdot \\ \cdot & & & & \cdot \\ b_{N1} & \cdot & \cdot & \cdot & b_{NM} \end{bmatrix} \begin{bmatrix} u_1 \\ u_2 \\ u_3 \\ \cdot \\ \cdot \\ \cdot \\ u_M \end{bmatrix}$$

which can then be restated in a matrix shorthand vector matrix form as

$$\dot{\mathbf{x}} = \mathbf{Ax} + \mathbf{Bu}$$

or, in general,

$$\dot{\mathbf{x}}(t) = \mathbf{A}(t)\mathbf{x}(t) + \mathbf{B}(t)\mathbf{u}(t)$$

Similarly, the output vector $\mathbf{y}$ of any system can be stated as follows:

$$\begin{bmatrix} y_1 \\ y_2 \\ y_3 \\ \cdot \\ \cdot \\ \cdot \\ y_P \end{bmatrix} = \begin{bmatrix} c_{11} & c_{12} & c_{13} & \cdot & \cdot & \cdot & c_{1N} \\ c_{21} & c_{22} & c_{23} & \cdot & \cdot & \cdot & c_{2N} \\ c_{31} & c_{32} & c_{33} & \cdot & \cdot & \cdot & c_{3N} \\ \cdot & & & \cdot & & & \cdot \\ \cdot & & & & \cdot & & \cdot \\ \cdot & & & & & \cdot & \cdot \\ c_{P1} & c_{P2} & c_{P3} & \cdot & \cdot & \cdot & c_{PN} \end{bmatrix} \begin{bmatrix} x_1 \\ x_2 \\ x_3 \\ \cdot \\ \cdot \\ \cdot \\ x_N \end{bmatrix} + \begin{bmatrix} d_{11} & \cdot & \cdot & \cdot & d_{1M} \\ d_{21} & & & & d_{2M} \\ d_{31} & & & & d_{3M} \\ \cdot & & & & \cdot \\ \cdot & & & & \cdot \\ \cdot & & & & \cdot \\ d_{P1} & \cdot & \cdot & \cdot & d_{PM} \end{bmatrix} \begin{bmatrix} u_1 \\ u_2 \\ u_3 \\ \cdot \\ \cdot \\ \cdot \\ u_M \end{bmatrix}$$

which, again, can be restated in shorthand vector matrix form as

$$\mathbf{y} = \mathbf{Cx} + \mathbf{Du}$$

For our particular example, we will only have *one input* and *one output* so the vector matrix equations reduce to

$$\dot{\mathbf{x}} = \mathbf{Ax} + \mathbf{B}u$$
$$y = \langle\mathbf{Cx}\rangle + du$$

where $u, y, \langle\mathbf{Cx}\rangle$, and d are scalars ($\langle\mathbf{Cx}\rangle$ represents the *inner* or *scalar* product).

Let's start by converting the previous frequency-domain transfer function into a time differential form, i.e.,

$$\frac{\text{Output}}{\text{Input}} = \frac{s(s + 2)^2}{s^3 + 2s^2 + 3s + 2}$$

First, cross-multiplying the Laplace transfer function, we obtain

$$\text{Output}\left[s^3 + 2s^2 + 3s + 2\right] = \text{Input}\left[s(s + 2)^2\right]$$

letting Output $\rightarrow y(t)$, Input $\rightarrow u(t)$ and realizing that $s \rightarrow {}^{d}\!/_{dt}$, then the previous expression becomes (assuming initial conditions are zero):

$$\frac{d^3y(t)}{dt^3} + 2\,\frac{d^2y(t)}{dt^2} + 3\,\frac{dy(t)}{dt} + 2y(t) = \frac{d^3u(t)}{dt^3} + 4\,\frac{d^2u(t)}{dt^2} + 4\,\frac{du(t)}{dt}$$

or in the shorthand dot notation,

$$\dddot{y} + 2\ddot{y} + 3\dot{y} + 2y = \dddot{u} + 4\ddot{u} + 4\ddot{u}$$

Restating this problem into the normal state space equation form, we get

$$\dot{x}_1 = x_2$$
$$\dot{x}_2 = x_3$$
$$\dot{x}_3 \equiv \dddot{y} = -2\ddot{y} - 3\dot{y} - 2y + \dddot{u} + 4\ddot{u} + 4\dot{u}$$

The problem with directly implementing this form of the simulation is the rather difficult aspect of handling input derivatives (i.e., $\dddot{u}$, $\ddot{u}$, $\dot{u}$). Derivatives, whether done in a numerical or symbolic simulation, create problems due to the noisy nature associated with these operators. Noisy operation means that, computationally, derivatives are very sensitive to slight variations in value. Therefore, any extraneous artifact not associated with the solution is amplified and will directly contribute to a lower quality result. Consequently, utilizing the following state flow diagram more easily facilitates the vector matrix setup for either a time- or frequency-domain analysis, regardless of the input derivative variable number or order.

Figure 5.4 shows the general simulation diagram formulation for a system that can be expressed as a rational polynomial in s or time-domain variable derivative of the form

$$\frac{y(s)}{u(s)} = \frac{a_n s^n + a_{n-1} s^{n-1} + \ldots + a_2 s^2 + a_1 s + a_0}{b_m s^m + b_{m-1} s^{m-1} + \ldots + b_2 s^2 + b_1 s + b_0}$$

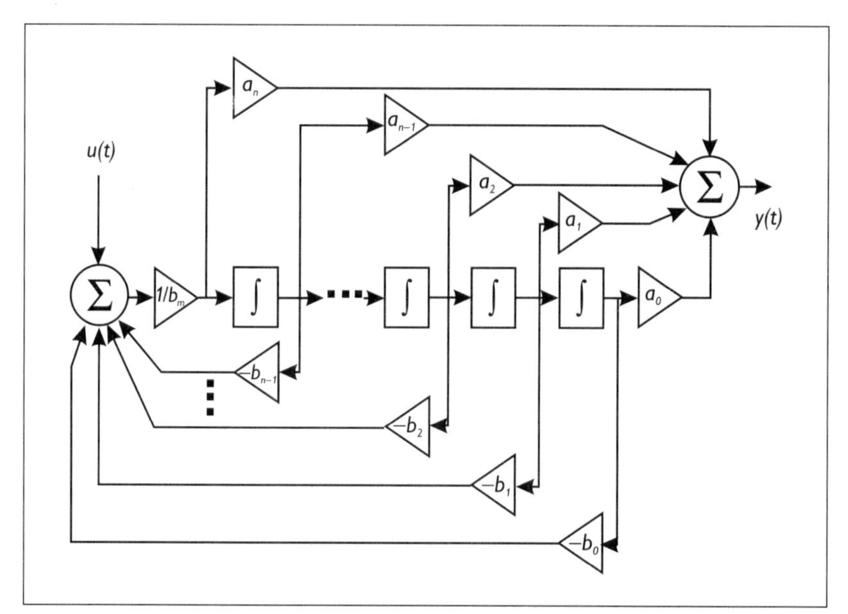

Figure 5.4
General simulation diagram.

Simulation diagram form for an arbitrary rational polynomial function. Hence, in the time-domain differential form,

$$\frac{y(t)}{u(t)} = \frac{a_n \frac{d^n}{dt^n} + a_{n-1} \frac{d^{n-1}}{dt^{n-1}} + \ldots + a_2 \frac{d^2}{dt^2} + a_1 \frac{d}{dt} + a_0}{b_m \frac{d^m}{dt^m} + b_{m-1} \frac{d^{m-1}}{dt^{m-1}} + \ldots + b_2 \frac{d^2}{dt^2} + b_1 \frac{d}{dt} + b_0}$$

for

$$m \geq n$$

This canonical simulation form is one of the easiest to implement provided one realizes that the state variables (i.e., $\dot{x}_1, \dot{x}_2, \dot{x}_3, \ldots, \dot{x}_{m-2}, \dot{x}_{m-1}$) are not directly representative of the different output (i.e.,y, $\dot{y}$, $\ddot{y}$, etc.) variable orders. However, one can easily obtain the solutions any of these y variables by directly taking the derivative of the output vector matrix equation. Therefore, plugging in our specific example into Figure 5.4 yields the simulation diagram of Figure 5.5.

From Figure 5.5, we can generate the previous example's state space matrices by inspection, thus the form

$$\dot{\mathbf{x}} = \mathbf{Ax} + \mathbf{B}u$$
$$y = \langle \mathbf{Cx} \rangle + du$$

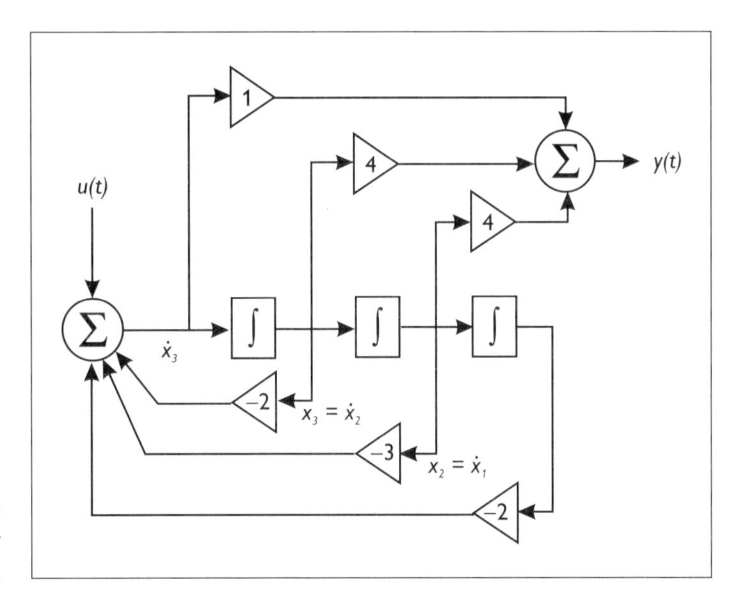

Figure 5.5 Simulation diagram for previously given example.

becomes

$$\begin{bmatrix} \dot{x}_1 \\ \dot{x}_2 \\ \dot{x}_3 \end{bmatrix} = \begin{bmatrix} 0 & 1 & 0 \\ 0 & 0 & 1 \\ -2 & -3 & -2 \end{bmatrix} \begin{bmatrix} x_1 \\ x_2 \\ x_3 \end{bmatrix} + \begin{bmatrix} 0 \\ 0 \\ 0 \end{bmatrix} u(t)$$

$$y(t) = [-2 \quad 1 \quad 2] \begin{bmatrix} x_1 \\ x_2 \\ x_3 \end{bmatrix} + [1] \; u(t)$$

$$= -2x_1 + x_2 + 2x_3 + u(t)$$

where

$$\mathbf{A} = \begin{bmatrix} 0 & 1 & 0 \\ 0 & 0 & 1 \\ -2 & -3 & -2 \end{bmatrix} \qquad \mathrm{B} = \begin{bmatrix} 0 \\ 0 \\ 0 \end{bmatrix} \equiv$$

$$\mathbf{C} = [-2 \quad 1 \quad 2] \qquad d = [1]$$

As the reader can see from the scalar output equation, the complete output time-domain solution for this system $y(t)$ is obtained by solving the column state vector for any given input function, $u(t)$:

$$\begin{bmatrix} x_1 \\ x_2 \\ x_3 \end{bmatrix} \equiv \mathbf{x} \text{ or } \mathbf{x}(t)$$

The state transition matrix

The time-domain solution for the state space approach uses the following formulation [2,3]:

$$\mathbf{x}(t) = \Phi(t - t_0)\mathbf{x}(t_0) + \int_{t_0}^{t} \Phi(t - \lambda)\mathbf{B}u(\lambda) \; d\lambda$$

where $\Phi(t - t_0)$ is the state transition matrix initiated at t_0 and is defined as

$$\Phi(t - t_0) \equiv e^{\mathbf{A}(t - t_0)}$$

Therefore, evaluating our specific transition matrix, and assuming $t_0 = 0$, we have the following transition matrix expression:

$$\Phi(t) \equiv e^{\mathbf{A}t} = e^{\begin{bmatrix} 0 & t & 0 \\ 0 & 0 & t \\ -2t & -3t & -2t \end{bmatrix}}$$

Cayley-Hamilton theorem

The Cayley-Hamilton theorem [2–4] says that a matrix solves its own characteristic equation (via the eigenvalues); hence, the transition or exponential matrix can be solved by implementing this theorem. Therefore, because we have a third-order system, the transition matrix becomes

$$e^{\mathbf{A}t} = \alpha_2(\mathbf{A}t)^2 + \alpha_1(\mathbf{A}t) + \alpha_0\mathbf{I}$$

where α_2, α_1, α_0 are scalars and represents a 3×3 identity matrix. Obviously, this matrix equation creates a 3×3 square matrix on both sides of the equation. Now we implement a scalar representation of the matrix equation since the Cayley-Hamilton theorem says a scalar form of eigenvalues (λ) will solve for the three unknown coefficients, α_2, α_1, α_0...

$$e^{\lambda_i t} = \alpha_2(\lambda_i t)^2 + \alpha_1(\lambda_i t) + \alpha_0$$

where there are three scalar equations to solve in a simultaneous fashion due to the order of the system. Hence,

$$e^{\lambda_1 t} = \alpha_2(\lambda_1 t)^2 + \alpha_1(\lambda_1 t) + \alpha_0$$

$$e^{\lambda_2 t} = \alpha_2(\lambda_2 t)^2 + \alpha_1(\lambda_2 t) + \alpha_0$$

$$e^{\lambda_3 t} = \alpha_2(\lambda_3 t)^2 + \alpha_1(\lambda_3 t) + \alpha_0$$

State space analysis with Maple

Now, let's determine the three (λ_1, λ_2, λ_3) eigenvalues via the following matrix formulation:

$$\det|\mathbf{A} - \lambda\mathbf{I}| = 0$$

Hence,

```
with (linalg):
A_Matrix := array ([[0,1,0],[0,0,1],[-2,-3,-2]]):
Identity := array ([[1,0,0],[0,1,0],[0,0,1]]):
Interim_1 := evalm (A_Matrix-(lambda)*Identity):
Interim_2 := det (Interim_1):
Interim_3 := solve (Interim_2=0,lambda):
Eigenvalue_1 := Interim_3[1];
Eigenvalue_2 := Interim_3[2];
Eigenvalue_3 := Interim_3[3];
```

$$\text{Eigenvalue\_1} := -1$$
$$\text{Eigenvalue\_2} := -\frac{1}{2} + \frac{1}{2} I\sqrt{7}$$
$$\text{Eigenvalue\_3} := -\frac{1}{2} - \frac{1}{2} I\sqrt{7}$$

or using Maple's `eigenvals` operand and abstracting the eigenvalues, we obtain the same result:

```
Eigenvalues := [eigenvals(A_Matrix)];
Eigenvalue_1 := op(1,Eigenvalues);
Eigenvalue_2 := op(2,Eigenvalues);
Eigenvalue_3 := op(3,Eigenvalues);
```

$$\text{Eigenvalues} := \left[-1, -\frac{1}{2} + \frac{1}{2} I\sqrt{7}, -\frac{1}{2} - \frac{1}{2} I\sqrt{7}\right]$$
$$\text{Eigenvalue\_1} := -1$$
$$\text{Eigenvalue\_2} := -\frac{1}{2} + \frac{1}{2} I\sqrt{7}$$
$$\text{Eigenvalue\_3} := -\frac{1}{2} - \frac{1}{2} I\sqrt{7}$$

Notice the *I* or imaginary component to our complex eigenvalues. Also, note that complex eigenvalues must appear as conjugate pairs (`Eigenvalue_2` and `Eigenvalue_3`), whereas eigenvalue 1 is real, as we have seen before in the frequency-domain approach.

Continuing with our state transition matrix computation, remember that

$$e^{\mathbf{A}t} = \alpha_2(\mathbf{A}t)^2 + \alpha_1(\mathbf{A}t) + \alpha_0\mathbf{I}$$

therefore via the Cayley-Hamilton theorem we generate the simultaneous equation set as follows:

$$\begin{aligned} e^{\lambda_1 t} &= \alpha_2(\lambda_1 t)^2 + \alpha_1(\lambda_1 t) + \alpha_0 \\ e^{\lambda_2 t} &= \alpha_2(\lambda_2 t)^2 + \alpha_1(\lambda_2 t) + \alpha_0 \\ e^{\lambda_3 t} &= \alpha_2(\lambda_3 t)^2 + \alpha_1(\lambda_3 t) + \alpha_0 \end{aligned}$$

or put into another form for Maple to solve

$$\begin{aligned} \text{Equation\_1} &= \alpha_2(\lambda_1 t)^2 + \alpha_1(\lambda_1 t) + \alpha_0 - e^{\lambda_1 t} \\ \text{Equation\_2} &= \alpha_2(\lambda_2 t)^2 + \alpha_1(\lambda_2 t) + \alpha_0 - e^{\lambda_2 t} \\ \text{Equation\_3} &= \alpha_2(\lambda_3 t)^2 + \alpha_1(\lambda_3 t) + \alpha_0 - e^{\lambda_3 t} \end{aligned}$$

This is clearly messy, thus,

```
Equation_1 := alfa2*(Eigenvalue_1*t)^2+alfa1*
 (Eigenvalue_1*t)+alfa0-exp(Eigenvalue_1 *t);
Equation_2 := alfa2*(Eigenvalue_2*t)^2+alfa1*
 (Eigenvalue_2*t)+alfa0-exp(Eigenvalue_2*t);
Equation_3 := alfa2*(Eigenvalue_3*t)^2+alfa1*
 (Eigenvalue_3*t)+alfa0-exp(Eigenvalue_3*t);
```

$$Equation\_1 := alfa2\ t^2 - alfa1\ t + alfa0 - e^{-t}$$

$$Equation\_2 := alfa2\left(-\frac{1}{2}+\frac{1}{2}I\sqrt{7}\right)^2 t^2 + alfa1\left(-\frac{1}{2}+\frac{1}{2}I\sqrt{7}\right)t + alfa0 - e^{((-1/2 + 1/2\,I\,\sqrt{7})\,t)}$$

$$Equation\_3 := alfa2\left(-\frac{1}{2}-\frac{1}{2}I\sqrt{7}\right)^2 t^2 + alfa1\left(-\frac{1}{2}-\frac{1}{2}I\sqrt{7}\right)t + alfa0 - e^{((-1/2 - 1/2\,I\,\sqrt{7})\,t)}$$

and then

```
Solutions := solve({Equation_1=0,Equation_2=0,
 Equation_3=0},{alfa2,alfa1,alfa0});
```

$$Solutions := \Bigg\{alfa1 = -\frac{1}{28}I\left(5\,e^{(I\sqrt{7}\,t)} + 2\,I\sqrt{e^{(I\sqrt{7}\,t)}}\sqrt{7}\,e^{(-t)}\sqrt{e^t} - I\sqrt{7}\,e^{(I\sqrt{7}\,t)} - 5 - I\sqrt{7}\right)\sqrt{7}\Big/\left(t\sqrt{e^{(I\sqrt{7}\,t)}}\sqrt{e^t}\right),\ alfa2 = -\frac{1}{448}\Big(-6\,I\sqrt{7}\,e^{(I\sqrt{7}\,t)} + 35\,e^{(-t)}\sqrt{e^{(I\sqrt{7}\,t)}}\sqrt{e^t} - 14\,e^{(I\sqrt{7}\,t)} - I\sqrt{7} - 21 + 7\,I\sqrt{e^{(I\sqrt{7}\,t)}}\sqrt{7}\,e^{(-t)}\sqrt{e^t}\Big)(-5 + I\sqrt{7})\Big/\left(t^2\sqrt{e^{(I\sqrt{7}\,t)}}\sqrt{e^t}\right),\ alfa0 = \frac{1}{7}\frac{-I\sqrt{7}\,e^{(I\sqrt{7}\,t)} + 7\,e^{(-t)}\sqrt{e^{(I\sqrt{7}\,t)}}\sqrt{e^t} + I\sqrt{7}}{\sqrt{e^{(I\sqrt{7}\,t)}}\sqrt{e^t}}\Bigg\}$$

Now abstract the roots and assign them to the appropriate variable:

```
XX := subs(Solutions,[alfa2,alfa1,alfa0]):
alfa2 := simplify(XX[1]);
alfa1 := simplify(XX[2]);
alfa0 := simplify(XX[3]);
```

$$\text{alfa2} := \left(4\,e^{\left(\frac{1}{2}t\,(-1+I\sqrt{7})\right)} + 3\,e^{\left(-\frac{1}{2}t\,(I\sqrt{7}+1)\right)} - I\sqrt{7}\,e^{\left(-\frac{1}{2}t\,(I\sqrt{7}+1)\right)} + I\,e^{(-t)}\sqrt{7} - 7\,e^{(-t)}\right)\left(-\frac{1}{112}I\sqrt{7} - \frac{1}{16}\right)\Big/ t^2$$

$$\text{alfa1} := \left(16\,e^{\left(-\frac{1}{2}t(I\sqrt{7}+1)\right)} + 5\,I\sqrt{7}\,e^{\left(\frac{1}{2}t(-1+I\sqrt{7})\right)} - 9\,e^{\left(\frac{1}{2}t(-1+I\sqrt{7})\right)} - 5\,Ie^{(-t)}\sqrt{7} - 7\,e^{(-t)}\right)\left(-\frac{1}{64} + \frac{5}{448}I\sqrt{7}\right)\Big/ t$$

$$\text{alfa0} := -\frac{1}{7}I\sqrt{7}\,e^{\left(\frac{1}{2}t(-1+I\sqrt{7})\right)} + e^{(-t)} + \frac{1}{7}I\sqrt{7}\,e^{\left(-\frac{1}{2}t(I\sqrt{7}+1)\right)}$$

Note the presence of the imaginary term, I. Unfortunately, Maple cannot automatically "see" certain trigonometric identities which would absorb the imaginary and real terms into trigonometric functions and, hence, would greatly ease our computations and simplify our results. The particular trigonometric identities of interest here are

$$\cos(\theta) = \frac{e^{I\theta} + e^{-I\theta}}{2}$$

$$\sin(\theta) = \frac{e^{I\theta} - e^{-I\theta}}{2I}$$

However, Maple can see these identities when asked to do so in a certain way. Let's reformulate the `alfa` coefficients by asking Maple to *combine* the exponentials and other terms in the coefficients into any applicable trigonometric forms with the `combine(expression,trig)` command. We will further ask Maple to *simplify* those trigonometric results before displaying them. The `evalc` command requires the combination process to use and be cognizant of complex forms in the trigonometric conversions.

```
alfa_0 := simplify(evalc(combine(alfa0,trig)));
alfa_1 := simplify(evalc(combine(alfa1,trig)));
alfa_2 := simplify(evalc(combine(alfa2,trig)));
```

$$\text{alfa\_0} := \frac{2}{7}\sqrt{7}\, e^{(-\frac{1}{2}t)} \sin\left(\frac{1}{2}\sqrt{7}\,t\right) + e^{(-t)}$$

$$\text{alfa\_1} := \frac{1}{14}\,\frac{-7\,e^{(-\frac{1}{2}t)}\cos\left(\frac{1}{2}\sqrt{7}\,t\right) + 5\sqrt{7}\,e^{(-\frac{1}{2}t)}\sin\left(\frac{1}{2}\sqrt{7}\,t\right) + 7\,e^{(-t)}}{t}$$

$$\text{alfa\_2} := \frac{1}{14}\,\frac{-7\,e^{(-\frac{1}{2}t)}\sin\left(\frac{1}{2}\sqrt{7}\,t\right) - 7\,e^{(-\frac{1}{2}t)}\cos\left(\frac{1}{2}\sqrt{7}\,t\right) + 7\,e^{(-t)}}{t^2}$$

Now we have the coefficients in terms that are necessary for substitution into the transition matrix. Therefore, solving for the transition matrix (assuming $t_0 = 0$):

$$\Phi(t) \equiv e^{\mathbf{A}t} = e^{\begin{bmatrix} 0 & t & 0 \\ 0 & 0 & t \\ -2t & -3t & -2t \end{bmatrix}} = \alpha_2(\mathbf{A}t)^2 + \alpha_1(\mathbf{A}t) + \alpha_0\mathbf{I}$$

by directly substituting the alpha coefficients (`alfa_0,alfa_1, alfa_2`) and performing the following substitution into the time variable t as

$$t = t - \zeta$$

we can set up the Transition_Matrix ($\Phi(t-\zeta)$ expression into the following (zero initial conditions and $t_0 = 0$) for integration. Also, substituting the input step function, $u(\zeta) = 1$, and the $\mathbf{B}$ matrix will also finalize the integrand product setup, hence, symbolically we perform the following:

$$\mathbf{x}(t) = \Phi(t - t_0)\mathbf{x}(t_0) + \int_{t_0}^{t} \Phi(t - \zeta)\mathbf{B}u(\zeta)\, d\zeta$$

for $t_0 = 0$ and $\mathbf{x}(t_0) = \mathbf{0}$ then becomes

$$\mathbf{x}(t) = \int_0^t \Phi(t - \zeta)\mathbf{B}u(\zeta)\ d\zeta$$

for $u(\zeta) = 1$ which becomes

$$\mathbf{x}(t) = \int_0^t \Phi(t - \zeta)\mathbf{B}\ d\zeta$$

Implementing this with the Maple commands,

```
Transition_Matrix := evalm(subs(t=t-zeta,alfa_2*
  (A_Matrix*t)^2+alfa_1*A_Matrix*t+alfa_0*Identity)):
B_Matrix := array([[0],[0],[1]]):
```

Now with the integrand component computed, we perform matrix multiplication of the `B_Matrix` and `Transition_Matrix` matrices, which completes the integrand expression

```
Integrand := evalm(Transition_Matrix&*B_Matrix);
```

Integrand :=

$$\left[\frac{1}{14}\sqrt{7}\ e^{(-\frac{1}{2}t + \frac{1}{2}\zeta)}\sin\left(\frac{1}{2}\sqrt{7}\ (t - \zeta)\right) - \frac{1}{2}\ e^{(-\frac{1}{2}t + \frac{1}{2}\zeta)}\cos\left(\frac{1}{2}\sqrt{7}\ (t - \zeta)\right) + \frac{1}{2}\ e^{(-t + \zeta)}\right]$$

$$\left[\frac{3}{14}\sqrt{7}\ e^{(-\frac{1}{2}t + \frac{1}{2}\zeta)}\sin\left(\frac{1}{2}\sqrt{7}\ (t - \zeta)\right) + \frac{1}{2}\ e^{(-\frac{1}{2}t + \frac{1}{2}\zeta)}\cos\left(\frac{1}{2}\sqrt{7}\ (t - \zeta)\right) - \frac{1}{2}\ e^{(-t + \zeta)}\right]$$

$$\left[-\frac{5}{14}\sqrt{7}\ e^{(-\frac{1}{2}t + \frac{1}{2}\zeta)}\sin\left(\frac{1}{2}\sqrt{7}\ (t - \zeta)\right) + \frac{1}{2}\ e^{(-\frac{1}{2}t + \frac{1}{2}\zeta)}\cos\left(\frac{1}{2}\sqrt{7}\ (t - \zeta)\right) + \frac{1}{2}\ e^{(-t + \zeta)}\right]$$

Obtain the individual state variables ($\mathbf{x}(t)$) results:

```
State_X1 := [int(Integrand[1,1],zeta=0..t)];
State_X2 := [int(Integrand[2,1],zeta=0..t)];
State_X3 := [int(Integrand[3,1],zeta=0..t)];
```

$$\text{State\_X1} := \left[\frac{1}{2}-\frac{1}{7}\sqrt{7}\,e^{(-\frac{1}{2}t)}\sin\left(\frac{1}{2}\sqrt{7}\,t\right)-\frac{1}{2}e^{(-t)}\right]$$

$$\text{State\_X2} := \left[-\frac{1}{2}e^{(-\frac{1}{2}t)}\cos\left(\frac{1}{2}\sqrt{7}\,t\right)+\frac{1}{14}\sqrt{7}\,e^{(-\frac{1}{2}t)}\sin\left(\frac{1}{2}\sqrt{7}\,t\right)+\frac{1}{2}e^{(-t)}\right]$$

$$\text{State\_X3} := \left[\frac{1}{2}e^{(-\frac{1}{2}t)}\cos\left(\frac{1}{2}\sqrt{7}\,t\right)+\frac{3}{14}\sqrt{7}\,e^{(-\frac{1}{2}t)}\sin\left(\frac{1}{2}\sqrt{7}\,t\right)-\frac{1}{2}e^{(-t)}\right]$$

Create the state variable vector (`State_Variable_Vector`) and convert the matrix or array into a vector for an inner product computation done later:

```
State_Variable_Vector := con-
vert(stack(State_X1,State_X2,State_X3),vector);
```

$$\text{State\_Variable\_Vector} := \left[\frac{1}{2}-\frac{1}{7}\sqrt{7}\,e^{(-\frac{1}{2}t)}\sin\left(\frac{1}{2}\sqrt{7}\,t\right)-\frac{1}{2}e^{(-t)}\right.$$
$$-\frac{1}{2}e^{(-\frac{1}{2}t)}\cos\left(\frac{1}{2}\sqrt{7}\,t\right)+\frac{1}{14}\sqrt{7}\,e^{(-\frac{1}{2}t)}\sin\left(\frac{1}{2}\sqrt{7}\,t\right)+\frac{1}{2}e^{(-t)}$$
$$\left.\frac{1}{2}e^{(-\frac{1}{2}t)}\cos\left(\frac{1}{2}\sqrt{7}\,t\right)+\frac{3}{14}\sqrt{7}\,e^{(-\frac{1}{2}t)}\sin\left(\frac{1}{2}\sqrt{7}\,t\right)-\frac{1}{2}e^{(-t)}\right]$$

Enter the **C** vector in the output equation:

```
C_Vector := vector([-2,1,2]);
```

$$\text{C\_Vector} := [-2 \quad 1 \quad 2]$$

and substitute into the output function by implementing the inner product operation of the `C_Vector` with the `State_Variable_Vector` plus addition of the input step function:

```
Output := innerprod(C_Vector,State_Variable_Vector)+1;
```

$$\text{Output} := \frac{11}{14}\sqrt{7}\,e^{(-1/2\,t)}\sin\left(\frac{1}{2}\sqrt{7}\,t\right) + \frac{1}{2}\,e^{(-t)} + \frac{1}{2}\,e^{(-1/2\,t)}\cos\left(\frac{1}{2}\sqrt{7}\,t\right)$$

As stated earlier, even though the state variables chosen, $x_1(t)$, $x_2,(t)$, $x_3(t)$, *do not* directly represent time derivative states of the output (as is true in some simulation diagram forms), the output equation, $y(t) = \langle \mathbf{Cx} \rangle + du(t)$, related these computed states and any inputs, to the final output result, `Output`.

Now we plot this result to compare with the previous section using the frequency-domain approach to get the solution:

```
with(plots):
plot(Output,t=0..10,color=black,axes=normal,
 labels=[Time, Response]);
```

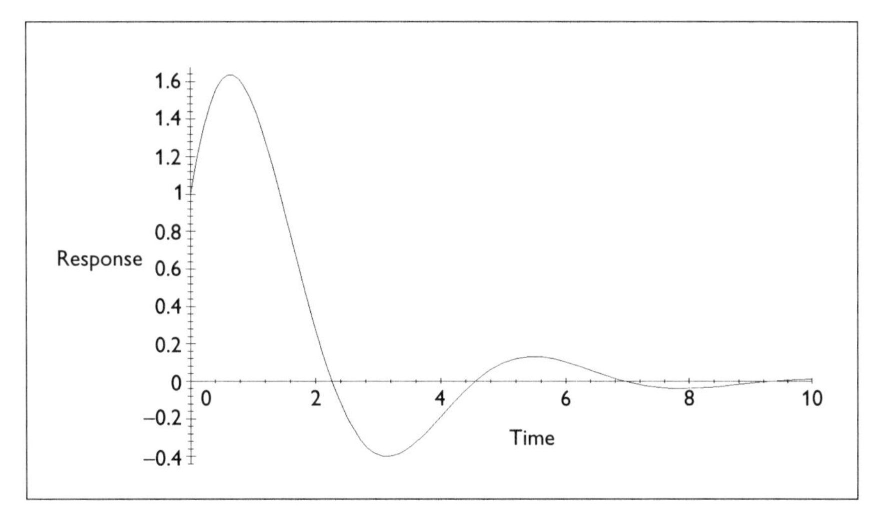

Figure 5.6 Output response for the third-order system example.

Figure 5.6 is identical to the frequency-domain output plot depicted in Figures 5.2 and 5.3. The difference with this approach is the more general mathematics involved with the system's dynamics. The state space approach is also useful when the coefficients are functions of time. This linear time-variant system cannot be handled by Laplace transform methods, hence, with the use of personal computers and Maple, one can perform some rather indepth dynamic analyses utilizing the state space technique just discussed (see MATRZANT.MWS file on enclosed diskette).

Conclusion

In this chapter, we have examined a simple template application of a third-order linear system for both frequency- and time-domain approaches. The fundamental difference was that the Laplace transform method was simpler to understand and mathematically set up, since all computations are algebraic in nature. However, if the system dynamics have nonconstant coefficients, then the state space or time-domain approach would have been the only method available to us for analysis. In either case, we had the ability to see the individual characteristic root effects. Though expressly discussed in the Laplace transform via partial fraction expansion, we did not get into the equivalent analysis via eigenvalue observation with the state space method. However, one obtains this individual root effect information on computation of the characteristic equation root solution in the state space approach. Since the characteristic equation has to be computed as a natural course of solving the control problem in the time domain (i.e., the eigenvalues), there was no need to examine specifically the individual root effect on the output response.

Maple has given the user a quick and exhaustive method of setting up and solving linear control problems of any order. In our template application, we used a documented third-order system, but it could have easily been a much higher order system without any change in either the mathematical or Maple syntax procedures.

As for nonlinear control problems, the standard procedure for setting up the solutions using describing or linearizing functions and numerical methods for obtaining specific solutions under a set of initial conditions is well documented [5,6] and easily implemented with Maple. The fundamental difference between linear and nonlinear solutions in Maple would be the increased number of procedures for numerical iteration to obtain bounded error solutions. However, the study of nonlinear systems with Maple was not the intention of the authors for this text at this time. Such systems require a much more rigorous mathematical base than we have presented for this section and is better left for another chapter.

References

[1] DeRusso, Roy, and Close, *State Variables for Engineers,* New York: John Wiley & Sons, 1965.

[2] Dorf, R., *Modern Control Systems,* Reading, MA: Addison-Wesley Publishing Co., 1967.

[3] Hirsch, M., and S. Smale, *Differential Equations, Dynamical Systems, and Linear Algebra,* New York: Academic Press, 1974.

[4] Saucedo, R., and E. Schiring, *Introduction to Continuous and Digital Control Systems,* New York: Macmillan Publishing Co., 1968.

[5] Ku, Y. H., *Analysis and Control of Nonlinear Systems,* The Ronald Press Co., 1958.

[6] Cunningham, W. J., *Introduction to Nonlinear Analysis,* New York: McGraw-Hill Book Company, 1958.

Chapter 6

Discrete control applications

DIGITAL CONTROL IS dependent on our being able to perform two fundamental operations, sampling and storage, both of which can be simulated in a Maple session. Through the process of sampling the signals present within a continuous system, we physically and mathematically transform the continuous system into a discrete or digital one. The effect of sampling a continuous control system can be rather dramatic as the following example illustrates. Using the Laplace operator s, define the transfer function of a dynamic system as follows:

$$\frac{1}{1+s^2}$$

which has the corresponding impulse response shown in Figure 6.1.

By simply sampling this conditionally stable system we transform it into an unconditionally stable one as the impulse response of the sampled system shows (Figure 6.2).

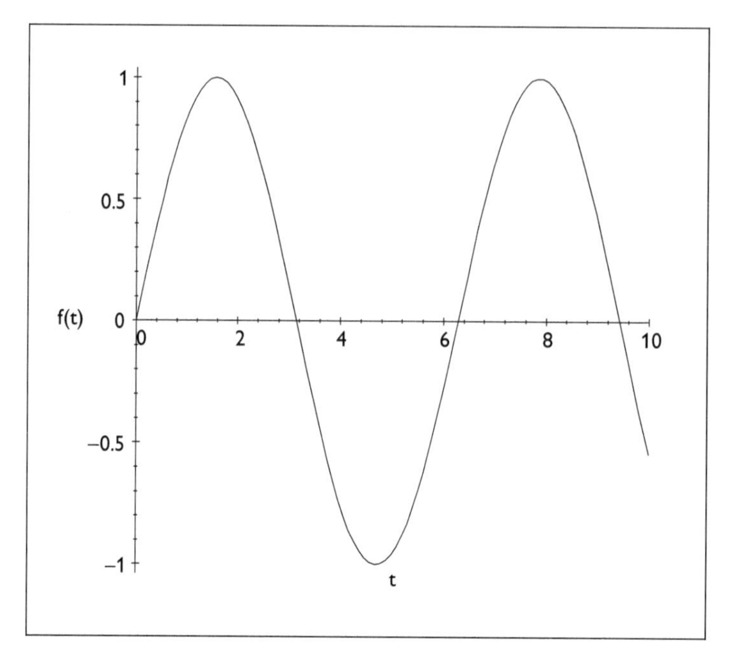

Figure 6.1 Continuous system.

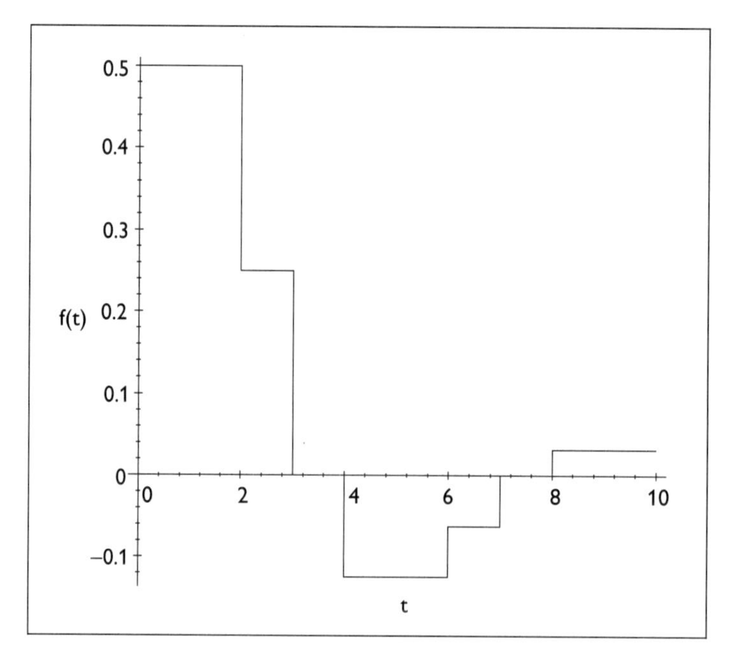

Figure 6.2 Sampled system.

The impulse response of the sampled system, in terms of the delay operator $\frac{1}{z}$, is

$$\frac{1}{2 + 2z^{-1} + z^{-2}}$$

The process of transforming a continuous system into a discrete one, computing the discrete system's time response using Maple, is discussed in this chapter.

First we see how, by using Maple, we can obtain the pulse transform function of a continuous system (converting from a continuous to a discrete system) and then apply a forcing function and compute the resulting time series. Then we see how Maple can help us form the state space matrices from a system's transfer function and then convert the matrices from one canonical form to another.

The pulse transfer function

The pulse transfer function is the digital version of the continuous system's transfer function and describes the digital system's behavior in terms of input and output pulses. Commonly, the input and output pulse trains are produced by sampling both the input and output signals repeatedly at a fixed rate, the sampling rate for the system. The idea of the pulse transfer function is conveyed in Figure 6.3 where $x(t)$ is the system input and $G(s)$ is the system's dynamics. You will notice that both starred variables, which are functions of time, and functions of z are used to label the same signals on the diagram. The starred variables represent sampled continuous signals, whereas the variables that are a function of z represent the continuous signals transformed into the discrete realm, via a transformation technique, and is the notation that we will be using in this chapter. Both sets of starred and z signals are equivalent and can be thought of as time series of weighted impulses. The connection between a continuous signal and its starred and z representation is shown in Figure 6.4.

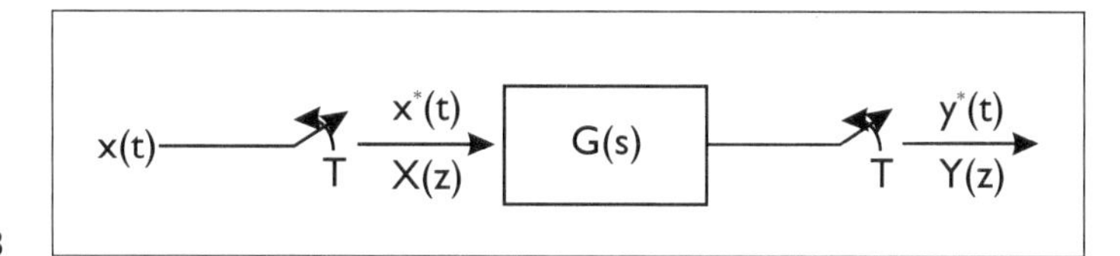

Figure 6.3

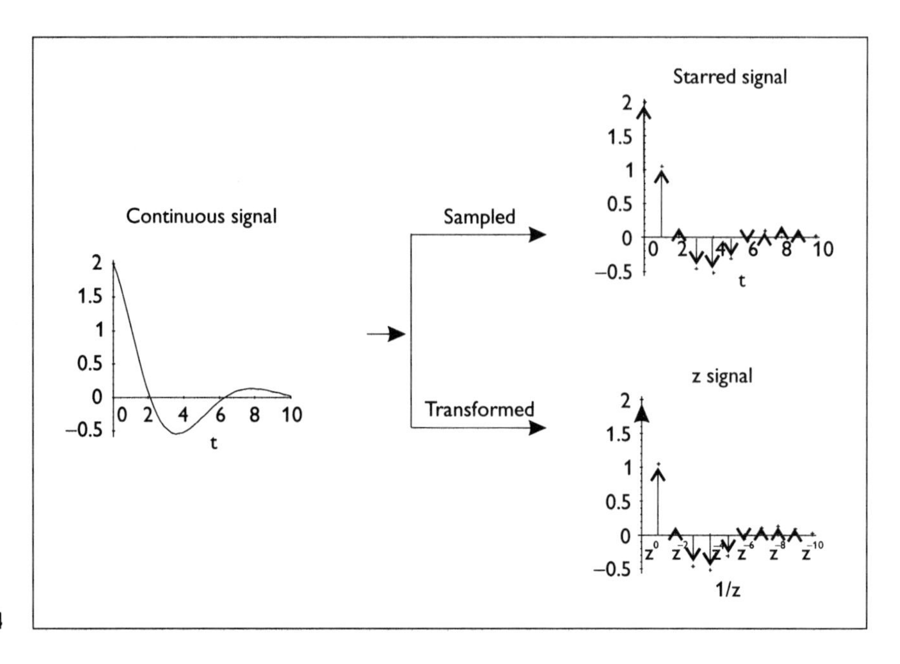

Figure 6.4

Transforming continuous signals

A number of techniques are available to us by which we can compute the Z-transform of a continuous signal. The most direct technique involves summing an infinite series of samples or weighted impulses; other techniques involve approximation by substitution and transforming from the s domain to the z domain. This final method is also known as the impulse or step invariant transform. We consider the direct approach first. A step of amplitude α can be described as the following series for positive n.

```
Step := n->alpha*z^(-n);
```

$$\mathrm{Step} := n \to \alpha\, z^{(-n)}$$

A unit step function, i.e., α=1, is shown in Figure 6.5. The first ten terms of the step function are easily calculated:

```
TERMS:=convert([seq(Step(n), n=0..10)], '+');
```

$$\mathrm{TERMS} := \alpha + \frac{\alpha}{z} + \frac{\alpha}{z^2} + \frac{\alpha}{z^3} + \frac{\alpha}{z^4} + \frac{\alpha}{z^5} + \frac{\alpha}{z^6} + \frac{\alpha}{z^7} + \frac{\alpha}{z^8} + \frac{\alpha}{z^9} + \frac{\alpha}{z^{10}}$$

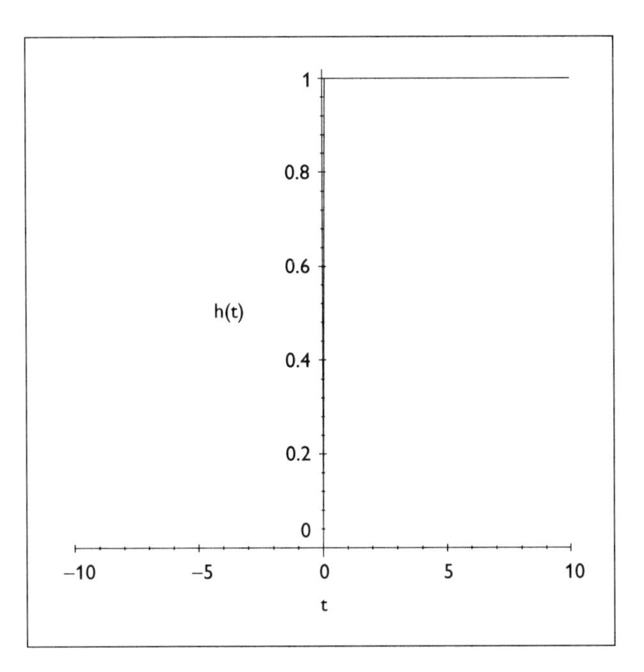

Figure 6.5
A unit step.

If we now sum our series from zero to infinity and a closed form exists, it will by definition be the Z-transform of the continuous step function $\alpha h(t)$.[1] The variable n needs to be reset because it currently has a value from the previous `seq` operation. Now we can perform the summation[2]

```
n:='n':
Sum(Step(n), n=0..infinity)=sum(alpha*z^(-n),
 n=0..infinity);
```

$$\sum_{n=0}^{\infty} \alpha\, z^{(-n)} = \frac{\alpha\, z}{-1 + z}$$

The right-hand side of this expression represents the closed-form solution, which we denote as `STEP`.

1. The unit step is commonly denoted by $h(t)$.

2. The function `Sum` is the inert form of the function `sum`. Maple returns inert function calls unevaluated.

```
STEP:=rhs(");
```

$$STEP := \frac{\alpha z}{-1 + z}$$

The convention when using the Z-transform is that the current sample is denoted by z^0; the next sample, the one taken after a delay of one sample period, is z^{-1}; the sample taken after two sample delays is z^{-2}; and so on. This means that a series of samples can be easily time shifted either by multiplying or dividing by the z operator (multiplying will shift the series forward in time whereas dividing will delay it). It is for this reason that the z operator is also known as the delay operator.

Next we find the Z-transform of the exponential sequence $Ae^{\tau t}$ where τ is the time constant:

```
Exp:=A*exp(tau*t);
```

$$Exp := A\ e^{(\tau t)}$$

As before we sum this expression for t = 0..infinity:

```
sum(Exp*z^(-t), t=0..infinity);
```

$$\sum_{t=0}^{\infty} A\ e^{(\alpha t)}\ z^{(-t)}$$

Unfortunately, this is not what we expect as Maple has returned an unevaluated form. We are still able to compute the Z-transform of the exponential sequence by performing a simple substitution, $r^t = e^{\tau t}$, and then performing the summation on this new expression:

```
TEMP:=subs(exp(tau*t)=r^t, Exp);
```

$$TEMP := A\ r^t$$

```
Z_Exp:=sum(TEMP*z^(-t), t=0..infinity);
```

$$Z\_Exp := -\frac{A\ z}{r - z}$$

Now we can perform the substitution, $r = e^{\tau T}$ where T is the sampling period, to obtain the general Z-transform of the exponential sequence:

```
Z_Exp:=normal(subs(r=exp(tau*T), Z_Exp));
```

$$Z\_Exp := \frac{A\ z}{-e^{(\tau\ T)} + z}$$

Using the Z-transform for the exponential sequence we can easily find the Z-transforms of a whole new class of functions, for example, trigonometric functions. Here we find the Z-transform of the function $\sin(\omega t)$:

```
ToExp := convert(sin(omega*t), exp);
```

$$ToExp := -\frac{1}{2}\ I\left(e^{(I\ \omega\ t)} - \frac{1}{e^{(I\ \omega\ t)}}\right)$$

Before we continue we need to change the expression `ToExp` slightly:

```
ToExp:=subs(1/exp(I*omega*t)=exp(-I*omega*t), ToExp);
```

$$ToExp := -\frac{1}{2}\ I\left(e^{(I\ \omega\ t)} - e^{(-I\ \omega\ t)}\right)$$

With reference to this expression, we see that τ in the general expression for the Z-transform of the exponential $Ae^{(\tau t)} = \ldots$, calculated earlier, is equal to $I\omega$ (the sign of τ equals the sign $I\omega$) and A is equal to unity, so substituting we get the following:

```
ZTF:=subs(exp(I*omega*t)=Z_Exp, tau=I*omega,
 exp(-I*omega*t)=Z_Exp, tau=-I*omega, A=1, ToExp);
```

$$ZTF := -\frac{1}{2} I \left(\frac{z}{-e^{(I\,\omega T)} + z} - \frac{z}{-e^{(-I\,\omega T)} + z} \right)$$

By representing the exponential terms as trigonometric functions we display the above expression ZTF in its usual form. Again, because we are using a CAS, the sequence of the operations and type of operation performed do not always match the sequence of operations that we would expect if we were solving the same problem by hand.

```
convert(ZTF, trig);
```

$$-\frac{1}{2} I \left(\frac{z}{-\cos(\omega\, T) - I\, \sin(\omega\, T) + z} - \frac{z}{-\cos(\omega\, T) + I\, \sin(\omega\, T) + z} \right)$$

```
normal(", expanded);
```

$$\frac{z\, \sin(\omega\, T)}{\cos(\omega\, T)^2 - 2\, \cos(\omega\, T)\, z + \sin(\omega\, T)^2 + z^2}$$

So, finally we have the Z-transform of $\sin(\omega t)$:

```
simplify(");
```

$$\frac{z\, \sin(\omega\, T)}{-2\, \cos(\omega\, T)\, z + 1 + z^2}$$

Of course, whenever using a computer algebra tool we should always exercise some caution as the final example of computing the Z-transform directly shows. Here we are trying to return the Z-transform of the unit ramp shown in Figure 6.6.

```
RAMP = sum(n*z^(-n), n=0..infinity);
```

$$\mathrm{RAMP} = 2\,\frac{\frac{1}{2}\,\frac{z^4\left(1-\frac{1}{z}\right)}{(-1+z)^3} - \frac{1}{2}\,z}{z^2}$$

```
simplify(");
```

$$\mathrm{RAMP} = \frac{-1 + 2\,z}{z(-1+z)^2}$$

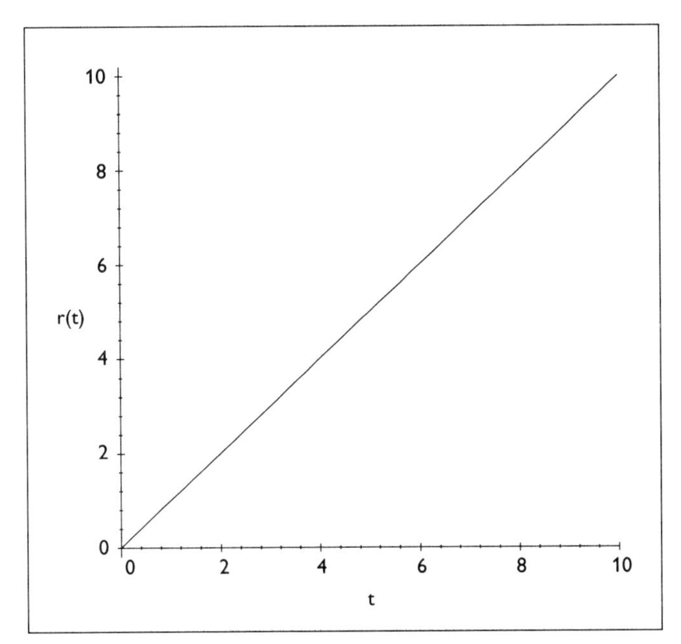

Figure 6.6
A unit ramp.

Unfortunately, Maple returns an incorrect answer when compared with the result returned by the built-in Z-transform function `ztrans`:

```
ztrans(t, t, z);
```

$$\frac{z}{(z-1)^2}$$

We confirm our suspicions by taking the inverse transform of the Z-transform of the ramp obtained using the direct method.

```
invztrans("", z, t);
```

$$\Delta(t)\ \mathrm{RAMP} = -\Delta(t-1) + t$$

The moral of the story is that care should always be exercised when using computers to manipulate complex expressions and answers should be checked wherever possible!

Impulse-invariant transformation

The method of substituting for the exponential terms, as in the *sin(ωt)* example given earlier, is more formally known as the impulse-invariant transformation and is commonly used when a continuous system needs to be quantized. Using this technique we compute the Z-transform of a test system by first obtaining the system's impulse response and expanding it using partial fractions and substituting for the exponential terms. This technique is easily demonstrated with a simple example. Consider the following continuous Laplace expression:

```
SYS:=4/(s^3+6.5*s^2+5.5*s);
```

$$\mathrm{SYS} := 4\,\frac{1}{s^3 + 6.5\,s^2 + 5.5\,s}$$

The first step is to transform the transfer function into partial fractions using `convert(..., parfrac,... .)`. Before we do this we must convert the floating-point numbers in the denominator of `SYS` into rationals:

```
SYSR:=convert(SYS, rational);
```

$$\mathrm{SYSR} := 4\,\frac{1}{s^3 + \frac{13}{2}\,s^2 + \frac{11}{2}\,s}$$

```
PF:=convert(SYSR, parfrac,s);
```

$$\mathrm{PF} := \frac{8}{11}\,\frac{1}{s} + \frac{32}{99}\,\frac{1}{2\,s + 11} - \frac{8}{9}\,\frac{1}{s + 1}$$

The partial fraction form of the transfer function is used because it ensures that when we take the inverse Laplace transform of it we get a result that consists of atomic elements; in this case, steps, impulses, and exponential terms. The Laplace transform pair is loaded from the `inttrans` package using `with` as follows:

```
with(inttrans, [laplace, invlaplace]):
IRESPONSE:=invlaplace(PF, s, t);
```

$$\mathrm{IRESPONSE} := \frac{8}{11} + \frac{16}{99}\,e^{(-11/2\,t)} - \frac{8}{9}\,e^{(-t)}$$

Now we can substitute for the step and the exponential terms using the transforms calculated earlier. We will do the substitutions one at a time starting with the step.

```
H[z]:=subs(alpha=select(type, IRESPONSE, numeric), STEP);
```

$$H_z := \frac{8}{11}\,\frac{z}{-1 + z}$$

```
Exps := select(has, IRESPONSE, exp);
```

$$\mathrm{Exps} := \frac{16}{99}\,e^{(-11/2\,t)} - \frac{8}{9}\,e^{(-t)}$$

Here we define a custom conversion routine ("teach" Maple's convert routine a new conversion type) to transform the exponential terms. The routine takes an expression containing an exponential and makes a copy (`a1`) of the numeric multiplier, if present, it makes a copy (`a2`) of the exponential and then finds the free variable (`a3`). The transformation is then made using `subs` and the contents of `a1, a2`, and `a3`.

```
'convert/toexp':=proc(a) local a1, a2, a3;
options 'Copyright Coded by Dr. Steve Adams 1995';
a1:=select(type, a, numeric);
a2:=op(select(type, a, function));
a3:=op(indets(a2));
subs(A=a1, alpha=a2, a3=T, A*z/(z-exp(alpha))):
end:
```

We apply the conversion routine to the remaining elements of the impulse response and compute the complete pulse transfer function H_z.

```
H[z]:=H[z] + map(convert, Exps, toexp);
```

$$H_z := \frac{8}{11}\frac{z}{-1+z} + \frac{16}{99}\frac{z}{z - e^{(-11/2\, T)}} - \frac{8}{9}\frac{z}{z - e^{(-T)}}$$

The Z-transform H_z still is general in that the sample period T is still present as a variable, which gives us a chance to see how the simple choice of sample period affects the impulse response of the system. The following sequence of Maple commands computes the time response of H_z, transforms the time response into function notation, and then generates a three-dimensional surface as the sample period T is varied:

```
h(t):=invztrans(H[z], z, t);
```

$$h(t) := \frac{1}{99}\,\frac{72\left(\left(e^{T}\right)^{t}\right)^{13/2} + 16\left(e^{T}\right)^{t} - 88\left(\left(e^{T}\right)^{t}\right)^{11/2}}{\left(\left(e^{T}\right)^{t}\right)^{13/2}}$$

```
h(t):=unapply(h(t), T, t);
```

$$h(t) := (T,\ t) \rightarrow \frac{1}{99}\,\frac{72\left(\left(e^{T}\right)^{t}\right)^{13/2} + 16\left(e^{T}\right)^{t} - 88\left(\left(e^{T}\right)^{t}\right)^{11/2}}{\left(\left(e^{T}\right)^{t}\right)^{13/2}}$$

```
plot3d(h('t')(T, t), t=0..3, T=0.01..1, labels=[''T'',
''t'', 'h(t)'], title='Effect Of T Size On System Impulse
Response', axes=BOXED, style=HIDDEN, color=BLACK, orienta-
tion=[-25, 70]);
```

Figure 6.7 shows how the effective time constant of the system increases as the sample period is reduced.

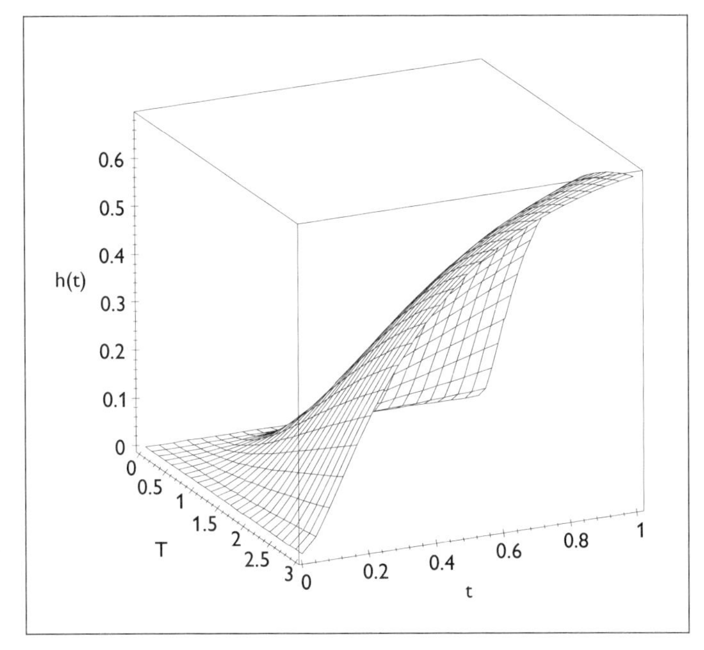

Figure 6.7 Effect of *T* size on system impulse response.

Substitution methods

Next we take a look at two of the substitution methods available for calculating the *Z*-transform of a continuous signal or system. The two techniques that we will look at are (1) using a numerical solution to the differential equation and (2) the bilinear transform. Both techniques provide a relationship between the Laplace operator s, which is equivalent to the differential operator $^{d}/_{dt}$, and the delay operator z, which means that we can transform between continuous and discrete functions.

The first approach relies on us being able to make the following approximation:

$$\frac{\partial}{\partial t}\,x(t) \approx \frac{x(n) - x(n-1)}{T}$$

This expression approximates the derivative of $x(t)$ with a finite difference, which is true for a sufficiently small sample period T. If we use the respective operator notation, $d/dt = s$, and the delay as $1/z$, in the above approximation, we get

```
Approx[1] := s=(1-z^(-1))/T;
```

$$\text{Approx}_1 := s = \frac{-\frac{1}{z} + 1}{T}$$

The bilinear transform on the other hand is derived by solving a simple first-order ordinary differential equation (ODE) of the form

```
ODE:=diff(y(t), t) + a*y(t) = b*u(t);
```

$$\text{ODE} := \left(\frac{\partial}{\partial t} y(t)\right) + a\ y(t) = b\ u(t)$$

using a popular numerical integration technique. The bilinear transform is in fact a conformal mapping, which translates the $j\omega$ axis of the s-plane onto the unit circle of the z-plane. If we integrate each side over the limits $(n-1)T$ to nT we get:

```
IntODE:=map(int,lhs(ODE), t=(n-1)*T..n*T)=int(rhs(ODE),
t=(n-1)*T..n*T);
```

$$\text{IntODE} := y(n\ T) - y(n\ T - T) + \int_{(n-1)T}^{n\ T} a\ y(t)\ dt = \int_{(n-1)T}^{n\ T} b\ u(t)\ dt$$

The unevaluated integrals can be approximated by applying the trapezoid rule (see `?student[trapezoid]`). First we isolate the integrals, using `select`, and then apply the trapezoid rule. The command `select(type, lhs(IntODE),specfunc(anything, int))` is used to isolate the functions of the form `int(anything, anything)` from the expression `lhs(IntODE)`.

```
TheInts:=[select(type, lhs(IntODE), specfunc(anything,
 int)),rhs(IntODE)];
```

$$\text{TheInts} := \left[\int_{(n-1)T}^{nT} a\ y(t)\ dt,\ \int_{(n-1)T}^{nT} b\ u(t)\ dt \right]$$

```
Y:=student[trapezoid](op(op(1, TheInts)), 1);
```

$$Y := \frac{1}{2}(n\,T-(n-1)T)$$
$$\left(a\,y((n-1)\,T)+2\left(\sum_{i=0}^{0} a\,y((n-1)\,T+i(n\,T-(n-1)T))\right)+a\,y(n\,T)\right)$$

Expanding and simplifying the above result we get

```
Y:=expand(simplify(Y));
```

$$Y := \frac{1}{2}\,T\,a\,y((n-1)\,T)+T\,a\left(\sum_{i=1}^{0} y(n\,T-T+i\,T)\right)$$
$$+\frac{1}{2}\,T\,a\,y(n\,T)$$

If we do the same with the second element of `TheInts` and combine the results we get the following:

```
U:=expand(simplify(student[trapezoid](op(op(2,
 TheInts)),1))):
```

```
RR:=convert([op(1..2, lhs(IntODE)), op(1,Y), op(3,Y)],
 `+`)=  convert([op(1, U), op(3, U)], `+`);
```

$$\text{RR} := y(n\,T)-y(n\,T-T)+\frac{1}{2}\,T\,a\,y((n-1)T)$$
$$+\frac{1}{2}\,T\,a\,y(n\,T)=\frac{1}{2}\,T\,b\,u((n-1)T)+\frac{1}{2}\,T\,b\,u(n\,T)$$

Taking the Z-transform of this difference equation by substituting for $x(nT) = x$ and $x((n-1)T) = xz^{-1}$, where x can be either y or b, we get

```
ZT:=subs( (n-1)*T = (n*T-T), (n*T-T)=1/z, n*T=1,RR);
```

$$ZT := y(1) - y\left(\frac{1}{z}\right) + \frac{1}{2}\, T\, a\, y\left(\frac{1}{z}\right) + \frac{1}{2}\, T\, a\, y(1)$$
$$= \frac{1}{2}\, T\, b\, u\left(\frac{1}{z}\right) + \frac{1}{2}\, T\, b\, u(1)$$

By convention transformed variables are represented as uppercase characters, for example, $Z(y(t)) \rightarrow Y(z)$ where $Z(y(t))$ is the Z-transform of $y(t)$. We can achieve this cosmetic change in the above transformed expression by first clearing the variables Y and U, which we used earlier, and then defining two functions as follows. When the expression `ZT` is used next Maple will automatically perform the simplification.

```
Y:='Y':U:='U':
y:=x->Y[z]*x:u:=x->U[z]*x:
```

Solving for $Y[z]/U[z]$, the system's transfer function, we get the following:

```
Discrete:=simplify(solve(ZT,Y[z]))/U[z];
```

$$\text{Discrete} := \frac{T\, b\, (1 + z)}{2\, z - 2 + T\, a + T\, a\, z}$$

If, by definition, the discrete and the continuous transfer functions are similar, we need to compare them in order to determine under what circumstances this is true. First, therefore, let us compute the s-domain representation of the ode

$$\frac{d}{dt}\, y(t) + ay(t) = bu(t)$$

The use of the Maple alias facility enables us to view the result in a concise form:

```
y:='y':u:='u':
alias(y[s]=laplace(y(t),t,s), b[s]=laplace(u(t), t,s)):
laplace(ODE, t, s);
```

$$y_s\ s - y(0) + a\ y_s = b\ b_s$$

Now solving the above for *y[s]/b[s]*, setting *y(0)=0*, and assuming that the system is initially at rest, we get the following continuous transfer function:

```
Continuous:=subs(y(0)=0,solve(", y[s]))/b[s];
```

$$\text{Continuous} := \frac{b}{s + a}$$

By comparing the `Discrete` and `Continuous` expressions we can see that for them to be equal the following must be true:

```
Approx[2]:=simplify(readlib(isolate)(Continuous=
 Discrete, s));
```

$$\text{Approx}_2 := s = 2\,\frac{-1 + z}{T(1 + z)}$$

This mapping is the bilinear transform. If we now apply this and the previous mapping to our test system with a sample period of $\frac{1}{5}$ we get:

```
tf[1]:=convert(simplify(subs(Approx[1], T=1/5, SYS)),
 rational);
```

$$tf_1 := \frac{8}{5}\,\frac{z^3}{-50 + 215\ z - 291\ z^2 + 126\ z^3}$$

```
tf[2]:=convert(simplify(subs(Approx[2], T=1/5, SYS)),
 rational);
```

$$tf_2 = \frac{4}{5}\,\frac{(1+z)^3}{(-1+z)\left(81 - 378\,z + 341\,z^2\right)}$$

Finally, we show the first transformation again so that the results of all three techniques can be easily compared:

```
tf[3]:=convert(simplify(subs(T=1/5,H[z])), rational);
```

$$tf_3 := \frac{8}{99}\,z\left(-11\,z\,e^{(-1/5)} + 2\,e^{(-11/10)}\,z + 9\,e^{(-13/10)} + 9\,z\right.$$
$$\left. + 2\,e^{(-1/5)} - 11\,e^{(-11/10)}\right)\Big/\left((-1+z)\left(z - e^{(-11/10)}\right)\left(z - e^{(-1/5)}\right)\right)$$

It is obvious that each approach results in a subtly different answer. The differences can be seen if we plot the impulse response obtained from each of the transfer functions. The following procedure generates a list of points that, when plotted, gives a staircase plot from a *Z*-transform and a time range. One thing to note in the procedure is the use of `eval` and the delay quotes. The delay quotes are used so that the sequence operator functions correctly and eval is used to force evaluation of the point pairs generated:

```
calc_response:=proc(X, Z, R)
options 'Copyright Coded by Dr. Steve Adams 1996';
local ans, pts, n, var;
var:= op(1, R);
ans:= invztrans(X,Z,var);
ans:=eval(['seq'([var, ans], R)]);
pts:=NULL;
for n to nops(ans)-1 do
 pts:=pts, ans[n], [op(1, ans[n+1]), op(2, ans[n])];
od;
[pts]:
end:
```

With reference to the transfer function tf_3,we can see that the degree of the numerator is the same as the denominator. This will result in Δ*s* being present in the corresponding time response and in Iris (Maple's graphical engine) being unable to plot expressions containing Δ*s*. The following function will remove any Δ*s* that appear:

```
Delta:=x->if x=0 then 1 else 0 fi:
```

Now we can plot and compare the three transfer functions (see Figure 6.8):

```
plot({plots[textplot]([0.1, 2.5, 'tf[1]']),
 calc_response(tf[1], z, 't'=0..15)):
plot({plots[textplot]([0.2, 4, 'tf[2]']),
 calc_response(tf[2], z, 't'=1..15)):
plot({plots[textplot]([16, 0.6, 'tf[3]']),
 calc_response(tf[3], z, 't'=0..15)}):
```

```
plots[display]({""", "", "}, title='Impulse Responses',
labels=[''t'','f(t)']);
```

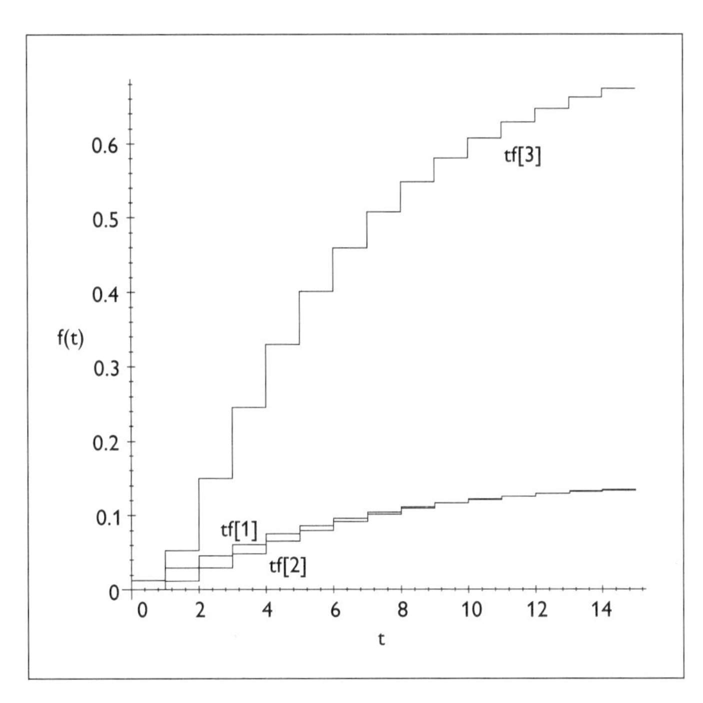

Figure 6.8
Impulse responses.

The three basic responses are similar except for the relative amounts of gain in each system.

Conclusion

As previously mentioned the pulse transfer functions obtained by different methods for the same continuous system are similar but not identical. What then are the advantages for choosing one method over another?

Impulse-invariant transform When the impulse-invariant transform is used, we are ensuring that the impulse response of the discrete system is identical to that of the continuous system, at least at the sample instances. The consequence of using this approach is the introduction of distortion due to aliasing. This is easily understood if the relationship between the frequency responses of the continuous and the discrete systems is examined. Unlike the frequency response of the continuous system, the frequency spectrum of the discrete system is repeated many times due to the fact that when the continuous spectrum is sampled with a period of *T* seconds the spectrum of the sampled signal is simply a scaled version of the continuous one repeated every *R* Hz. The repetition frequency *R* is equal to *1/T* Hz and the scale factor is *1/T*. The sampling period must be set sufficiently high so that the repeated spectra do not overlap, thus eliminating any chance of aliasing and mismatch in the discrete system's impulse response (see Figure 6.9). This is not possible in practical systems although it is possible to approximate such conditions.

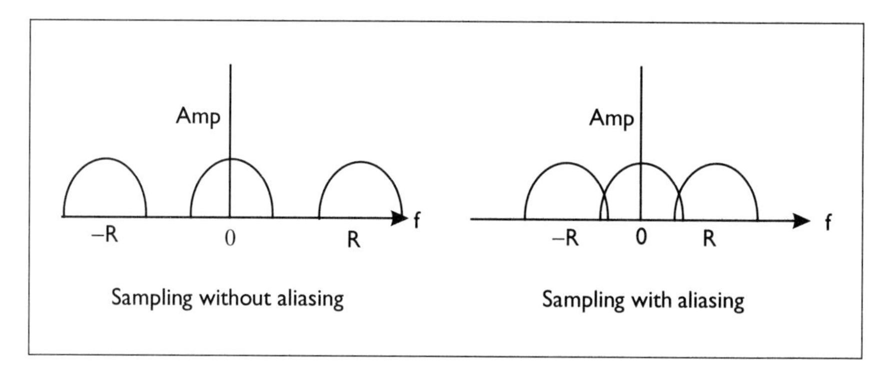

Figure 6.9

Another point of interest is the actual mapping of the *s*-plane onto the *z*-plane. This mapping is as follows: The left-hand side of the *s*-plane maps onto the interior of the unit circle centered on the origin of the *z*-plane, whereas the right-hand side of the *s*-plane maps to the exterior as shown in Figure 6.10.

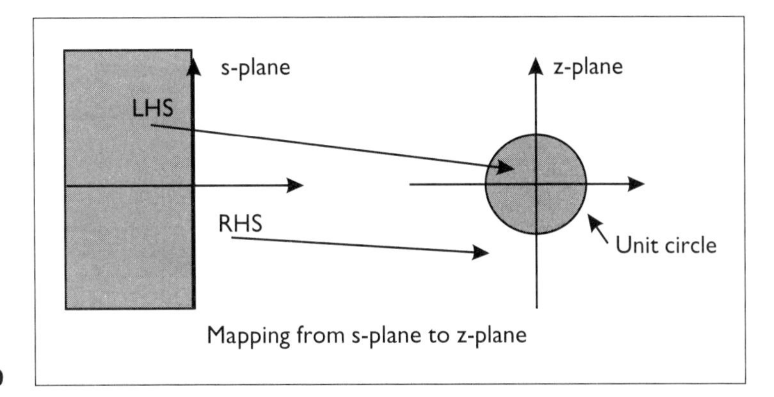

Figure 6.10

Despite the nonlinear relationship of the overall mapping, the relationship between the continuous frequency and the corresponding discrete frequency is linear, which means that the shape of the frequency response is preserved and hence the identical (at the sample instances) impulse responses of the continuous and discrete systems.

Numerical approximation This transformation method, although easy, gives less than ideal results. In order to achieve accurate transformation from the continuous to the discrete worlds, very high sample rates are necessary. This tends to result in inefficient designs in every area except for low-pass digital filters. The mapping provided by this method is like the previous one, nonlinear. Unlike the previous one, however, the relationship linking the continuous frequency spectrum to the discrete one is also nonlinear. Whereas the invariant impulse response transform mapped the s-plane to either the inside or outside of the unit circle, this transform maps the left-hand side of the s-plane onto the interior of a circle of radius

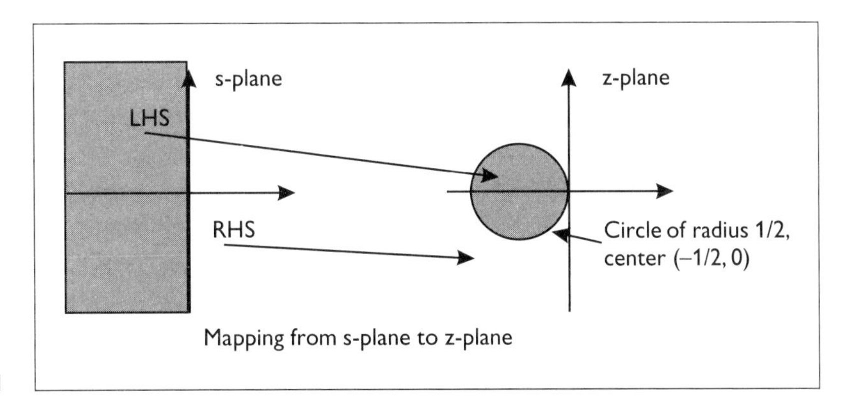

Figure 6.11

As before, the right-hand side maps to the outside of this circle. Although different than the earlier one, this mapping does preserve stability but without mapping the $j\omega$-axis into the unit circle in the z-plane.

Bilinear transform This transform method has these advantages over the previous two: Aliasing is avoided, it is efficient, and the dc gain of the system is preserved ($s = 0 \rightarrow z = 1$) and is valid for any order of system since any n^{th}-order system can be represented as n first-order systems. Although the mapping provided by this technique is, as we would expect, nonlinear, the frequency mapping between the s- and the z-planes is one to one and aliasing has been avoided at the cost of distorting the frequency axis. Like the first mapping, the bilinear transform maps the stable poles and zeros to the interior of the unit circle centered at the origin of the z-plane and the unstable ones to its exterior.

Calculating the time response

The computation of a system's time response (at the sample instants) is, in effect, the inverse transformation from the discrete realm to the continuous. This process can be performed in one of three basic ways: Solve the recurrence relationship formed by the Z-transform, polynomial or synthetic long division, or the direct method.

The recurrence relationship

A recurrence relationship is one that describes the current value of a sequence in terms of its history and in some cases its future. For example, the exponential filter describes the following recurrence relationship: $y_{(n)} = r_{(n)}e^{\alpha} + (1 - e^{\alpha})y_{(n-1)}$ where the current output $y_{(n)}$ is a weighted combination of the current filter input $r_{(n)}$ and the previous filter output $y_{(n-1)}$. Recurrence relationships are easily generated from a system's discrete transfer function and solved, after which the corresponding time sequence can be computed. The following sequence of Maple commands demonstrates how this is done using the transfer function

$$\frac{1}{1 - 3z^{-1} - 0.2z^{-2}},$$

which is equal to $\frac{Y(z)}{U(z)}$, the ratio of the input signal to output signal.

```
SYS:=1/(1-3/z-0.2/z^2);
```

$$SYS := \frac{1}{1 - 3\,\frac{1}{z} - .2\,\frac{1}{z^2}}$$

Using `numer` and `denom` we first isolate the numerator and the denominator:

```
BOT:=denom(SYS);
```

$$BOT := z^2 - 3\,z - .2$$

```
TOP:=numer(SYS);
```

$$TOP := z^2$$

We can see from the two expressions that Maple rationalizes and simplifies `SYS` to

$$\frac{z^2}{z^2 - 3z - 0.2}$$

prior to computing the numerator and the denominator. The next stage is to manipulate these expressions by dividing them both by z raised to the degree of the numerator, in this case z^2, which is equal to `TOP`, to transform all of the forward time shifts into time delays. We will assume that the system's forcing function $U(z)$ is a unit step. Note, however, that the forcing function can be any sequence, as discussed in the next section.

```
LHS:=expand(BOT/TOP);
```

$$LHS := 1 - 3\,\frac{1}{z} - .2\,\frac{1}{z^2}$$

```
RHS:=TOP/TOP;
```

$$RHS := 1$$

We are now at the stage where we can convert the *Z*-transform into a recurrence relationship that can be solved by replacing terms in *z* with terms of the form $\Psi(n - \beta)$ where Ψ is the signal of interest, β an integer denoting the number of delays associated with the term, and *n* the current time count. The custom conversion routine defined below enables us to convert the polynomial in $1/z$ into its equivalent recurrence relationship. The routine takes the expression to be converted, the name of the signal, the *from* variable and the *to* variable as parameters.

```
'convert/toRR':=proc(x, y, z, t) local a1,a2;
option 'Copyright Coded by Dr. Steve Adams 1995';
a1:=select(type, x, numeric);
a2:=degree(x, z);
a1*y(t+a2);
end:
```

We now translate `LHS` with the signal name `Y`, the variable `z`, and the variable `n` as follows. The map function enables us to convert each term of the expression `LHS` in a single operation and is equivalent to `convert(op(i, LHS), toRR, Y, z, n) for i=1..nops(LHS)`.

```
RR := map(convert, LHS, toRR, Y, z, n);
```

$$RR := Y(n) - 3\ Y(n - 1) - .2\ Y(n - 2)$$

Now using `rsolve`, the Maple recurrence relationship solver, we can solve RR for `Y(n)` with the initial conditions `Y(0)=3` and `Y(1)=0`:

```
SEQ:=rsolve({RR=RHS, Y(0)=3, Y(1)=0}, Y(n));
```

$$SEQ := \frac{3}{7}\frac{(-47 + 21\sqrt{5})\sqrt{5}\left(-2\dfrac{1}{-7\sqrt{5} + 15}\right)^n}{-7\sqrt{5} + 15} + \frac{3}{7}\frac{(47 + 21\sqrt{5})\sqrt{5}\left(-2\dfrac{1}{15 + 7\sqrt{5}}\right)^n}{15 + 7\sqrt{5}} - \frac{5}{11}$$

$$+\frac{10}{77}\frac{(-16+7\sqrt{5})\sqrt{5}\left(-2\frac{1}{-7\sqrt{5}+15}\right)^{n}}{-7\sqrt{5}+15}$$

$$+\frac{10}{77}\frac{(16+7\sqrt{5})\sqrt{5}\left(-2\frac{1}{15+7\sqrt{5}}\right)^{n}}{15+7\sqrt{5}}$$

We plot this function, which is valid at the sample instances, by transforming it to function notation, generating the sequence of samples, converting the samples to a staircase plot, and then displaying the result (Figure 6.12):

```
TIME:=unapply(SEQ, n):
THE_SEQ:=[seq([T, TIME(T)], T=[0,1,2,3,4,5])]:
STAIR:=NULL:
for n to nops(THE_SEQ)-1 do
STAIR:=STAIR, THE_SEQ[n], [op(1, THE_SEQ[n+1]),
 op(2, THE_SEQ[n])];
od:
plot([STAIR], labels=['t','f(t)'], title=
 'Impulse Response');
```

We can deduce from the plot of Figure 6.12 that this particular system is unstable. This is further confirmed by looking at the poles of the system.

```
solve(BOT, {z});
```

$$\{z = 3.065247585\},\ \{z = -.065247585\}$$

One of the poles lies outside the unit circle, indicating the system is unstable.

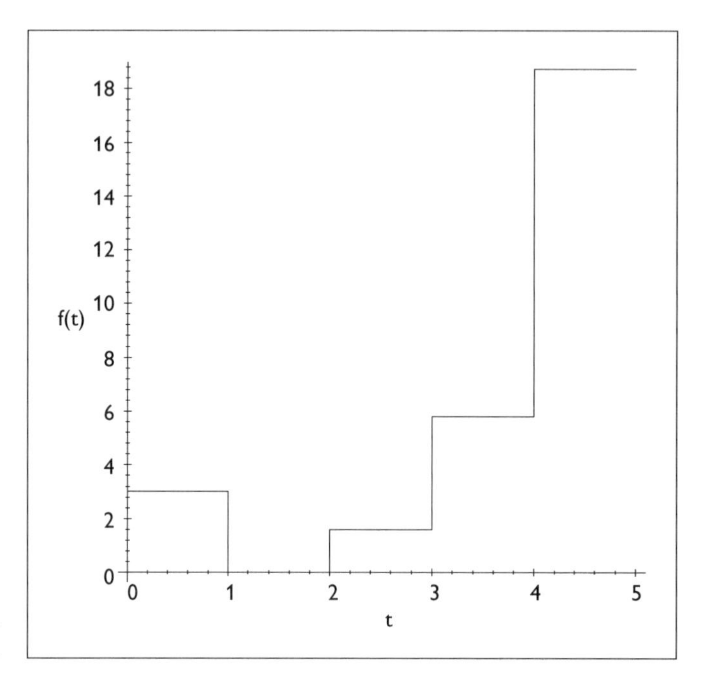

Figure 6.12
Impulse response.

The direct method

The direct method is more general than the recurrence relationship method because it does not rely on a closed-form solution being available. It is also well suited to computer implementation as well as being able to incorporate random input sequences. As before we use SYS as our example pulse transfer function. The basic principle for computing the output sequence is as follows, the discrete transfer function $G(z)$ is equal to the quotient $Y(z)/U(z)$, which can be manipulated to yield $Y_n + aY_{n-1} + bY_{n-2} + \ldots = U_n + U_{n-1} + \ldots$ Further manipulation yields the desired equation for the current output as a function of the current input and the previous inputs and outputs: $Y_n = -aY_{n\,1} - bY_{n-2} + \ldots + U_n - U_{n-1} + \ldots$

```
SYS;
```

$$\frac{1}{1 - 3\,\frac{1}{z} - .2\,\frac{1}{z^2}}$$

As before we need to operate on the numerator and denominator of this expression separately to isolate the sample weights.

```
BOT:=denom(SYS);
```

$$BOT := z^2 - 3z - .2$$

Next we determine the degree of the polynomial `BOT` by determining the free variable and then using `degree`.

```
VAR:=indets(SYS);
```

$$VAR := \{z\}$$

```
TO:=degree(BOT, VAR);
```

$$TO := 2$$

Using this as the upper bound, we form a list of ordered coefficients in an explicit fashion instead of using `coeffs` because this command will not return zero for any missing terms.

```
YWEIGHTS:=[seq(coeff(BOT, op(VAR), n), n=0..TO)];
```

$$YWEIGHTS := [-.2, -3, 1]$$

Reading this list from right to left we have the weights to be applied to the current output $Y_{(n)}$, the previous output $Y_{(n-1)}$, and the output prior to that $Y_{(n-2)}$, respectively.

Although trivial in this example, repeating the same sequence of operations would return a list of weights associated with current and past values of the input sequence. In this example the list is, by inspection, `[1]`:

```
UWEIGHTS:=[1];
```

$$UWEIGHTS := [1]$$

Now that we have the weightings and the respective delays, we can compute the output sequence from a given input sequence and a set of initial conditions. Here we assume that the system is initially at rest:

```
ICS:=[0, 0];
```

$$ICS := [0,\ 0]$$

The first output is calculated, assuming that the first input sample is one, as follows:

```
OUTPUT:=zip((x, y)-> -x*y, ICS, op(1..2, YWEIGHTS)) +
 zip((x,y)-> x*y, UWEIGHTS, [1]);
```

$$OUTPUT := [0,\ 0] + [1]$$

The elements of the lists are summed to return Y_n:

```
OUTPUT:=map(convert,OUTPUT, `+`);
```

$$OUTPUT := 1$$

Before we can repeat the process and compute the next value in the output sequence we need to update the list containing the previous output values, in our case the variable ICS:

```
ICS:=[ICS[1], OUTPUT];
```

$$ICS := [0,\ 1]$$

Before we plot the system output for a forcing function of a bipolar square wave with a peak-to-peak amplitude of two (Figure 6.13), we must remind ourselves that our test system is unstable in its current configuration (see the previous section).

By placing a 10:1 attenuator in the forward path we stabilize the system (Figure 6.14). This particular modification is easily accomplished by simply dividing the $Y_{(n)}$ weighting in the calling sequence as shown here:

```
plot(output_sequence([-2, -3, 1/10], [1], [0, 0],
 [1, 1, 1, -1, -1, -1, 1, 1, 1, -1, -1, -1]),
 title='System Output', labels=[`t`,'o/p']);
```

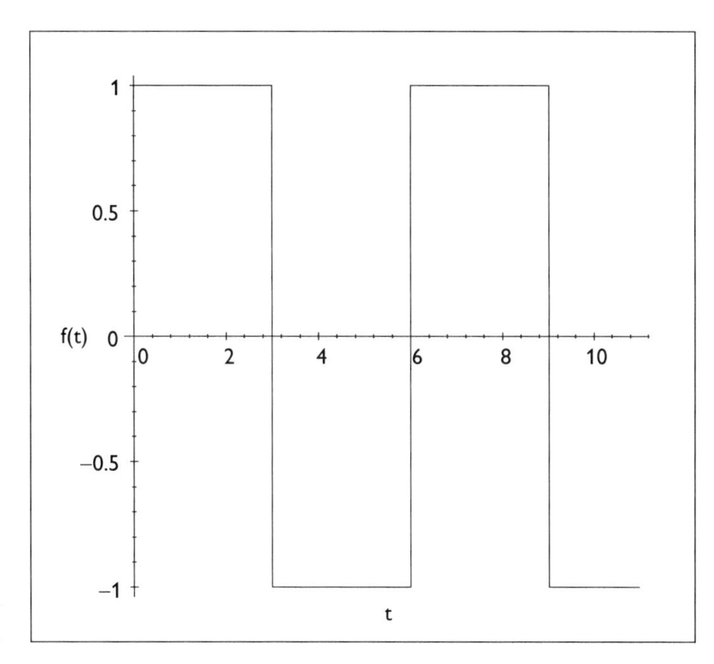

Figure 6.13
Forcing function.

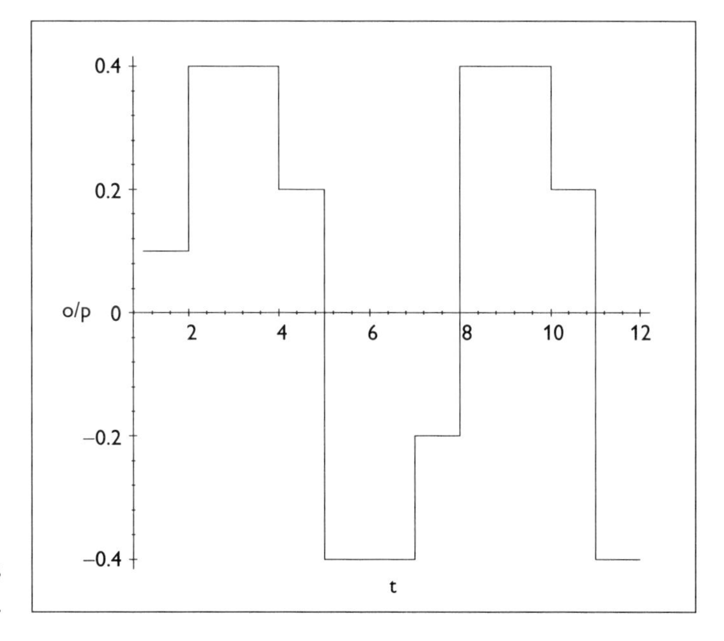

Figure 6.14
System output.

This entire process has been automated in the function `output_sequence` found on the program disk.

```
output_sequence := proc(YWEIGHTS, UWEIGHTS, ics, IP)
local OUTPUT, ICS, n, temp;
ICS:=ics;
OUTPUT:=NULL:
for n to nops(IP) do
temp:=op(3, YWEIGHTS)*map(convert, zip((x, y)-> -x*y, ICS,
[op(1..2, YWEIGHTS)]) +
 zip((x,y)-> x*y, UWEIGHTS, [IP[n]]), '+');
OUTPUT:=OUTPUT, [n,temp];
ICS:=[ICS[1], IP[n]];
od;
temp:=[OUTPUT];
OUTPUT:=NULL:
for n to nops(temp)-1 do
OUTPUT:=OUTPUT, temp[n], [op(1, temp[n+1]), op(2,
 temp[n])];
od;
[OUTPUT];
end:
```

State space equations and their canonical forms

One of Maple's primary uses to the control engineer is as a fast, efficient, and accurate manipulator of equations, expressions, and matrices using the host of functions found in the `linalg` package. In this section, we use Maple to convert discrete transfer functions into their state space forms and generate the many useful canonical forms.[3]

Transfer function to state space (the controllable canonical form)

Historically, dynamic systems have been described using differential equations. Now, however, state space descriptions using matrices are common. Using state matrices a dynamic system can now be described in the following manner:

3. Although discrete systems are being used in this example, the techniques are equally applicable to continuous systems.

$$\dot{\boldsymbol{x}} = \boldsymbol{Ax} + \boldsymbol{Bu}$$
$$\boldsymbol{y} = \boldsymbol{Cx} + \boldsymbol{Du}$$

where $\dot{\boldsymbol{x}}$, $\boldsymbol{x}$, $\boldsymbol{A}$, $\boldsymbol{B}$, $\boldsymbol{u}$, $\boldsymbol{y}$, $\boldsymbol{C}$, and **D** are all matrices. The system states are held in the **x** matrix, the **A** matrix contains the system coefficients, the **B** matrix contains the input gains, **u** is the input matrix, **y** is the system output, the **C** matrix contains the output gains, and **D** is known as the disturbance matrix.

Using the pulse transfer function of a discrete system it is a relatively simple process to generate the corresponding state space matrices. Although this looks like a complex process, do not forget that we are using Maple, which does all of the housekeeping for us, ensuring that no terms are missed and no signs are dropped. Omitting terms and dropping signs is all too easy when we are dealing with high-order complex systems containing both numbers and symbols.

In the following example, we derive the state space representation of a third-order discrete system with symbolic coefficients:

```
SYS:=z^3/(a*z^3 + b*z^2+c*z+d);
```

$$SYS := \frac{z^3}{a\,z^3 + b\,z^2 + c\,z + d}$$

We have deliberately selected a system where the degrees of the numerator and the denominator are equal, which means that we must first divide out the transfer function by converting it to a continued fraction using `convert(..., confrac, ...)`:

```
D_SYS:=convert(SYS, confrac, z);
```

$$D\_SYS := \frac{1}{a} - b\Bigg/\left(a\left(z - \frac{-b^2 + c\,a}{a\,b} - \frac{b\,d - c^2}{b^2\left(z + \frac{c(2\,b\,d - c^2)}{b(b\,d - c^2)} + \frac{d^3\,b}{(b\,d - c^2)^2\left(z - \frac{d\,c}{b\,d - c^2}\right)}\right)}\right)\right)$$

The first term is the disturbance matrix **D** [we use `Dist` because `D` is a Maple system name (the differential operator) and is protected]:

```
Dist:=linalg[matrix](1,1, [[op(1, D_SYS)]]);
```

$$\text{Dist} := \begin{bmatrix} \frac{1}{a} \end{bmatrix}$$

The first row of the **A** matrix is made up of the coefficients of the numerator of the remaining polynomial `D_SYS`:

```
POLY:=normal(op(2, D_SYS), expanded);
```

$$\text{POLY} := \frac{-b\, z^2 - d - c\, z}{a^2\, z^3 + a\, b\, z^2 + a\, c\, z + d\, a}$$

For convenience we really want the coefficient of the leading term of the denominator to be unity and we must allow for this when we form the matrices:

```
DIV:=lcoeff(denom(POLY), z);
```

$$\text{DIV} := a^2$$

By picking off each coefficient of the denominator in turn, starting with the highest-but-one term in z, negating it, and dividing by the leading coefficient (`DIV`), we can construct the **A** matrix:

```
TO:=degree(denom(POLY), z)-1;
```

$$\text{TO} := 2$$

```
FIRST:=[seq(-coeff(expand(denom(POLY)), z, TO-n)/DIV,
 n=0..TO)];
```

$$\text{FIRST} := \left[-\frac{b}{a}, -\frac{c}{a}, -\frac{d}{a} \right]$$

The other two rows of the **A** matrix are [1, 0, 0] and [0, 1, 0], so the final **A** matrix becomes:

```
A:=linalg[matrix](3,3, [ FIRST, [1,0,0],[0,1,0]]);
```

$$A := \begin{bmatrix} -\frac{b}{a} & -\frac{c}{a} & -\frac{d}{a} \\ 1 & 0 & 0 \\ 0 & 1 & 0 \end{bmatrix}$$

The **C** matrix is just the negated numerator coefficients of the transfer function divided by `DIV` starting with the highest-term-but-one in z and continuing through the lowest term in z:

```
C:= linalg[matrix](1,3, [[seq(-coeff(expand(numer(POLY)), z,
TO-n)/DIV, n=0..TO)] ]);
```

$$C := \begin{bmatrix} \frac{b}{a^2} & \frac{c}{a^2} & \frac{d}{a^2} \end{bmatrix}$$

Finally, because the system is a single-input/single-output system, the **B** matrix is

```
B:=linalg[matrix](3,1, [[1], [0], [0]]);
```

$$B := \begin{bmatrix} 1 \\ 0 \\ 0 \end{bmatrix}$$

These state matrices are said to be in the *controllable canonical form* and are well suited to the design of state variable feedback controllers.

Jordan canonical form

In many designs it is advantageous to decouple the system through the diagonalization of the **A** matrix and the application of a transformation matrix. If the system has distinct eigenvalues, then the matrix transformation $\mathbf{x} = \mathbf{Pz}$, where **P** is a Vandermonde matrix (see

`?linalg[vandermonde]`) formed from the eigenvalues, is possible. Maple has the built-in function `jordan` with which to perform the transformation. For the sake of brevity we set values for the a, b, c, and d to one, two, three, and four, respectively, and then diagonalize the **A** matrix.

```
NEWA:=linalg[jordan](subs(a=1, b=2, c=3, d=4, eval(A)),
TRANS);
```

$$\mathrm{NEWA} := \left[\frac{1}{2}\left(\frac{35}{27}+\frac{5}{9}\sqrt{6}\right)^{1/3} - \frac{5}{18}\frac{1}{\left(\frac{35}{27}+\frac{5}{9}\sqrt{6}\right)^{1/3}} - \frac{2}{3} - \frac{1}{2}I\sqrt{3}\left(-\left(\frac{35}{27}+\frac{5}{9}\sqrt{6}\right)^{1/3} - \frac{5}{9}\frac{1}{\left(\frac{35}{27}+\frac{5}{9}\sqrt{6}\right)^{1/3}}\right), 0, 0\right]$$

$$\left[0, \frac{1}{2}\left(\frac{35}{27}+\frac{5}{9}\sqrt{6}\right)^{1/3} - \frac{5}{18}\frac{1}{\left(\frac{35}{27}+\frac{5}{9}\sqrt{6}\right)^{1/3}} - \frac{2}{3} + \frac{1}{2}I\sqrt{3}\left(-\left(\frac{35}{27}+\frac{5}{9}\sqrt{6}\right)^{1/3} - \frac{5}{9}\frac{1}{\left(\frac{35}{27}+\frac{5}{9}\sqrt{6}\right)^{1/3}}\right), 0\right]$$

$$\left[0, 0, -\left(\frac{35}{27}+\frac{5}{9}\sqrt{6}\right)^{1/3} + \frac{5}{9}\frac{1}{\left(\frac{35}{27}+\frac{5}{9}\sqrt{6}\right)^{1/3}} - \frac{2}{3}\right]$$

Simplifying this we get

```
map(evalf, NEWA);
```

$$\begin{bmatrix} -.1746854042 + 1.546868888\ I, & 0, & 0 \\ 0, & -.1746854042 - 1.546868888\ I, & 0 \\ 0, & 0, & -1.650629192 \end{bmatrix}$$

The `jordan` function returns a transition matrix, specified by the last argument passed to the function, which in our case is tagged `TRANS`.

```
evalf(eval(TRANS), 4);
```

$$\begin{bmatrix} 1. & 1.825 + 1.547\ I & .2891 + 2.554\ I \\ 1. & 1.825 - 1.547\ I & .2891 - 2.554\ I \\ 1. & .3487 & 2.423 \end{bmatrix}$$

The inverse of this transition matrix can then be used to transform the **B**, **C**, and **Dist** matrices.

```
'B':=evalf(evalm(linalg[inverse](eval(TRANS)) &* B));
```

$$B := \begin{bmatrix} .2019853944 + .3056525967\ I \\ .1805460653 - .1509657080\ I \\ -.1093801481 - .1043649861\ I \end{bmatrix}$$

```
'C':=evalf(evalm(subs(a=1, b=2, c=3, d=4, eval(C)) &* linalg
 [inverse](TRANS)));
```

$$C := [.5080883927 - .250518744\ I,\ .5080883927 + .2590518744\ I,\ .9838232148]$$

Observable canonical form

The dual form of the controllable canonical form is known as the *observable canonical form*. Like the controllable form the coefficients of the transfer function appear directly in the state matrices. The controllable form matrices can be transformed into the observable form simply as follows: $A \rightarrow A^T$, $B \rightarrow C^T$, and $C \rightarrow B^T$; hence, we get:

```
A:=linalg[transpose](A);
```

$$A := \begin{bmatrix} -\frac{b}{a} & 1 & 0 \\ -\frac{c}{a} & 0 & 1 \\ -\frac{d}{a} & 0 & 0 \end{bmatrix}$$

```
temp:=eval(C):
```

```
C:=linalg[transpose](B);
```

$$C := \begin{bmatrix} 1 & 0 & 0 \end{bmatrix}$$

and

```
B:=linalg[transpose](temp);
```

$$B := \begin{bmatrix} \frac{b}{a^2} \\ \frac{c}{a^2} \\ \frac{d}{a^2} \end{bmatrix}$$

Chapter 7

Discrete data processing

MOST COMMON DISCRETE data processing applications fall into one of two general categories: digital signal processing and image processing. In many instances the operations required are similar if not identical. Although in this discussion we will concentrate mainly on image processing tools, many of the tools developed are equally applicable to signal processing. Because we will be using plots to generate our test data and to display the processed results, we will start with a brief overview of the Maple plotting routines and structures.

Maple plots

The most commonly used plotting functions are `plot` and `plot3d` and they are immediately accessible. In addition to these, two additional plotting packages are available in Maple: `plots` and `plottools`. The `plots` package contains additional data visualization tools, and the plottools packages contains a set of graphics primitives and graphics manipulation routines such as rotation and translation functions.

The plot structure

All maple plotting functions produce either a PLOT or a PLOT3D data structure describing the image to be displayed. Both types of plot structure have the same basic form: PLOT(plot_object1, plot_object2, .., plot_objectn, plot_options) or PLOT3D(plot_object1, plot_object2, .., plot_objectn, plot_options). For example, here are plot structures for the plots in Figures 7.1 and 7.2:

```
PLOT(CURVES([[0,0],[2,2],[1,-1],[1,3],[3,4]]), TEXT([1.1,3],
 'Strange Curve',ALIGNABOVE, ALIGNRIGHT, FONT(TIMES,
 BOLDITALIC,15)));
```

```
PLOT3D(MESH([[[0,0,0], [1,0,0]], [[0,1,2], [1,1,2]],
 [[0,2,0], [1,2,0]]]),POLYGONS([[0, 0,.5], [1, 0, .5],
[1,2,.5], [0, 2, .5]]), ORIENTATION(150, 70),
 TITLE('A-FRAME'), COLOUR(RGB,0,0,0));
```

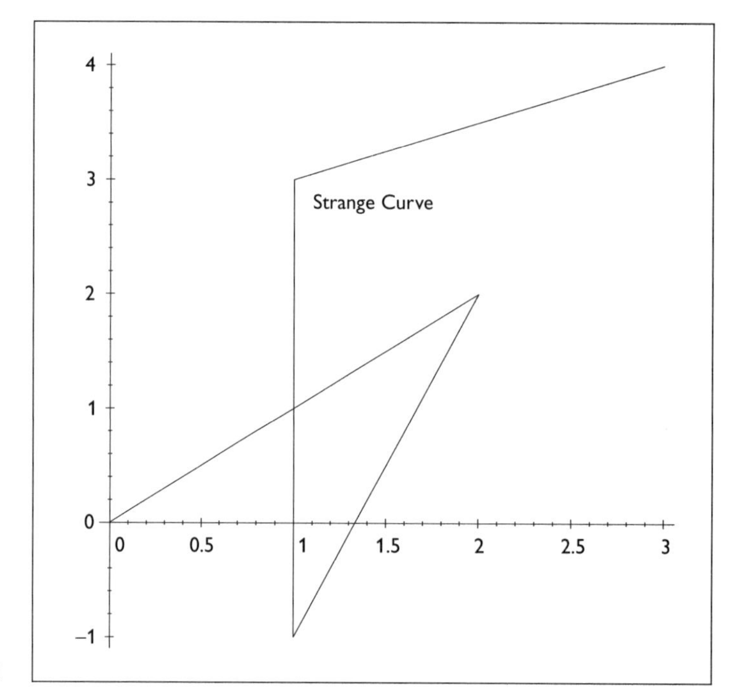

Figure 7.1

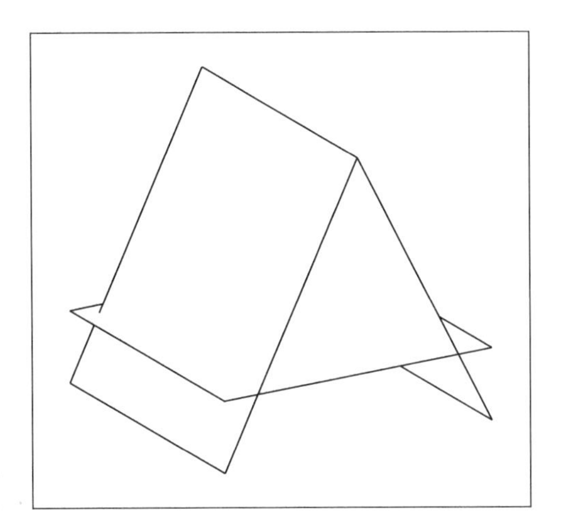

Figure 7.2
A_FRAME.

The objects $\texttt{plot\_object}_\texttt{i}$ can take a number of different forms but they conform to the same basic syntax throughout, namely, `object_type( data )`. The `object_type` describes how the data are to be drawn and can be one of the following: `CURVES`, `POINTS`, `POLYGONS`, `TEXT`, `GRID`, or `MESH`. The objects `POINTS`, `POLYGONS`, and `TEXT` can be either two- or three-dimensional depending on whether the data supplied to each are in the form of pairs or triplets. The `plot_options` are used to set the style of the resulting plot and allow us to set such things as color, axes styles, view ranges, font styles, point symbols, and so on. The table in Appendix B shows the correspondence between the various plot data structures and the user-defined plot options. Unless otherwise stated, the data structures and their corresponding plot options are equally applicable to two- and three-dimensional plot structures.

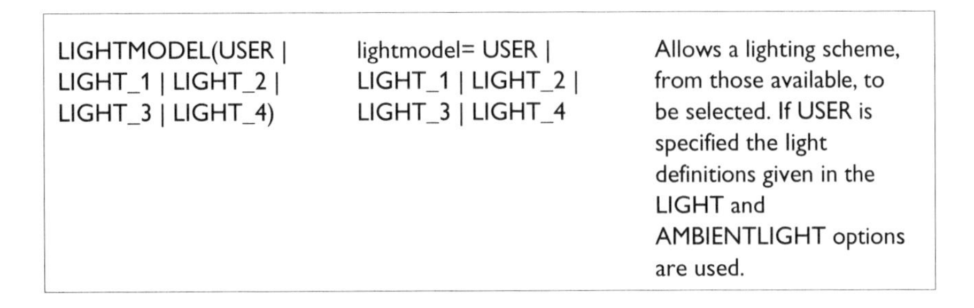

LIGHTMODEL(USER | LIGHT_1 | LIGHT_2 | LIGHT_3 | LIGHT_4)	lightmodel= USER | LIGHT_1 | LIGHT_2 | LIGHT_3 | LIGHT_4	Allows a lighting scheme, from those available, to be selected. If USER is specified the light definitions given in the LIGHT and AMBIENTLIGHT options are used.

Before we move on, it is worth mentioning that the plot data structures are in reality unevaluated function calls that are evaluated when they are passed to IRIS, Maple's graphics interface. An unevaluated function can have many uses in Maple; in this case, it is just a wrapper placed around some plot data as a convenient way of storing the data and context in a way that it can be passed to and processed by IRIS. For more information see `?plot[structure]` and `plots[options]`.

Image conversion

In this first section, we develop a set of conversion tools: `togreyscale`, `tofalsecolor`, `normalize`, and `histogram`. These tools enable us to convert images' color formats from RGB color to monochrome and from monochrome to RGB color, normalize monochrome color data to the range 0 .. 1, where 0 is black and 1 is white, and return a histogram of a data set, respectively. In this particular case we develop our own procedure in deference to Maple's own histogram function `(stats[statplots, histogram])` because there is no exact match between the functionality required and that provided. The mismatch is because Maple's histogram procedure has been designed and implemented as a statistical tool, not as a DSP tool.

To develop and apply these tools, we need to be able to both retrieve and manipulate a plot's color information and then reattach it. In a Maple plot the color information can be specified in three ways: an `RGB` color specification, an `HSV` color specification, or a `HUE` color specification. The `RGB` specification uses three floating-point values, each between 0 and 1, for each of the primary colors. `HUE`, on the other hand, uses a single floating-point value, also between 0 and 1, to select the color, whereas the `HSV` specification requires three floating-point values between 1 and 0: one for color, one for saturation, and one for brightness. The `HUE` definition cycles through the spectrum with the 0–1 transition being the join. Hence, `COLOUR(HUE, 0)` equals `COLOUR(HUE, 1)` equals black (Figure 7.3).

NOTE that because of Maple's Canadian origins, many spellings are UK-English, not American-English.

```
PLOT(POLYGONS([[0,0],[2,0],[2,2],[0,2]]), COLOUR(RGB, 0, 0,
 0), TITLE('A Black Square'));
```

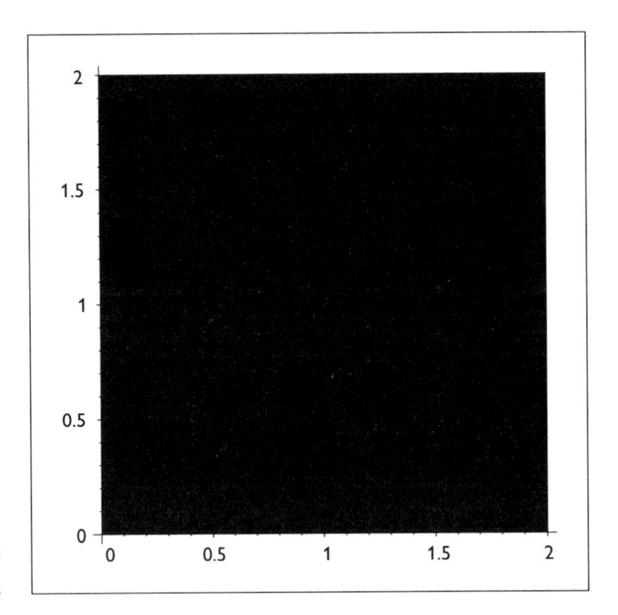

Figure 7.3
A black square.

The following plot structure created by explicitly assigning color values using a color function shows how the color information is attached to a plot object. We use the RGB format so our color function has to provide values for each color. Although we have taken care to scale the color function's output to lie within the range 0 .. 1 this is not strictly necessary because Maple normalizes the color data prior to rendering it.

```
COLORFUNC:=[x/3, (3-y)/3, x*y/9];
```

$$\mathrm{COLORFUNC} := \left[\frac{1}{3}\,x,\; 1 - \frac{1}{3}\,y,\; \frac{1}{9}\,x\,y\right]$$

Using this color function we compute plot structure. By assigning it to a variable, we force Maple to display the data and not the plot.

```
TESTPLOT:=plot3d(sin(x*y), x=0..3, y=0..3, color=COLORFUNC,
 numpoints=9);
```

TESTPLOT := PLOT3D(GRID(0 .. 3., 0 .. 3., [[0, 0, 0, 0],
[0, .8414709848078965, .9092974268256817, .1411200080598672],
[0, .9092974268256817, −.7568024953079282, −.2794154981989259],
[0, .1411200080598672, −.2794154981989259, .4121184852417566]],
COLOR(*RGB*, 0, 1., 0, 0, .6666666666666667, 0, 0,

.3333333333333334, 0, 0, 0, 0, .3333333333333333, 1., 0,
.3333333333333333, .6666666666666667, .1111111111111111,
.3333333333333333, .3333333333333334, .2222222222222222,
.3333333333333333, 0, .3333333333333333, .6666666666666666, 1., 0,
.6666666666666666, .6666666666666667, .2222222222222222,
.6666666666666666, .3333333333333334, .4444444444444444,
.6666666666666666, 0, .666666666666666, 1., 1., 0, 1.,
.6666666666666667, .3333333333333333, 1., .3333333333333334,
.6666666666666666, 1., 0, 1.)), *AXESLABELS*(*x*, *y*,), *TITLE*(),
STYLE(*PATCH*))

With reference to the preceding Maple output, we can see that Maple inserted the plot's color information after the three-dimensional surface definition. We can, therefore, isolate the color information by selecting the correct portion of the plot structure. In this particular example, the information we are interested in is surrounded by the wrapper `GRID` and can be picked out by `op`. The `op` function allows us to select operands from within data structures, in the current release,[1] `op's` functionality has been extended to operate on nested data structures implicitly. The call `op([4, 1], ...)` returns the first operand of the fourth operand of the plot structure, namely, the color information.

```
DATA:=op([4, 1], TESTPLOT);
```

DATA := COLOR(*RGB*, 0, 1., 0, 0, .6666666666666667, 0, 0,
.3333333333333334, 0, 0, 0, 0, .333333333333333, 1., 0,
.3333333333333333, .6666666666666667, .1111111111111111,
.3333333333333333, .3333333333333334, .2222222222222222,
.3333333333333333, 0, .3333333333333333, .6666666666666666, 1., 0,
.6666666666666666, .6666666666666667, .2222222222222222,
.6666666666666666, .3333333333333334, .4444444444444444,
.6666666666666666, 0,.6666666666666666, 1., 1., 0, 1.,
.6666666666666667, .3333333333333333, 1., .3333333333333334,
.666666666666666, 1., 0, 1.)

1. This syntax is valid for releases greater than Maple Vr3. For earlier releases, nested ops must be used; i.e., op (4, op(1, ...)).

We can see that the color information is presented as a sequence of data points, which must be taken three at a time to provide the actual color. If we take a closer look at the data, we can see that there are more data points than there appear to be grid points, i.e., `numpoints` equals 9 while there are 16 data points. This is because Maple generates a grid that is numbered from zero to $\sqrt{\text{numpoints}}$, hence the extra points. Here we read in a previously created test image and then plot the three original planes. Now that we can isolate the color information, the next step is to isolate the individual color planes—red, green, and blue. By way of an example, we remove the red color plane from the image data using a for loop as shown:

```
RDATA:=NULL:
for n from 2 to nops(DATA) by 3 do
 RDATA:=RDATA, op(n, DATA);
od:
[RDATA];
```

[0, 0, 0, 0, .3333333333333333, .3333333333333333, .3333333333333333, .3333333333333333, .6666666666666666, .6666666666666666, .6666666666666666, .6666666666666666, 1., 1., 1., 1.]

The list of red color values has no spatial information (i.e., where in the plot grid it applies) associated with it so we must recreate this manually. This is a simple task of converting the list into a square matrix whose ij^{th} element corresponds with the $(i-1)(j-1)^{\text{th}}$ red intensity in the image grid. The offset is necessary because Maple matrix indices must start at one.

```
REDMAT:= linalg[matrix](sqrt(nops([RDATA])),
 sqrt(nops([RDATA])),[RDATA]);
```

REDMAT :=
[0, 0, 0, 0]
.3333333333333333 , .3333333333333333 , .3333333333333333 ,
.3333333333333333]
[.6666666666666666 , .6666666666666666 , .6666666666666666 ,
.6666666666666666]
[1., 1., 1., 1.]

The intensity of the image's red component can be viewed with `matrixplot`, which is found in the `plots` package (Figure 7.4):

```
plots[matrixplot](REDMAT, axes= FRAME,
labels=['x','y','Int'],
title='Image's red plane', style=HIDDEN, color=BLACK);
```

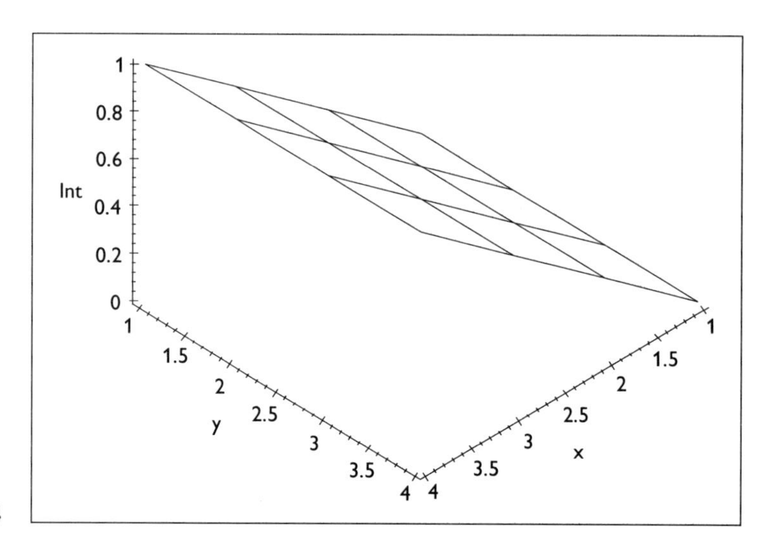

Figure 7.4

The Maple code just developed has been encapsulated in the following two procedures. The first procedure takes two arguments, the color data and the color plane required, and returns the selected color information as a list. Within the body of the procedure, the selected color plane is converted into an index offset into the data structure by using a table. The color information is then stepped through and the appropriate data are removed using a `for` loop. The second procedure takes the color information list and returns a graphical representation of it using `matrixplot`:

```
GET_COLOR:=proc(data, colour)
 options 'Copyright Coded by Dr. Steve Adams 1995';
 local colordat, temp, n;
 colordat:=table([RED=1, GREEN=2, BLUE=3]);
 temp:=NULL;
 for n from 1+colordat[colour] to nops(data) by 3 do
  temp:=temp, op(n, data);
 od;
 [temp];
end:
```

```
COLOR_PLOT:=proc(data, Title)
 options 'Copyright Coded by Dr. Steve Adams 1996';
 local count;
 count:=sqrt(nops(data));
 plots[matrixplot](linalg[matrix](count, count, data),
 style=WIREFRAME, color=BLACK, axes=FRAME, title=Title,
 labels=['x','y','Int']):
end:
```

Applying these functions to a previously obtained color (stored in the file `colour.dat`) plot to return the three color planes—red, blue, and green—and plotting the results as a graphics array we get the following:

```
read('colour.dat'):
```

```
P1:=COLOR_PLOT(GET_COLOR(DATA, RED), 'Red Plane'):
```

```
P2:=COLOR_PLOT(GET_COLOR(DATA, GREEN), 'Green Plane'):
```

```
P3:=COLOR_PLOT(GET_COLOR(DATA, BLUE), 'Blue Plane');
```

```
plots[display](array(1..1, 1..3, [P1, P2, P3])):
```

Figures 7.5, 7.6, and 7.7 show the intensities of the three primary colors (red, blue, and green, respectively) at each point on a three-dimensional surface. The actual surface is irrelevant at this point because we are only interested in its color map.

Togreyscale

A common digital image processing transformation is to change an image's color map into a greyscale one by forcing the red, green, and blue components of the RGB data structure to be equal. This approach can drastically reduce the size required to store an image. In this example we use, as a starting point, the color information stored earlier in the variable `DATA`.

Maple supports an extensive set of conversion routines that can be easily extended by the user. The procedure `convert` uses `helper` routines to manipulate data passed to it. These "helper" routines all have the same form of procedure name: `convert/helper_name` and are all invoked in the same way: `convert(data, helper_name)`. Hence, in this particular case we create the helper routine `convert/togreyscale`, shown next, to convert an image's color information

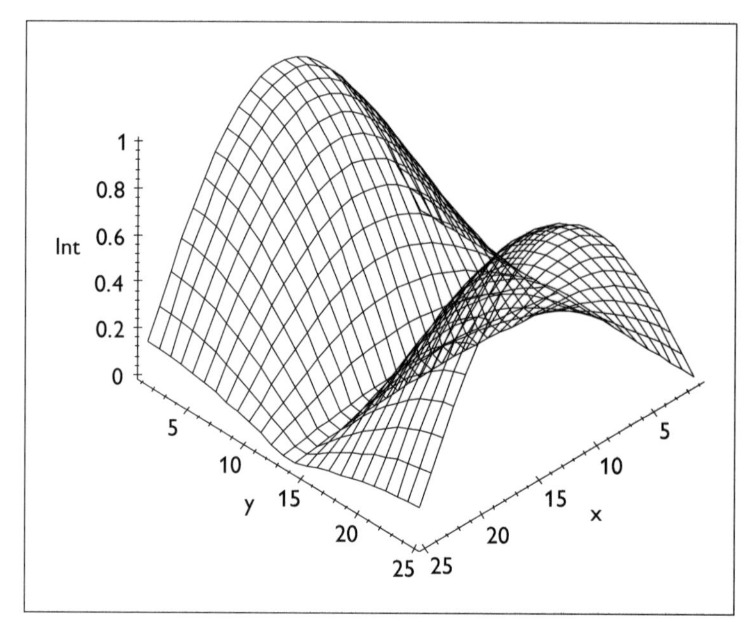

Figure 7.5
Red plane.

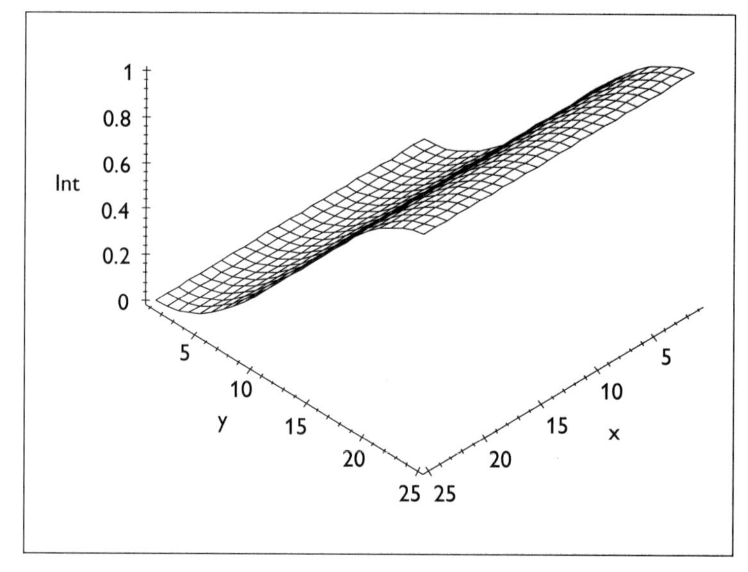

Figure 7.6
Green plane.

into greyscale information that is invoked with `convert(data, togreyscale)`. The core of this routine is a `for` loop that allows us to step through the data backwards in steps of three. Each triplet is averaged to compute the greyscale value.

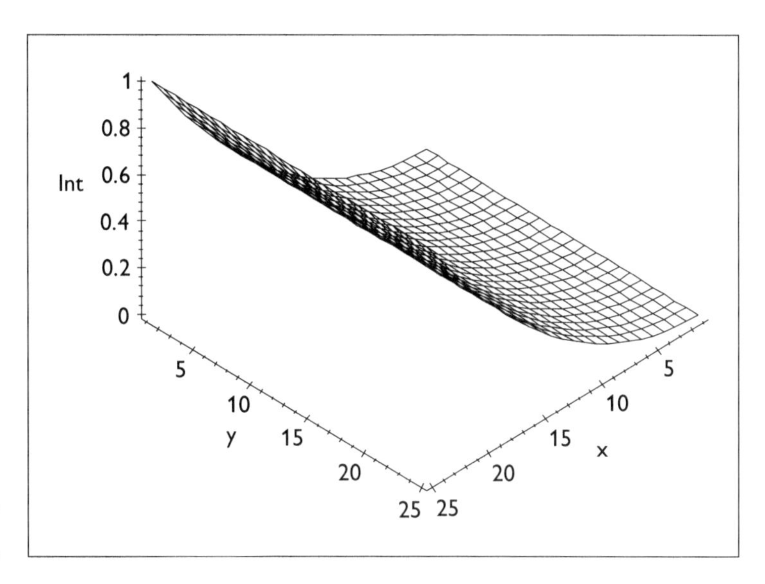

Figure 7.7
Blue plane.

```
'convert/togreyscale':=proc(data)
 options 'Copyright coded by Dr. Steve Adams 1995';
 local n, START, new_data, level;
 new_data:=NULL;
 START:=nops(data)-1;
 for n from START-1 to 1 by -3 do
  level:= convert([op(n..n+2, data)], '+')/3;
  new_data:=level$3, new_data;
 od;
 COLOUR(RGB, new_data);
end:
```

Converting the color data to greyscale data and plotting the three individual color planes enables us to compare the new `color` data with the original data shown earlier.

```
NEW_DATA:=convert(DATA, togreyscale):
```

As above we plot each color plane of the new greyscale image for comparison:

```
> P11:=COLOR_PLOT(GET_COLOR(NEW_DATA, RED), 'New Red Plane'):

> P21:=COLOR_PLOT(GET_COLOR(NEW_DATA, GREEN),
  'New Green Plane'):

> P31:=COLOR_PLOT(GET_COLOR(NEW_DATA, BLUE), '
  New Blue Plane'):

> plots[display](array(1..1, 1..3, [P11, P21, P31]));
```

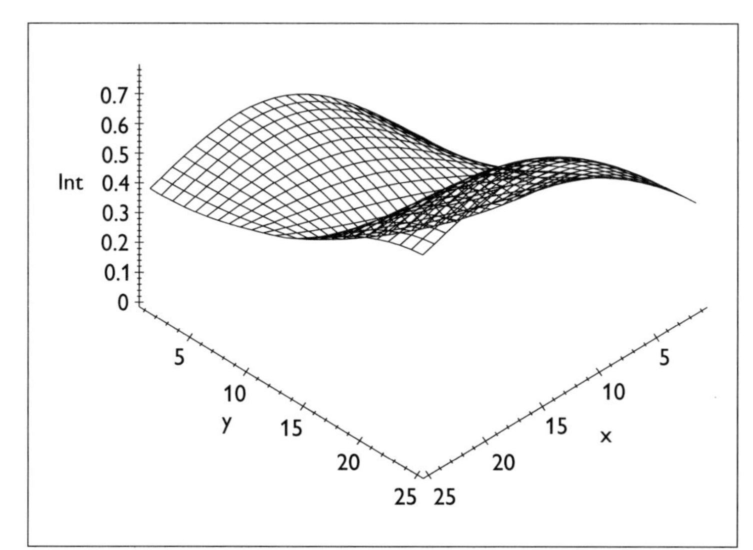

Figure 7.8
New red plane.

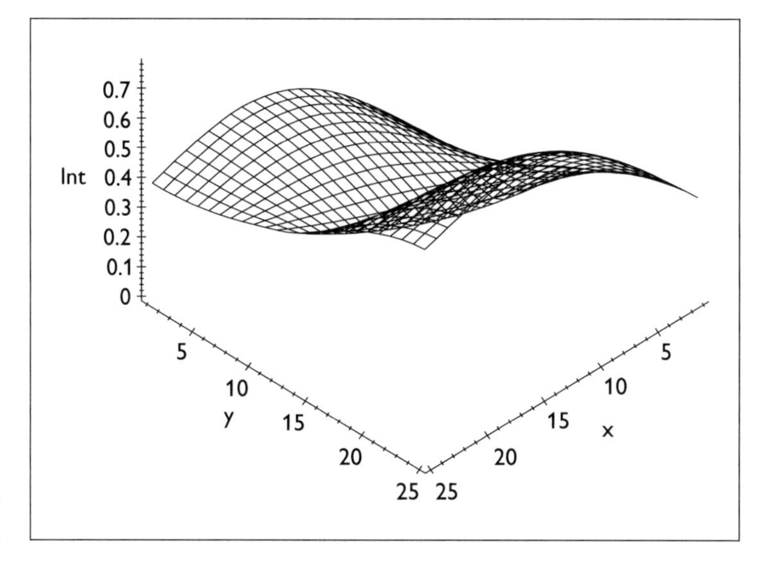

Figure 7.9
New green plane.

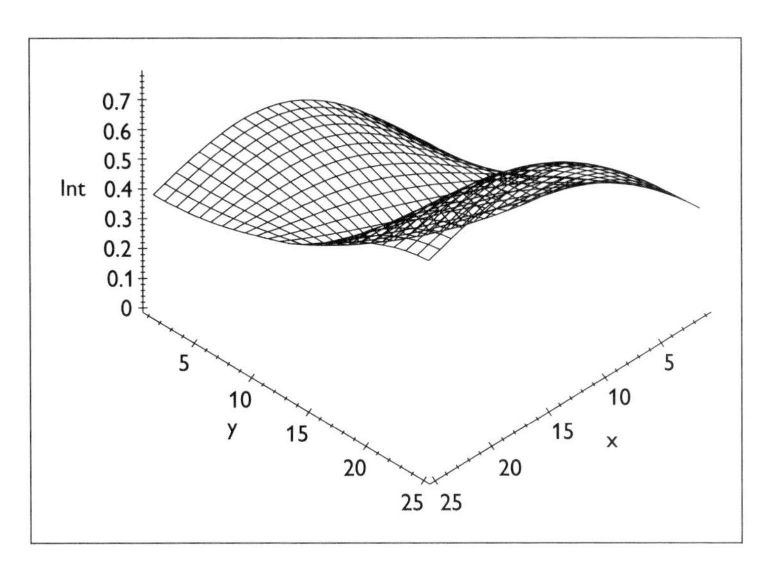

Figure 7.10
New blue plane.

With reference to Figures 7.8, 7.9, and 7.10, we can see that all of the color intensities have the same value for any given point on the image, which means that the color map is now a greyscale one. The final stage in this process is to substitute the greyscale information into the original plot structure:

```
NEW_PLOT:=subs(DATA=NEW_DATA, TESTPLOT):
```

The new greyscale image can now be viewed, which is left as an exercise for the reader.

Before we move to our next topic, normalization, we conclude this section by outlining how Maple generates greyscale images. Unless otherwise specified by explicitly defining a color map, Maple produces an image object with the default color map applied automatically by IRIS. The default color map, usually a hue value dependent on the *x-y-z* coordinate of the point in question, can be overridden by setting the user option shading to be equal to a valid map, currently, `XYZ`, `XY`, `Z`, `ZGREYSCALE`, `ZHUE`, and `NONE`. Depending on which color map is selected, IRIS computes the color at each point on the surface as it renders it. If we want to display a greyscale image, we simply set shading as shown: `shading= ZGREYSCALE`. This has the same effect as setting the color map explicitly with the red, green, and blue components equal.

Normalize

If we refer to the greyscale information displayed in the last section, we see that the intensity values do not cover all of the intensity values available, namely, 0 .. 1. It can be seen that the intensity values only range between 0.2 to 0.7 resulting in the contrast of the image being compressed. A simple, but effective, image processing technique is contrast adjustment is one in which the original intensity values are mapped onto the entire supportable range of intensity values as shown in Figure 7.11.

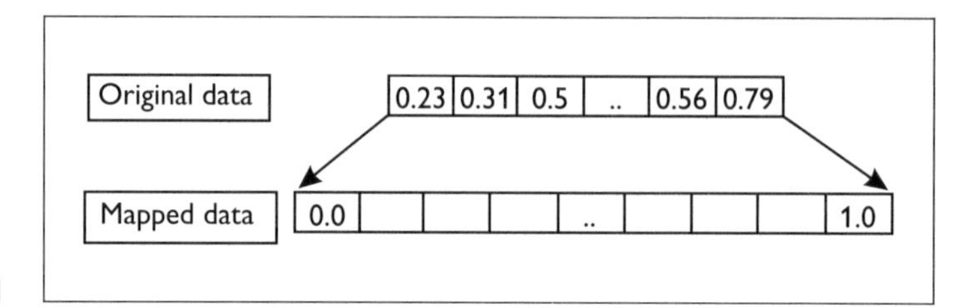

Figure 7.11

The actual mapping process that we will perform is linear, in which we will first shift the data and then stretch it to cover the range 0 .. 1. But first we need to isolate the original data and, because all of the color planes contain the same information in the greyscale image, we will simply isolate the red plane data:

```
DATA_ONLY:=GET_COLOR(NEW_DATA, RED):
```

Before we adjust the contrast we will take a look at a histogram of the original data. Before we can do this, however, we need to manipulate the raw data to take into account that we are dealing with a discrete system. In a discrete system the number of valid levels is determined by the bit length or depth of the imaging system. In our example, we will use a bit depth of eight giving 256 (0 .. 255) valid levels:

```
BIT_DEPTH:=8;
```

$$BIT\_DEPTH := 8$$

Here we look at the first five elements of the data:

```
DATA_ONLY[1..5];
```

$$[0, .001306928340, .005207319179, .01164030823, .02050551083]$$

Now we quantize the original data to the bit depth of the system using `map` and `round` as shown next. The function `round` rounds down a numeric value to the nearest integer:

```
QUANTIZED:=map(round,map((x,y)-x*y, DATA_ONLY,
2^BIT_DEPTH-1)):
```

Here we look at the first five elements of the quantized data:

```
QUANTIZED[1..15];
```

$$[0, 0, 1, 3, 5, 8, 11, 15, 20, 24, 29, 34, 40, 45, 50]$$

Now we are ready to sort the quantized data into bins, one unit wide, and plot them to produce the required histogram. Filling the bins is done very simply by using a table and a `for` loop. First we create 256 empty bins as shown:

```
BINS:=table([seq(n=0, n=0..2^BIT_DEPTH-1)]):
```

Now we step through the sorted data incrementing the bin counts as appropriate:

```
for n in QUANTIZED do
BINS[n]:=BINS[n]+1;
od:
```

Before we can plot the data we have to convert the table entries, which is in essence unsorted, into a sorted list of data pairs: `[bin number, entries]`.

```
pts := [seq([n, BINS[n]], n=0..2^BIT_DEPTH-1)]:
```

In Figure 7.12, we plot the histogram of the original data quantized to eight bits:

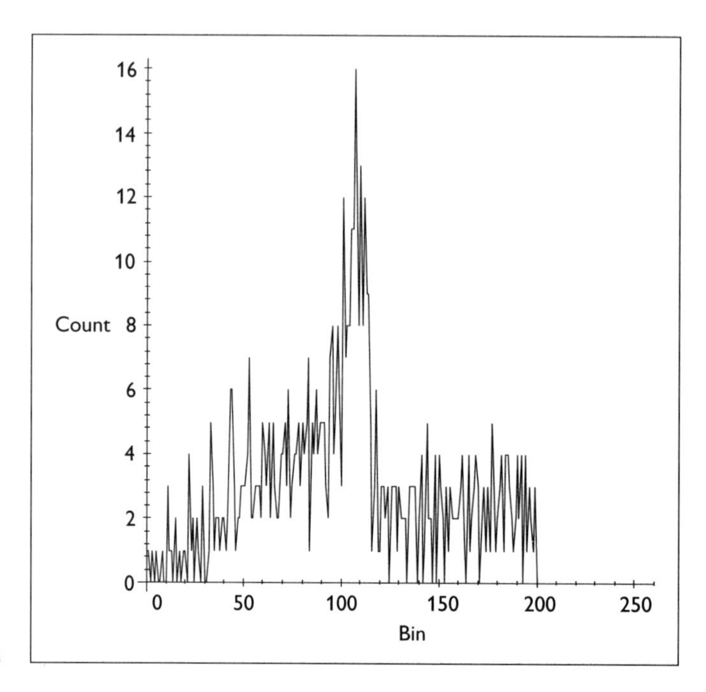

Figure 7.12

```
plot(pts, labels=['Bin','Count']);
```

The histogram plot, although valid, needs some cosmetic adjustment because Maple by default plots a list by connecting all of the points in sequence. What we want is a sequence of vertical lines the length of which is proportional to the number of the entries in the bin. We can achieve this very easily by scanning the point pairs searching for nonzero abscissa values. Once a nonzero value is discovered, two additional points are inserted, one on either side, which forces Maple to plot a line starting and ending at the point `[ bin number, 0]`. For example, given the point sequence `[0, 0], [1, 10], [2, 4]` we would generate the new point sequence `[0, 0],` **`[1, 0]`**`, [1, 10],` **`[1, 0]`**`,` **`[2, 0]`**`, [2, 4],` **`[2, 0]`**, where the additional point pairs are emboldened for clarity. This process is demonstrated in Figure 7.13. The following Maple function performs this task:

```
FUNC:=x->if x[2]<>0 then [x[1], 0], x, [x[1], 0] else x fi;
```

```
FUNC :=proc(x)
options operator,arrow;
if x[2] <>0 then [x[1],0],x,[x[1],0] else x fi
end
```

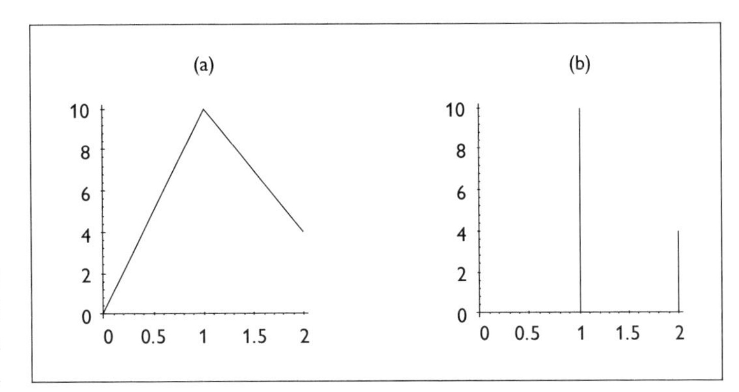

Figure 7.13 (a) Original, and (b) with extra points.

Here we test our new function to check that it operates as we expect:

```
FUNC([1,1]);
```

$$[1, 0], [1, 1], [1, 0]$$

```
FUNC([1,0]);
```

$$[1, 0]$$

Now we are ready to replot the histogram (Figure 7.14):

```
plot(map(FUNC, pts), title='Histogram Plot',
 labels=['Bins','Count']);
```

The process just explained to generate a histogram plot of a set of data has been automated in the function `histogramplot` found in the program disk.

The histogram confirms our original observation that the contrast of the original image is compressed because there are no intensity values in the range 200 .. 250 present in the original plot. The original data are therefore adjusted for contrast by first resetting the origin and then stretched the translated data so that it covers the supported range 0 .. 1. This resets the origin:

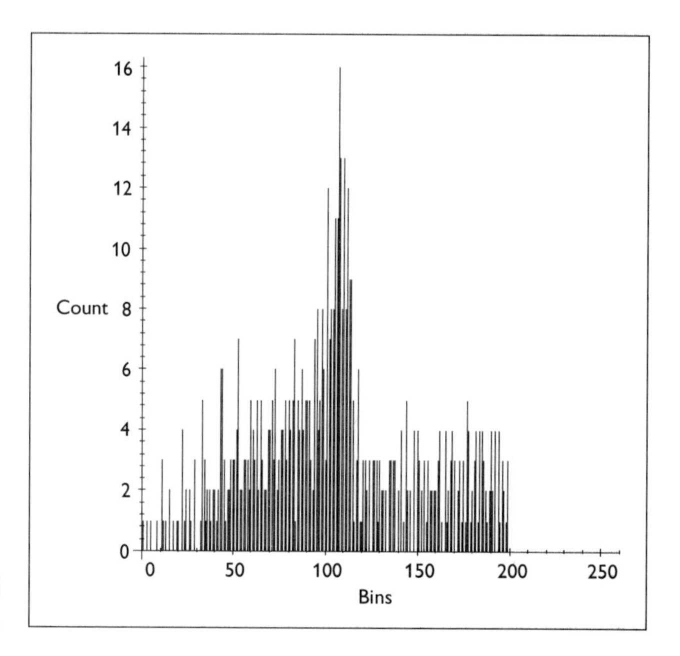

Figure 7.14
Histogram plot.

```
LOW:=min(op(DATA_ONLY));
```

$$LOW := 0$$

```
SHIFTED := map((x,y)->x-y, DATA_ONLY, LOW):
```

Now we stretch the shifted data:

```
HIGH:=max(op(SHIFTED));
```

$$HIGH := .7819887006$$

```
STRETCHED := map((x, y)->x/y, SHIFTED, HIGH):
```

Here we look at the first and last five elements of the adjusted data set:

```
[op(op(1..5,sort(STRETCHED))), `..`, op(op(nops(STRETCHED)
 -5..nops(STRETCHED), sort(STRETCHED)))];
```

[0, .001671288011, .006659072152, .01488551973, .02622225975, ..,
.9865825261, .9871622880, .9948650055, .9959871150, .9994963150,
1.000000000]

As before we first quantize the data to eight bits and then using `histogramplot` we plot a histogram of the adjusted data (Figure 7.15):

```
QUANTIZED:=map(round,map((x,y)->x*y, STRETCHED,
 2^BIT_DEPTH-1)):
```

```
histogramplot(QUANTIZED, 8, title='Normalized Histogram
 Plot');
```

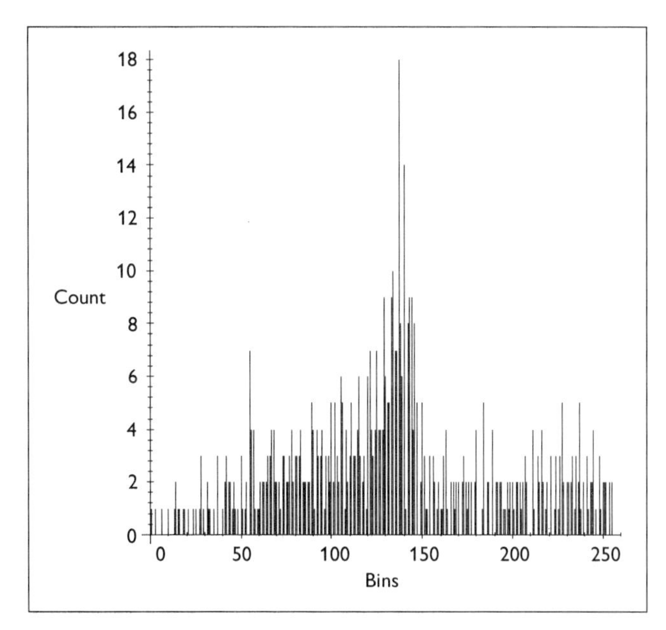

Figure 7.15 Normalized histogram.

With reference to the histogram of Figure 7.15 we can see that the image's contrast has been successfully stretched, but the overall shape of its intensity profile has remained unchanged. The number of instances of various intensity values occurring in the contrast adjusted image will be different than the unadjusted image as a result of the mapping and quantization process.

The normalize function can be found on the program disk.

Tofalsecolor

In many applications the analysis of an image is simplified by adding false color, as the next example demonstrates. We wish to investigate the three-dimensional image shown in Figure 7.16 by applying contours of equal color intensity.

```
plot3d( exp(-sin(x^2)-cos(y^2)), x=-2..2, y=-2..2,
 labels=['x','y','z'], style=CONTOUR, color=BLACK );
```

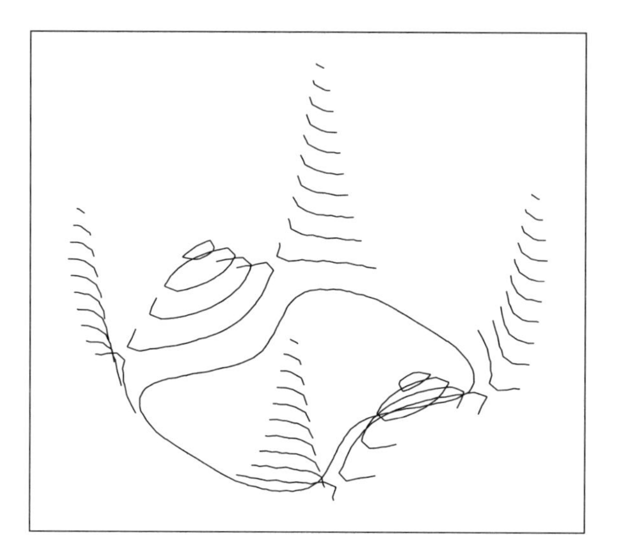

Figure 7.16

With the image in its original form we are unable to deduce much regarding the dominant plateau region of the surface. One possible approach is to apply a mapping function to the data, analogous to applying false color, and plotting the result (Figure 7.17):

```
MAP:=log;
```

$$MAP := \log$$

```
plot3d(MAP(exp(-sin(x^2)-cos(y^2))), x=-2..2, y=-2..2,
 labels=['x','y','z'], style=CONTOUR, color=BLACK );
```

By manipulating the data in this way we can 'see' information buried in the data.

Using Maple we can add false color to the adjusted greyscale data generated earlier (stored in the variable `NEW_DATA`) using mapping functions. These mapping functions are normally functions of intensity and can return any value in the range 0 .. 1. The example functions that we will be using as defaults, one for each plane, are as follows:

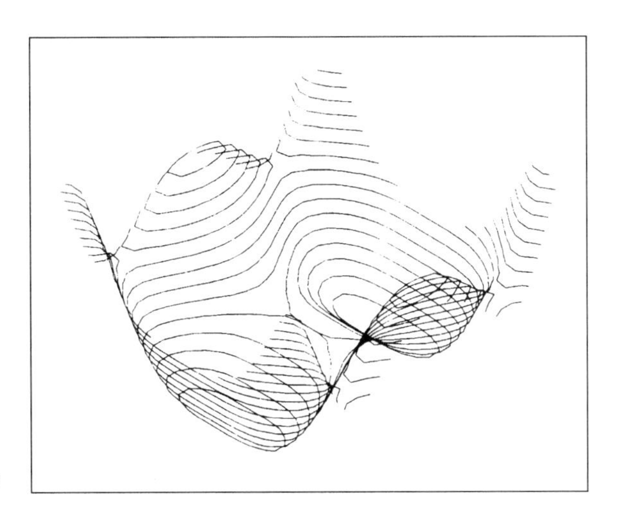

Figure 7.17

```
a:=t->((1-cos(t*2*Pi))^2)/4;
b:=t->(1-cos(t*Pi))/2;
c:=t->(cos(t*Pi)+1)/2;
```

$$a := t \rightarrow \frac{1}{4}\,(1 - \cos(2\ t\ \pi))^2$$

$$b := t \rightarrow \frac{1}{2} - \frac{1}{2}\cos(t\ \pi)$$

$$c := t \rightarrow \frac{1}{2}\cos(t\ \pi) + \frac{1}{2}$$

Here we plot the mapping functions as a function of intensity over the range 0 .. 1 (Figure 7.18):

```
plots[display]([plots[textplot]([[0.1, 0.9, `a`],[0.5, 0.9,
`b`],[ 0.9, 0.9, `c`]]),plot({a(t), b(t), c(t)}, t=0..1,
title='False Color Functions', color=BLACK,
labels=[`Old`,'New'])]);
```

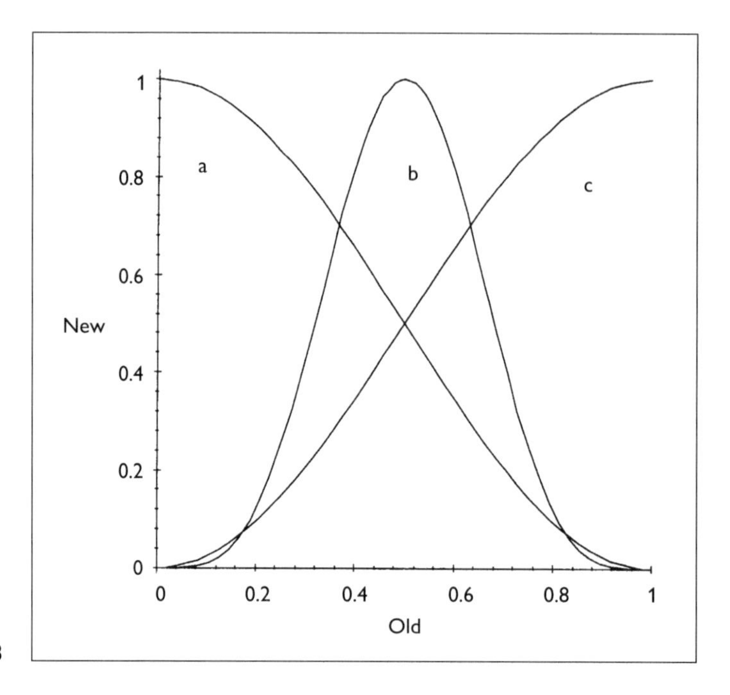

Figure 7.18

The process of applying false color is outlined in the next example. First we get the color data for the other two planes (we have already normalized the red color plane data and stored it in the variable `STRETCHED`) in our test image and normalize it:

```
DATA2:=normalize(GET_COLOR(NEW_DATA, GREEN)):
DATA3:=normalize(GET_COLOR(NEW_DATA, BLUE))):
```

Using this and the previously obtained data (`STRETCHED`) we apply the false color mapping functions in order to generate three new color planes:

```
NEW_COLOR1:=map(a, STRETCHED):
NEW_COLOR2:=map(b, DATA2):
NEW_COLOR3:=map(c, DATA3):
```

The three sets of false color information need to be knitted together in the correct order so the complete `COLOUR` data structure can be built. We do this in two stages:

```
TEMP:=zip((x,y)->[x,y], NEW_COLOR1, NEW_COLOR2):
 COLOUR(RGB, op(map(op,zip((x,y)->[op(x), y], TEMP,
 NEW_COLOR3)))):
```

Again this is left to the reader to do as an exercise. Before moving on, we will look at the first false color plane data as a three-dimensional surface (Figure 7.19):

```
plots[matrixplot](linalg[matrix](25,25,NEW_COLOR1),
 title='False Color Plane 1', labels=['x','y','Int'],
 color=BLACK, style=HIDDEN, axes=FRAME);
```

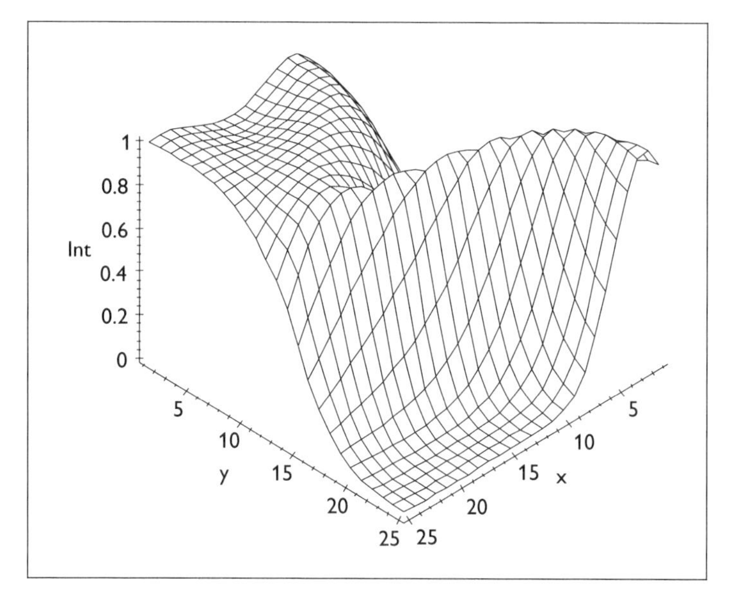

Figure 7.19
False color plane 1.

The process of applying false color to greyscale images has been extended and encapsulated in the procedure `'convert/tofalsecolor'` which is available on the data disk.

Conclusion

In the previous section we demonstrated how Maple can be used effectively as a tool to manipulate and investigate discrete images. We have also shown how Maple can help us to develop prototype algorithms for digital image processing techniques by first developing the algorithm interactively and then implementing the algorithm as a procedure. The three tools developed in this manner were the conversion routines `togreyscale` and tofalsecolor, the utility functions normalize and GET_COLOR, and the plotting routines COLOR_PLOT and histogramplot.

Linear filters

Another common discrete data processing application is that of time series data filtering. With Maple's list and array structures, we can manipulate time series data very effectively. In this section we use Maple to develop routines to implement five of the more common filtering techniques, namely, first-order differencing, moving average, moving median, and exponential filtering.

Differencing

Differencing is a technique commonly applied to time series data in disciplines as diverse as finance to image processing. For example, recently, the stock market appears to be relentlessly increasing and therefore it may appear to be a good long-term investment. However, if the average increases in the average price are less than the prevailing rate of inflation, then it would not be such an attractive proposition. We can use differencing techniques to determine the underlying trend, if any, concealed within the data. In this particular implementation we operate on a list data structure using the for-next construct. First we read the series data into the current Maple session. The data file is in a Maple-friendly form (i.e., the file contains text that conforms to the Maple syntax) so the data are automatically assigned to the variable SERIES_DATA. The time series data consists of a list with 256 entries, a portion of which, along with a plot of the data (Figure 7.20), is given here:

```
> read(`series.dat`):
```

```
> [op(op(1..10,SERIES_DATA)), `..`, op(op(246..255,
  SERIES_DATA))];
```

$$[2, 0, 1, 0, 1, 0, 0, 1, 0, 0, .., 4, 1, 0, 3, 1, 2, 2, 2, 0, 2]$$

```
> plots[listplot](SERIES_DATA, title='Original Data',
  labels=[`t`,'amp']);
```

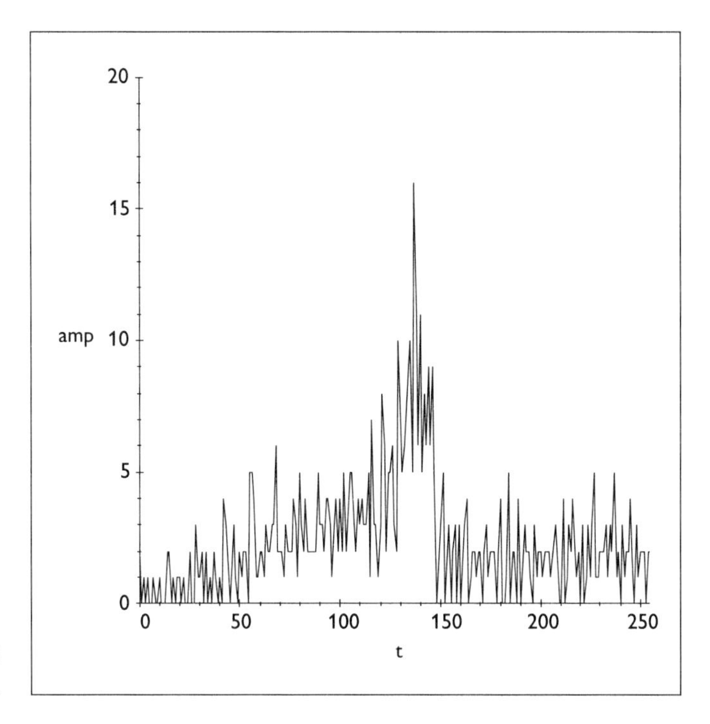

Figure 7.20
Original data.

Using a sample of the entire data set we will develop the difference algorithm:

```
SAMPLE:=SERIES_DATA[1..10];
```

$$\text{SAMPLE} := [2, 0, 1, 0, 1, 0, 0, 1, 0, 0]$$

The algorithm used is as follows: The ith filtered output is the difference between the ith and the $(i+1)^{\text{th}}$ inputs. We compute the filter's output sequence using the `for-next` construct shown:

```
temp:=NULL:
for n to nops(SAMPLE)-1 do
 the_diff:=SAMPLE[n+1] - SAMPLE[n]:
 temp := temp, the_diff:
 print(temp);
od:
[temp];
```

$$-2$$
$$-2,\ 1$$
$$-2,\ 1,\ -1$$
$$-2,\ 1,\ -1,\ 1$$
$$-2,\ 1,\ -1,\ 1,\ -1$$
$$-2,\ 1,\ -1,\ 1,\ -1,\ 0$$
$$-2,\ 1,\ -1,\ 1,\ -1,\ 0,\ 1$$
$$-2,\ 1,\ -1,\ 1,\ -1,\ 0,\ 1,\ -1$$
$$-2,\ 1,\ -1,\ 1,\ -1,\ 0,\ 1,\ -1,\ 0$$
$$[\ -2,\ 1,\ -1,\ 1,\ -1,\ 0,\ 1,\ -1,\ 0\ ]$$

We can see that the manipulated data do not contain the first and last data points of the original list and, hence, each time the difference filter is used the length of the available data set is reduced by two as we lose the first and the last data points from the list. The function `difference`, found on the program disk, takes a list of samples and returns a list of their differences. Here we use this function to return a list of differences from the original data and plot it (Figure 7.21) using `listplot`:

```
DIFFERENCED:=difference(SERIES_DATA):

plots[listplot]([DIFFERENCED], title='Difference Filter
 Output',labels=['t','delta']);
```

The output from the difference filter shows the underlying trends present. With reference to both the original and the filtered data, we can see that the test data exhibit periods of uniform growth and decline on top of which periods of volatile activity are impressed. By using the difference filter it is easy to highlight the periods of high volatility that are directly applicable to, for example, the finance industry.

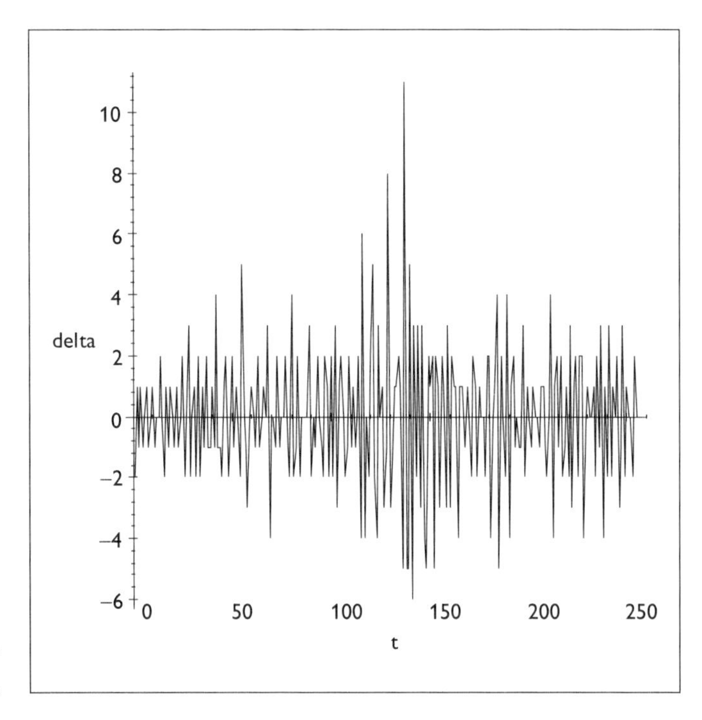

Figure 7.21 Difference filter output.

Moving average

Another common filtering technique that is effective at reducing the noise component of a data set is that of returning the moving average of a data set. Unfortunately this type of filtering has the tendency to blur edges because it mimics a low-pass filtering operation. The moving average algorithm, like the differencing algorithm, is relatively simple to implement in Maple through the manipulation of matrices. Put simply, a moving average filter takes a list of samples and produces a new one, the elements of which are the average of a windowed version of the original. This process is depicted in Figure 7.22, where we take a three-element window and pass it the data to generate the new filtered data. The first element of the filtered data is the same as the original because only a single element is covered by the filter window. The second element is the average of the first two of the original data as the window has now moved to cover them. The third entry in the filtered data set is the average of the first three elements of the original data set and so on. As before we use `SAMPLE` as our test data set with which to develop our filter.

```
WINDOW:=3;
```

$$WINDOW := 3$$

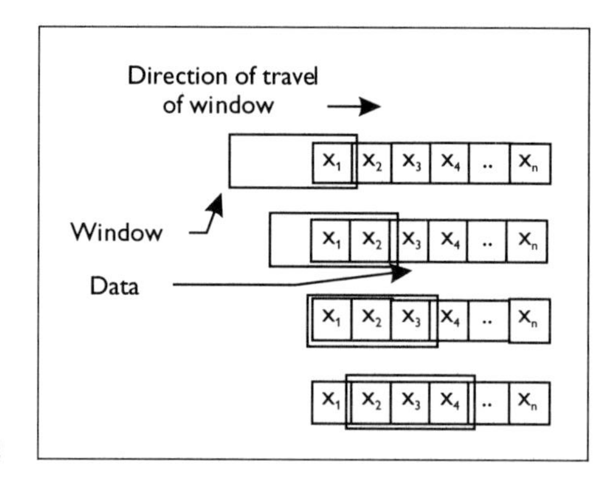

Figure 7.22

Create a matrix of the data that is larger by `2(WINDOW-1)` and pad it with zeros at both the beginning and the end. This makes it easier to sweep the window over the data at its extremities.

```
ZEROS:=0$WINDOW-1;
```

$$ZEROS := 0, 0$$

```
MAT:=linalg[matrix](1, nops(SAMPLE)+(WINDOW-1)*2, [ZEROS,
op(SAMPLE), ZEROS]);
```

$$MAT := [0\ 0\ 2\ 0\ 1\ 0\ 1\ 0\ 0\ 1\ 0\ 0\ 0\ 0]$$

Using the `for` loop shown next, we sweep the window over the preceding matrix and compute the next element in the filtered data set by first removing the windowed elements, converting the submatrix to a list of lists, isolating the sublist, and then finding the average of its entries. The result is appended to the previously calculated results.

```
FILTERED:=NULL:
for n to nops(SAMPLE) do
 temp:=linalg[submatrix](MAT, 1..1, n..n+2):
 temp:=convert(temp, listlist):
 temp:=convert(op(temp), `+`)/WINDOW:
 FILTERED:=FILTERED, temp:
 print(FILTERED);
od:
[FILTERED];
```

$$\frac{2}{3}$$

$$\frac{2}{3}, \frac{2}{3}$$

$$\frac{2}{3}, \frac{2}{3}, 1$$

$$\frac{2}{3}, \frac{2}{3}, 1, \frac{1}{3}$$

$$\frac{2}{3}, \frac{2}{3}, 1, \frac{1}{3}, \frac{2}{3}$$

$$\frac{2}{3}, \frac{2}{3}, 1, \frac{1}{3}, \frac{2}{3}, \frac{1}{3}$$

$$\frac{2}{3}, \frac{2}{3}, 1, \frac{1}{3}, \frac{2}{3}, \frac{1}{3}, \frac{1}{3}$$

$$\frac{2}{3}, \frac{2}{3}, 1, \frac{1}{3}, \frac{2}{3}, \frac{1}{3}, \frac{1}{3}, \frac{1}{3}$$

$$\frac{2}{3}, \frac{2}{3}, 1, \frac{1}{3}, \frac{2}{3}, \frac{1}{3}, \frac{1}{3}, \frac{1}{3}, \frac{1}{3}$$

$$\frac{2}{3}, \frac{2}{3}, 1, \frac{1}{3}, \frac{2}{3}, \frac{1}{3}, \frac{1}{3}, \frac{1}{3}, \frac{1}{3}, \frac{1}{3}$$

$$\left[\frac{2}{3}, \frac{2}{3}, 1, \frac{1}{3}, \frac{2}{3}, \frac{1}{3}, \frac{1}{3}, \frac{1}{3}, \frac{1}{3}, \frac{1}{3}\right]$$

The function `moving_ave`, which is an extension of the `for` loop shown previously, is found on the program disk and can be used to filter a list of sample points. In addition to the data list, the window length must also be specified. Here we apply `moving_ave` to the data list used in the previous example (Figure 7.23):

```
FILTERED:=moving_ave(SERIES_DATA,3):
plots[listplot]([FILTERED], title='Moving Average Filter
 Output', lables=['t','amp']);
```

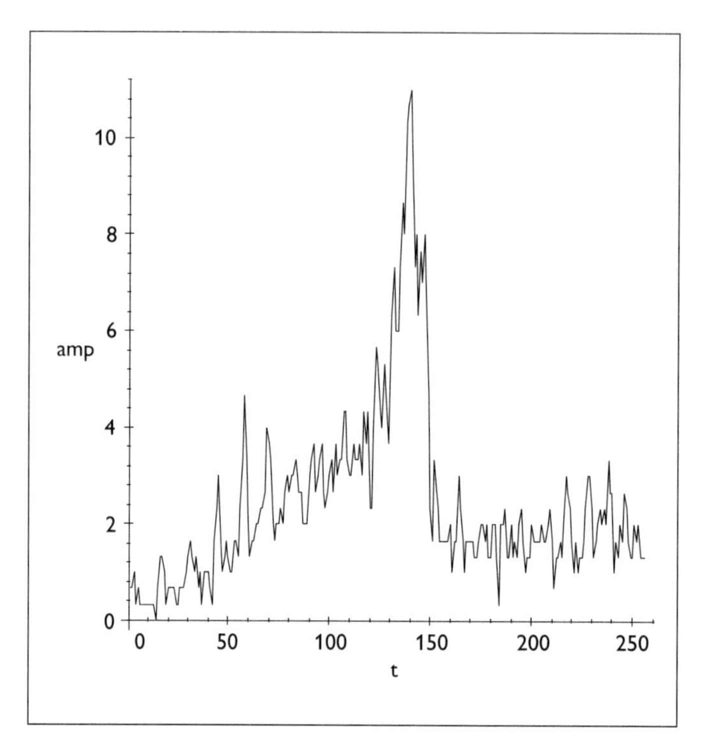

Figure 7.23 Moving average filter output.

Now we can investigate the effect on the filtered data of altering the window size:

```
MANY_TIMES:=SERIES_DATA:
for n to 3 do
MANY_TIMES:=MANY_TIMES, moving_ave(op(n, [MANY_TIMES]),
 3*n)
od:
```

Using `listplot` we can plot the output from each iteration (Figures 7.24 through 7.27). By placing the resulting plots into an array structure we can compare the output for each iteration easily. By using a table we can easily place a relevant title on each plot.

```
TITLES:=table([1='Original Data', 2='Window = 3', 3='Window
 = 6', 4='Window = 9']):
temp:=[seq(plots[listplot]([MANY_TIMES[n]], title=TITLES[n]
 view=[0..255, 0..20]), n=1..4)]:
plots[display](array(1..2,1..2, [temp[1..2], temp[3..4]])):
```

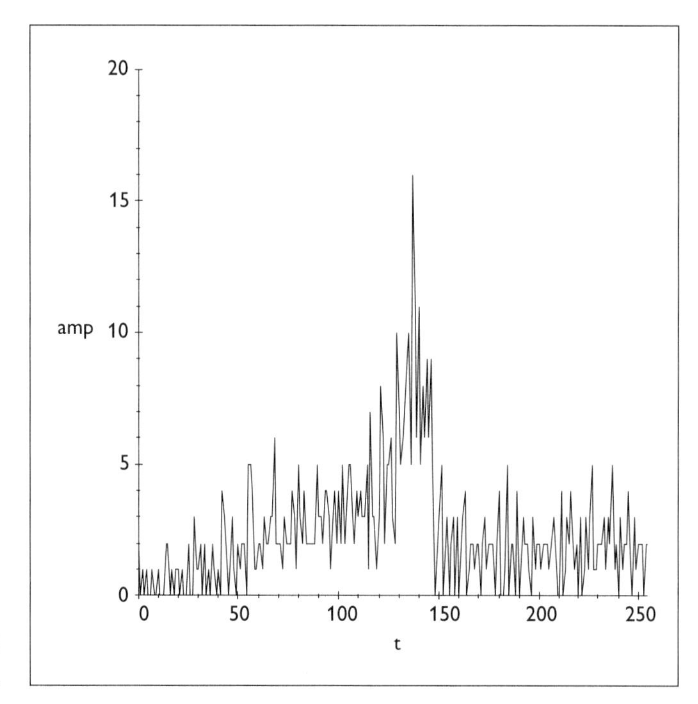

Figure 7.24
Original data.

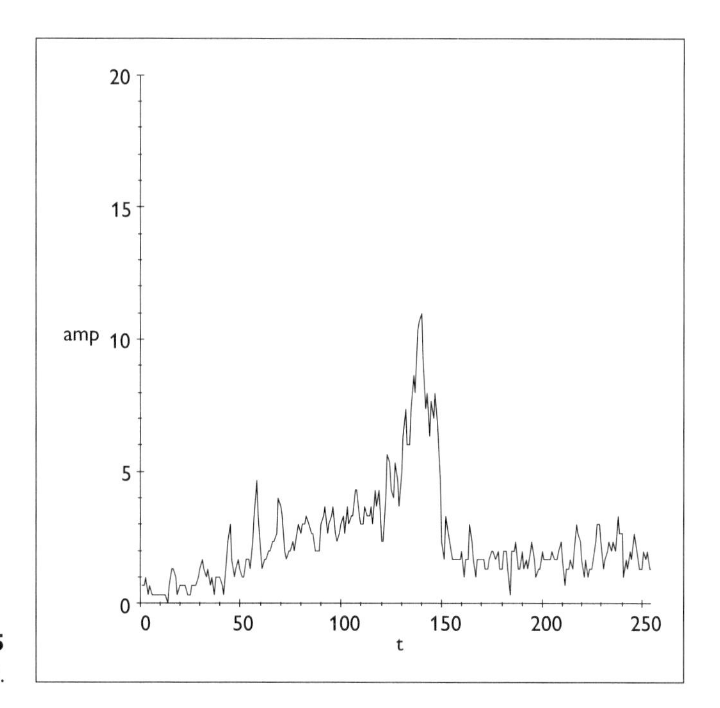

Figure 7.25
Window = 3.

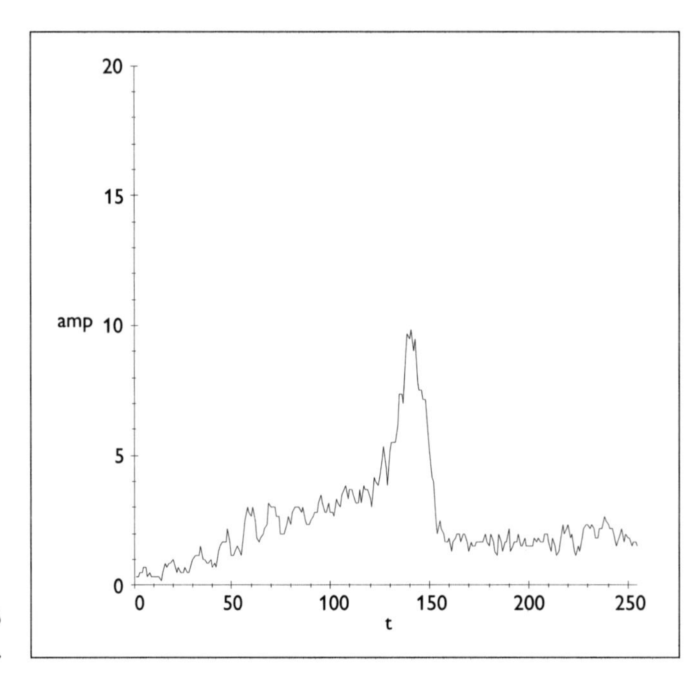

Figure 7.26
Window = 6.

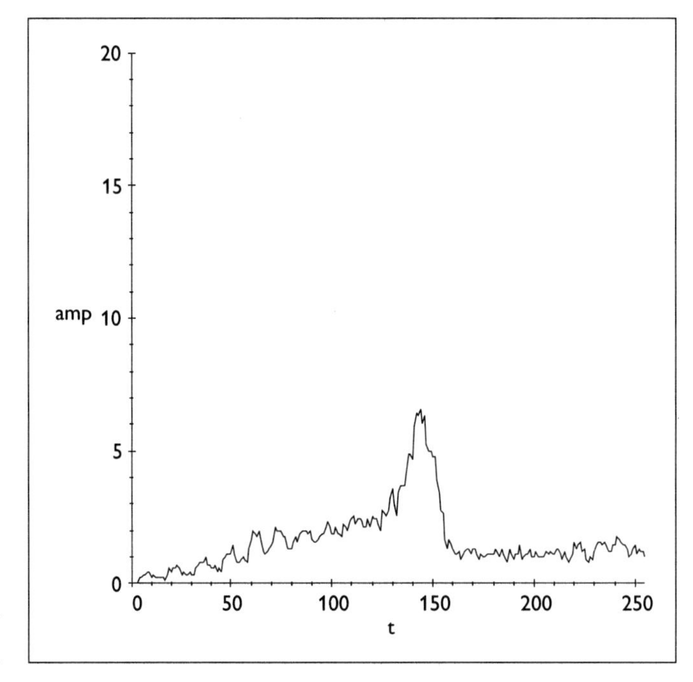

Figure 7.27
Window = 9.

The moving average filter is an easy filter to implement that possesses good noise reduction qualities. With reference to Figures 7.24 through

7.27, we can see that there is a trade-off between the amount of smoothing achieved, determined by the window size and the retention of information in areas of high volatility or rapid change. As the window size is increased, more elements are averaged, which results in increased noise rejection but individual peaks in the data tend to be blurred. The reduction in signal amplitude is an artifact of the averaging process that can be avoided through normalization techniques.

Moving median

The moving median filter is similar to the moving average filter in terms of operation (computing the new data from the old data by sweeping a window over the old data). but instead of using the average value, we use the median value of the windowed data. So, for example, the median of the data list `SAMPLE` is:

```
stats[describe, median](SAMPLE);
```

$$0$$

We can see more clearly how this answer was arrived at by first sorting `SAMPLE` and then taking the middle entry:

```
sort(SAMPLE);
```

$$[0, 0, 0, 0, 0, 0, 1, 1, 1, 2]$$

```
op(floor(nops(SAMPLE)/2), ");
```

$$0$$

Next we repeat the filtering process while sweeping the window across the data:

```
WINDOW:=3:
MAT:=linalg[matrix](1, nops(SAMPLE), SAMPLE):
FILTERED:=op(1..WINDOW-1, SAMPLE):
```

```
for n from WINDOW to nops(SAMPLE)-WINDOW+1 do
 temp:=linalg[submatrix](MAT, 1..1, n..n+WINDOW-1):
```

```
  temp:=convert(temp, listlist):
  temp:=stats[describe, median](op(temp)):
  FILTERED:=FILTERED, temp:
  print(FILTERED);
od:
```

$$2,\ 0,\ 1$$
$$2,\ 0,\ 1,\ 0$$
$$2,\ 0,\ 1,\ 0,\ 0$$
$$2,\ 0,\ 1,\ 0,\ 0,\ 0$$
$$2,\ 0,\ 1,\ 0,\ 0,\ 0,\ 0$$
$$2,\ 0,\ 1,\ 0,\ 0,\ 0,\ 0,\ 0$$

Now we can compare the original with the filtered data:

```
FILTERED:=FILTERED, op(nops(SAMPLE)-WINDOW+2..nops
(SAMPLE), SAMPLE);
```

$$\mathrm{FILTERED} := 2, 0, 1, 0, 0, 0, 0, 0, 0, 0$$

```
SAMPLE;
```

$$[2, 0, 1, 0, 1, 0, 0, 1, 0, 0]$$

When applying a median filter, we have the option of applying it until the filtered data are identical to the unfiltered data. At this point no more filtering is possible so we stop. In Maple we would implement this as follows. First we save the original data for the current iteration:

```
OLD:=[FILTERED]:
FILTERED:=moving_median([FILTERED],3);
```

$$\mathrm{FILTERED} := [2, 0, 0, 0, 0, 0, 0, 0, 0, 0]$$

Are the filtered data equal to the original?

```
linalg[iszero](linalg[matrix](1, nops(OLD), zip((x,y)->x-y,
 OLD,FILTERED)));
```

$$false$$

This time around they are not, so we must continue:

```
OLD:=FILTERED:
FILTERED:=moving_median(FILTERED,3);
```

$$FILTERED := [2, 0, 0, 0, 0, 0, 0, 0, 0, 0]$$

We continue to repeat the operations in the loop until the output data from an iteration are equal to the prior iteration. Although it is not necessary to continue in this particular way, the following code can be used to filter the data repeatedly and compare the filter's output with its input. If the filter's input and output are different, then the filtering operation ceases; if not, it continues:

```
while not linalg[iszero](linalg[matrix](1, nops(OLD),
 zip((x,y)->x-y, OLD, FILTERED))) do
OLD:=FILTERED:
FILTERED:=moving_median(FILTERED,3);
od;
```

Here we use the function `moving_median` first in a single-shot mode and then with repeated application. The mode of operation is set with a third, optional, argument. If this argument is omitted, the single-shot mode of operation is used by default. The difference in the filter's output is obvious.

```
moving_median(SAMPLE, 3, repeated=false);
```

$$[2, 0, 1, 0, 0, 0, 0, 0, 0, 0]$$

```
moving_median(SAMPLE, 3, repeated=true);
```

$$[2, 0, 0, 0, 0, 0, 0, 0, 0, 0]$$

Here we test the moving median filter on our test data and show the results in Figure 7.28:

```
FILTERED:=moving_median(SERIES_DATA,3):
plots[listplot]([FILTERED], title='Moving Median Filter
 Output');
```

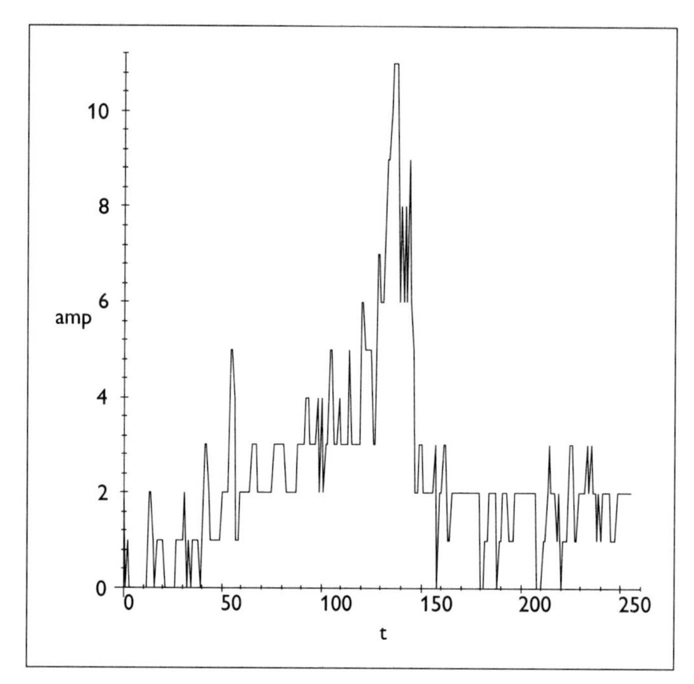

Figure 7.28 Moving median filter output.

Now we can look at the effect on the filtered data as we pass it through the filter more than once:

```
MANY_TIMES:=SERIES_DATA:
for n to 3 do
MANY_TIMES:=MANY_TIMES, moving_median(op(n, [MANY_TIMES]),
3)
od:
```

Using `listplot` we can plot the output from each iteration (Figures 7.29 through 7.32). By placing the resulting plots into an array structure, we can compare the output for each iteration easily. By using a table we can easily place a relevant title on each plot.

```
TITLES:=table([1='Original Data', 2='First Iteration',
 3='Second Iteration', 4='Third Iteration']):
temp:=[seq, plots[listplot]([MANY_TIMES[n]],
 title=TITLES[n] , view=[0..255, 0..20]), n=1..4)]:
plots[display](array(1..2, 1..2, [temp[1..2], temp[3..4]])):
```

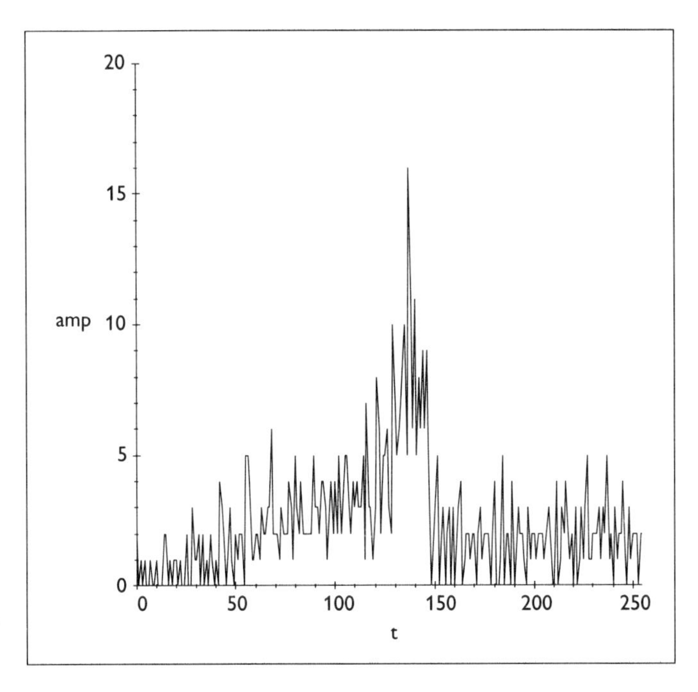

Figure 7.29
Original data.

The moving median filter is another easy filter to implement that possesses good noise reduction qualities. Like the moving average filter discussed earlier, there is a trade-off between the amount of smoothing achieved, determined by the window size, and the retention of information in areas of high volatility or rapid change. The major advantage of the moving median filter over the moving average is its ability to reject noise while retaining individual peaks in the data. The reduction in signal amplitude, which is a consequence of the filtering process, is not as dramatic as in the moving average filter's case and can be avoided through normalization techniques.

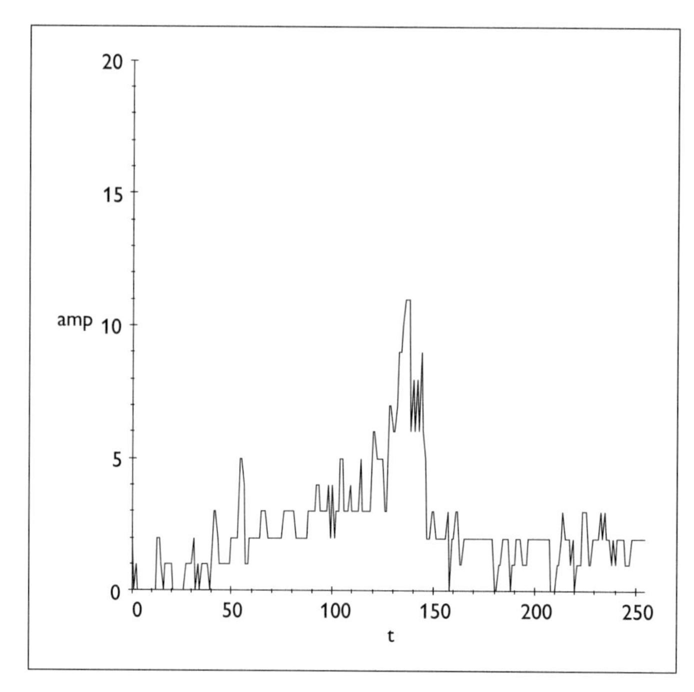

Figure 7.30
First iteration.

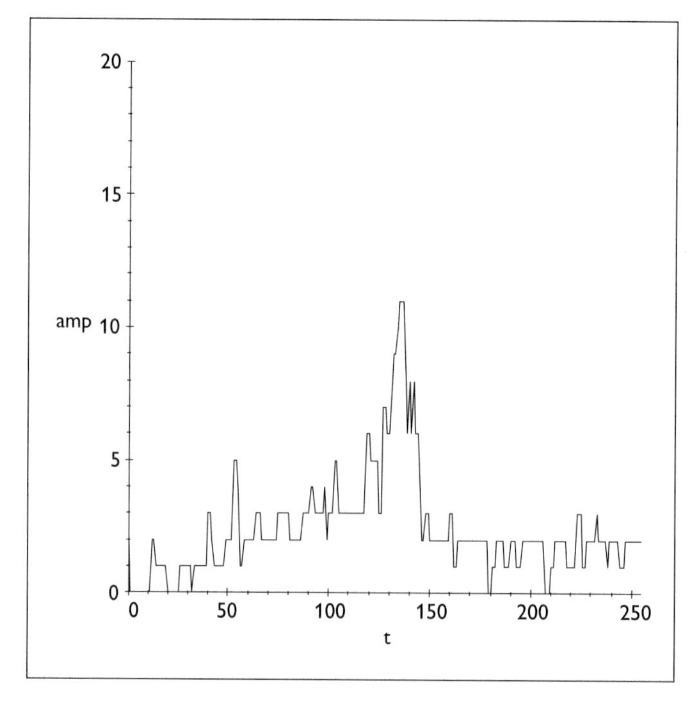

Figure 7.31
Second iteration.

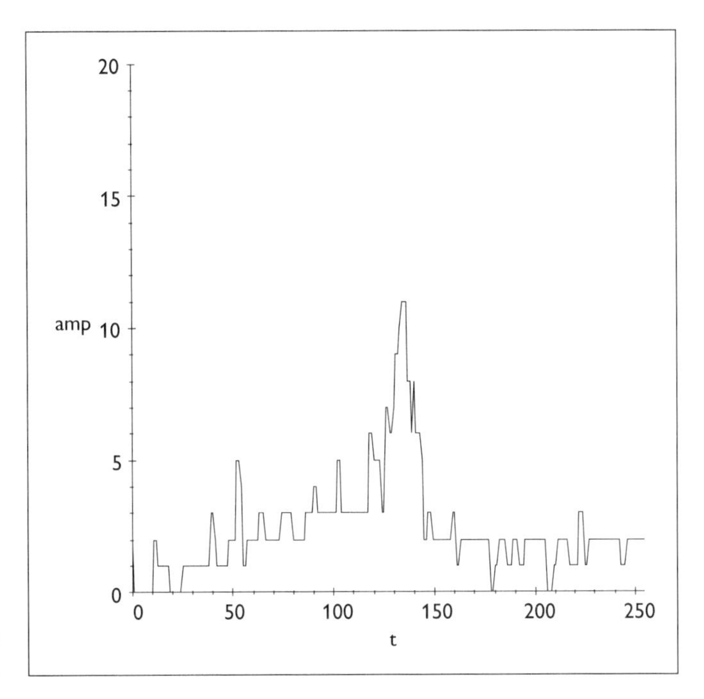

Figure 7.32
Third iteration.

Exponential filtering

The filters that we have looked at so far have all operated on the data, giving each element in the data set an equal weight. The filters have also all been without memory, that is to say, once a filtered sample has been computed it plays no further role in the filtering function. The exponential filter is different on both counts: Previously calculated filter outputs are used in the calculation of the current output, and previous outputs have a weight associated with them that decreases exponentially the further back in time we go. The first task is to build the type of filter that we want, in this case a simple exponential filter:

```
filter := y(t)/r(t)=exp(a*t);
```

$$\text{filter} := \frac{y(t)}{r(t)} = e^{(a\,t)}$$

Using the Z-transform function found in the integral transforms package `inttrans` we transform the continuous filter into its discrete form. We are using the normal conventions in terms of t and z.

```
with(inttrans):
```

Before we actually perform this filtering operation we develop a simple function that performs cross multiplication:

```
cross_multiply:= x->numer(lhs(x))*denom(rhs(x)) =
 numer(rhs(x))*denom(lhs(x));
```

cross_multiply :=
$$x \rightarrow \operatorname{numer}(\operatorname{lhs}(x))\,\operatorname{denom}(\operatorname{rhs}(x)) = \operatorname{numer}(\operatorname{rhs}(x))\,\operatorname{denom}(\operatorname{lhs}(x))$$

Applying this to the filter transfer function we get

```
Filter:=cross_multiply(Y(z)/R(z)= ztrans(rhs(filter),
 t, z));
```

$$\text{Filter} := Y(z)\left(z - e^{a}\right) = z\,R(z)$$

The filter response in terms of z can now be manipulated. Here we multiply each side by z to ensure that the input $R(z)$ is always the current sample.

```
Filter:=expand(Filter/z);
```

$$\text{Filter} := Y(z) - \frac{Y(z)\,e^{a}}{z} = R(z)$$

Using Maple's alias function we simplify the printed form of the transfer function:

```
alias(Y=Y(z), R=R(z));
```

$$I, Y, R$$

Next we get the coefficients of $Y(z)$ and $R(z)$, respectively. These will be the weightings that are applied to the filter samples:

```
map(coeffs, Filter, {Y, R});
```

$$1 - \frac{e^a}{z} = 1$$

```
coefflist:=map(convert,",list);
```

$$\text{coefflist} := \left[1, -\frac{e^a}{z}\right] = [1]$$

The next stage is to recreate the filter transfer function in terms of the current output, the previous outputs, and the current input. We define these as $y0, y1$, and $r0$, respectively.

```
left:=zip((x,y)->x*y,lhs(coefflist), ['y.n'$(n=0..1)]);
```

$$\text{left} := \left[y0, -\frac{e^a y1}{z}\right]$$

```
right:=zip((x,y)->x*y,rhs(coefflist), ['r.n'$n=0..0]);
```

$$\text{right} := [r0]$$

```
new_filter:=subs(z=1, readlib(isolate)(convert(left, '+') =
convert(right, '+'), y0));
```

$$\text{new\_filter} := y0 = r0 + e^a y1$$

The first thing that we notice is that the filter has gain. We can eliminate this by adjusting the weighting of the filter input r_0. It is common when designing digital filters of this nature to ensure that the filter coefficients add to unity. In this case we need to multiply $r0$ by A and $y1$ by $1 - A$, where A is e^a and is known as the filter weight.

```
new_filter:=subs(r0=r0*A, exp(a)=(1-A), A=exp(a),
new_filter);
```

$$new\_filter := y0 + r0\ e^{a} + \left(1 - e^{a}\right) y1$$

Now we can use `unapply` to convert the transfer function into function notation:

```
NEW_FILTER := unapply(rhs(new_filter), r0, y1, a);
```

$$NEW\_FILTER := (r0, y1, a) \rightarrow r0\ e^{a} + \left(1 - e^{a}\right) y1$$

Finally, we test the filtering function.

```
NEW_FILTER(1,0,-0.5);
```

$$.6065306597$$

```
NEW_FILTER(0, ", -0.5);
```

$$.2386512185$$

So far so good. Now using a loop we compute and plot (Figure 7.33) the filter's impulse response:

```
LAST:=0:ANS:=NULL:R:=1:
for n from 0 to 10 do
 LAST:=NEW_FILTER(R, LAST, -0.5):
 ANS:=ANS, [n, LAST]:
 if n>=0 then R:=0 fi:
od:
plot([ANS], title='Impulse Response', labels=['t','amp']);
```

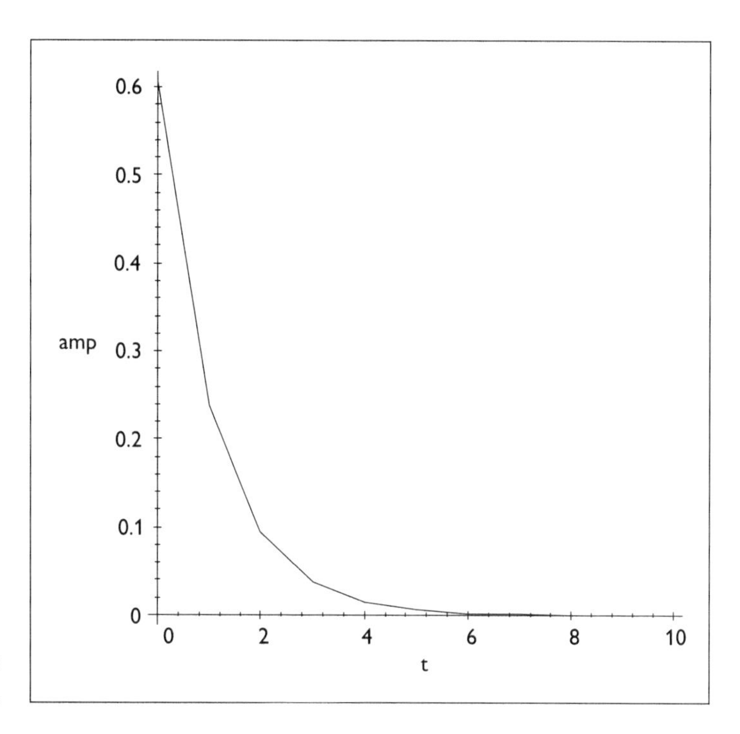

Figure 7.33
Impulse response.

Here we apply the new filter to our test data. We are using the function `exp_filter` found on the program disk.

```
exp_filter(SAMPLE, -0.7);
```

[.9931706076, .4999766797, .7482809121, .3766956080, .6862194089, .3454529353, .1739060845, .5841321825, .2940607252, .1480344906]

By applying the `exp_filter` to our test data, we can exponentially smooth it. In the first case we use a filter weight of −0.7 and plot (Figure 7.34) the filter output using `listplot` as shown:

```
exp_filter(SERIES_DATA, -0.7):
plots[listplot](["], title='Output from Exponential
 Filter', labels=['t','amp']);
```

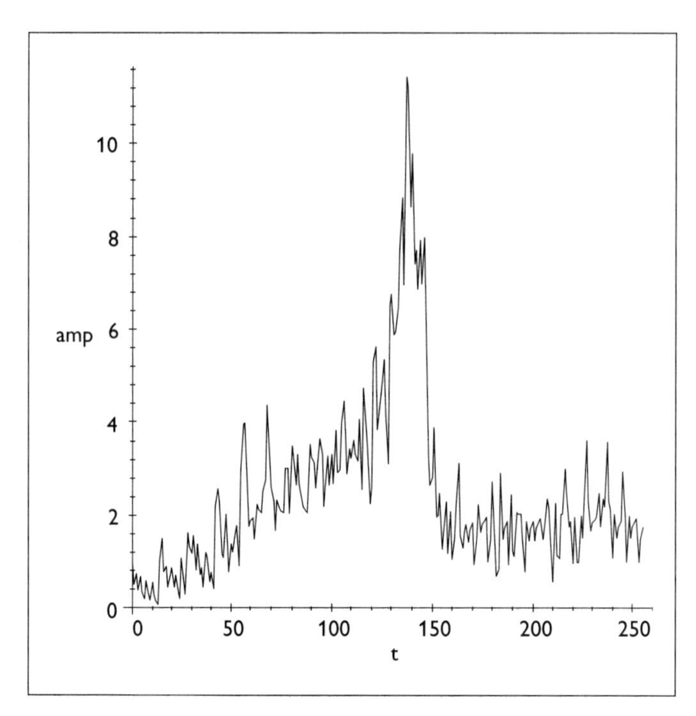

Figure 7.34 Output from exponential filter.

As with the previous types, it is beneficial to compare the exponential filters response to changing filter weights. The next few lines of Maple code produce plots of the original and filtered data, with filter weights −0.7, −0.3, and −0.1, and displays them all in a single graphics array (Figures 7.35 through 7.38):

```
MANY_TIMES:=READINGS:WEIGHTS:=[-0.7, -0.3, -0.1]:
for n to 3 do
MANY_TIMES:=MANY_TIMES,exp_filter(READINGS, WEIGHTS[n]):
od:
```

```
TITLES:=table([1='Original Data', 2='Filter Weight = -0.7',
 3='Filter Weight = -0.3', 4='Filter Weight = -0.1']):
temp:=[seq(plots[listplot]([MANY_TIMES[n]],
 title=TITLES[n], view=[0..255, 0..20]), n=1..4)]:
plots[display](array(1..2,1..2, [temp[1..2], temp[3..4]])):
```

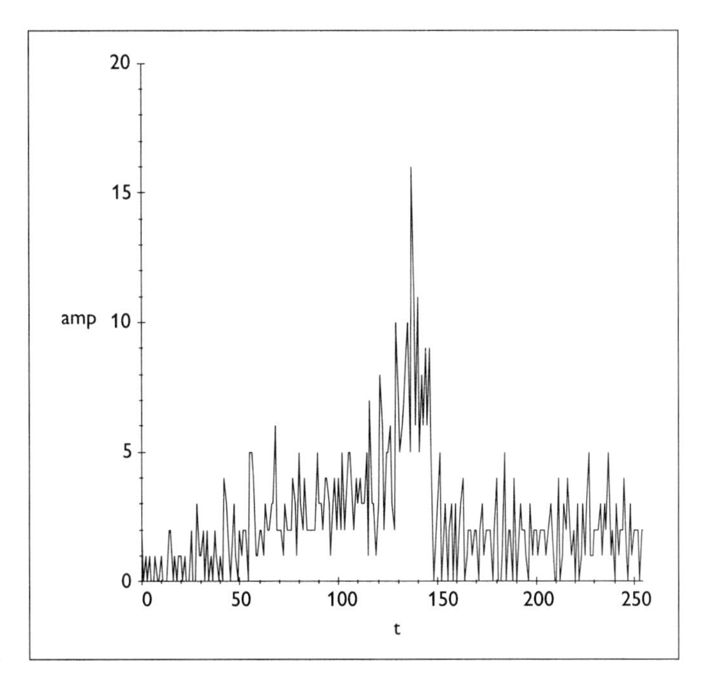

Figure 7.35
Original data.

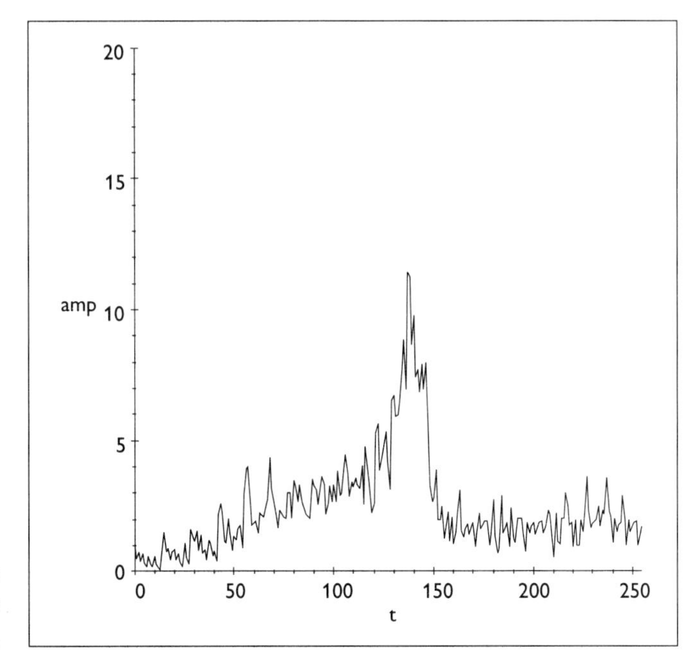

Figure 7.36
Filter weight = –0.7.

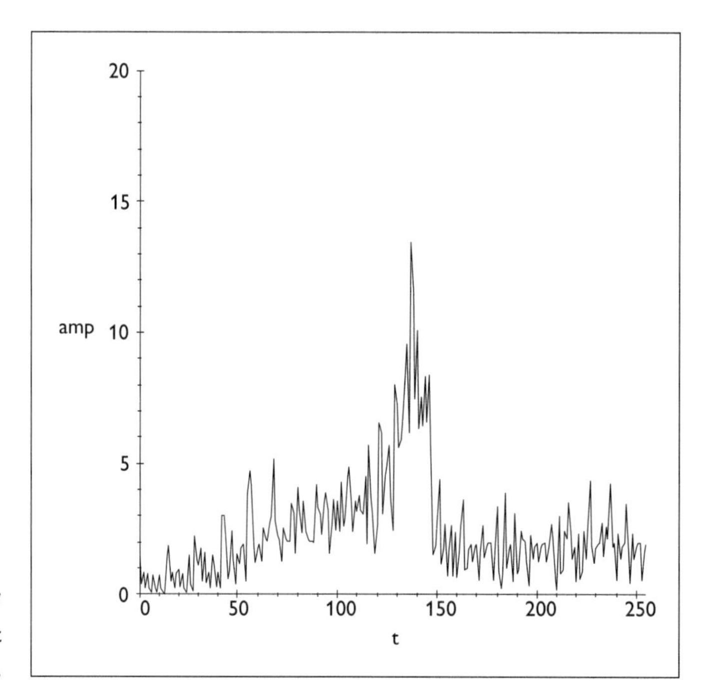

Figure 7.37 Filter weight = –0.3.

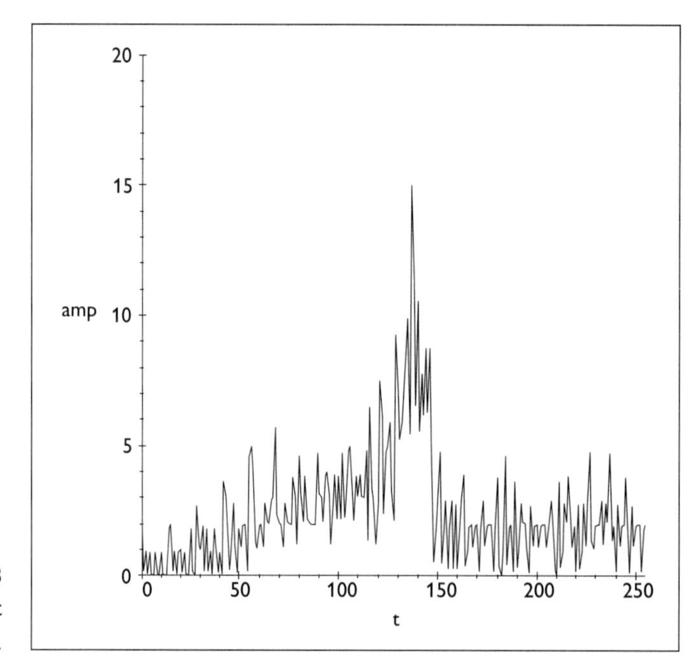

Figure 7.38 Filter weight = –0.1.

The exponential filter is unique among the filter configurations discussed in this section in that it contains hysteresis (i.e., there is a memory component in its implementation which means that the next filter output is affected by previous filter outputs). With reference to Figures 7.35 through 7.38, we can see that the exponential filter has good noise reduction qualities while being able to retain data that are changing rapidly. The level of noise reduction and the corresponding ability to track fast changing data are determined by the filter weight—the higher the weight the greater the noise reduction. The reduction in signal amplitude is due to the filter process and can be avoided through the application of normalization techniques.

Conclusion

The final graphics show the original series data along with the outputs from the difference, moving average, moving median, and the exponential filters (Figure 7.39 through 7.47).

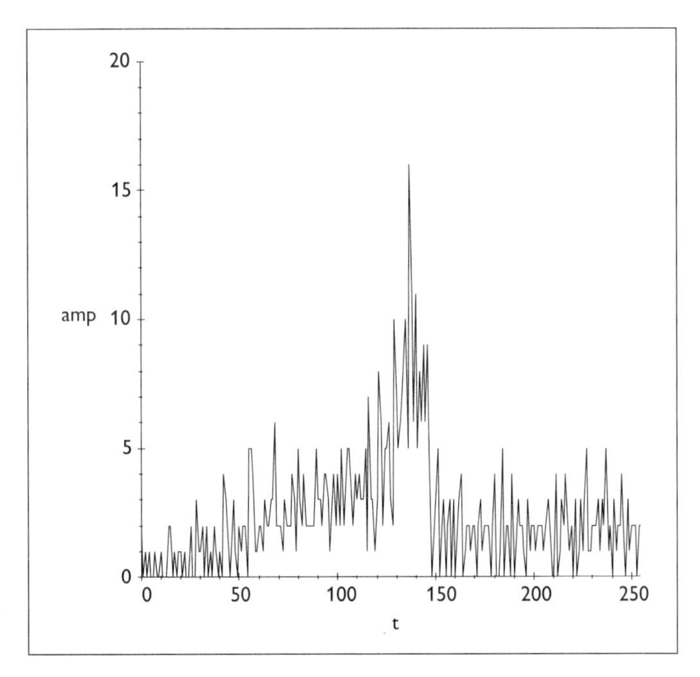

Figure 7.39
Original data.

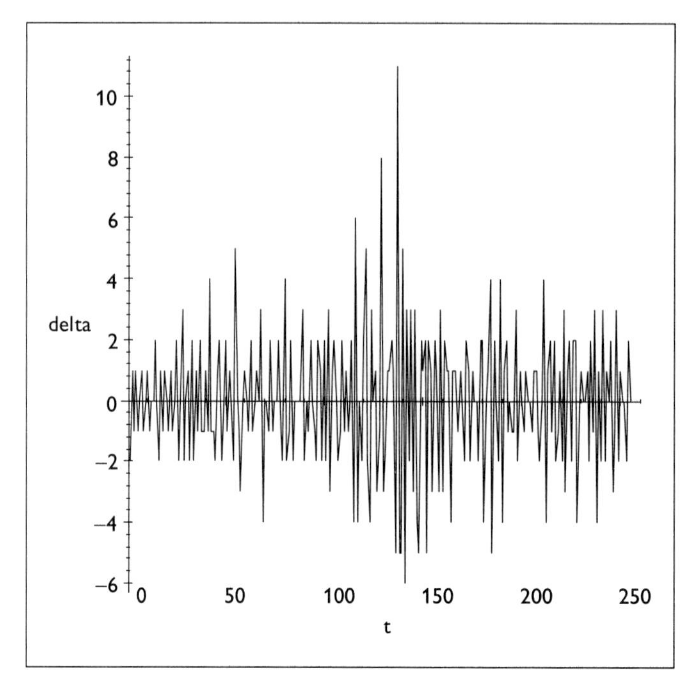

Figure 7.40 Difference filter output.

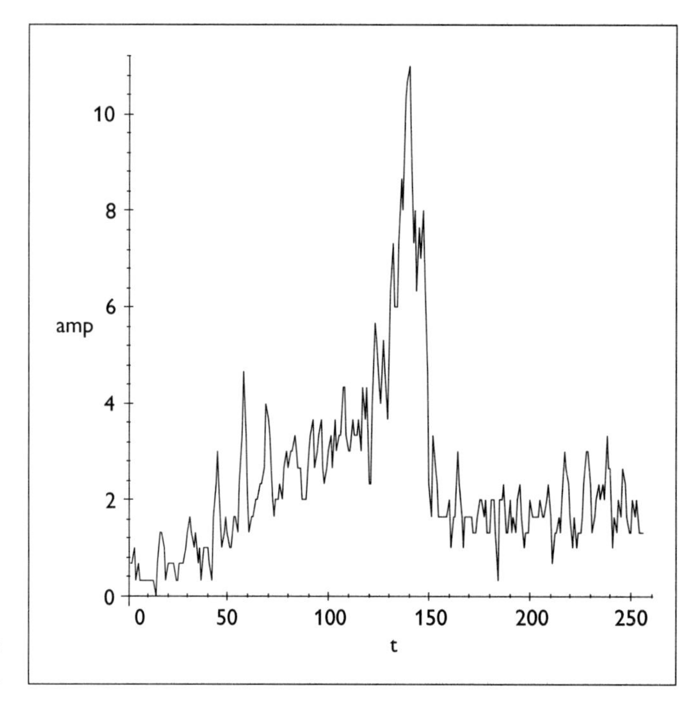

Figure 7.41 Moving average filter output.

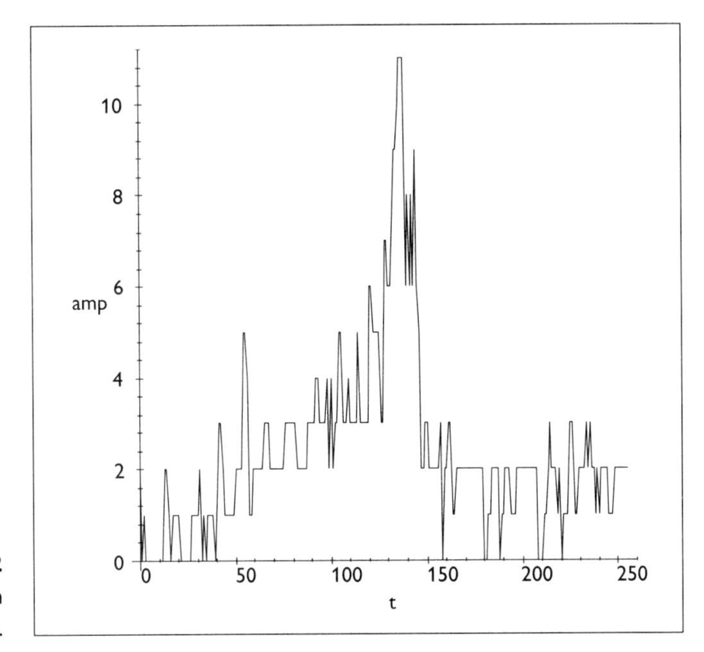

Figure 7.42 Moving median filter output.

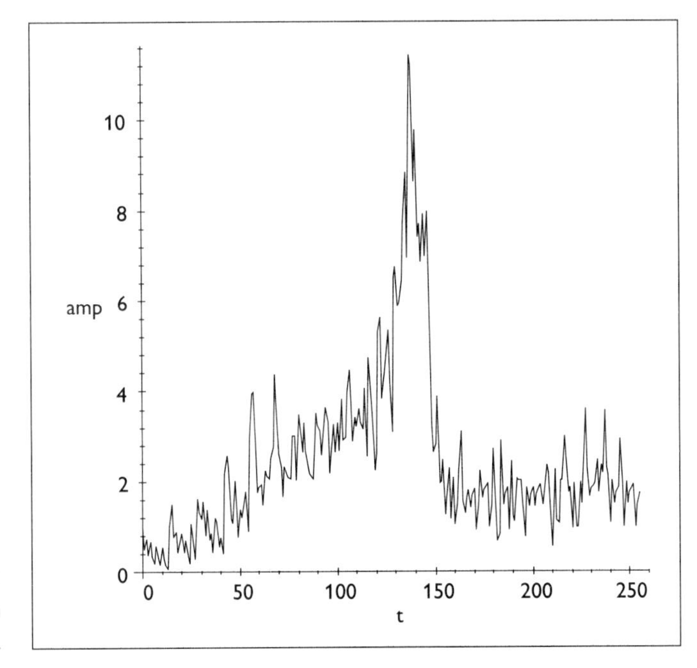

Figure 7.43 Output from exponential filter.

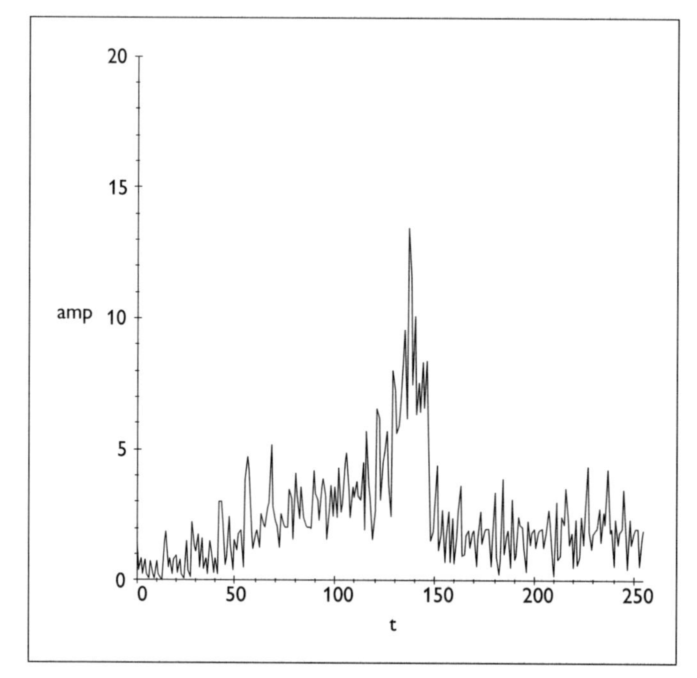

Figure 7.44
Filter weight
= –0.3.

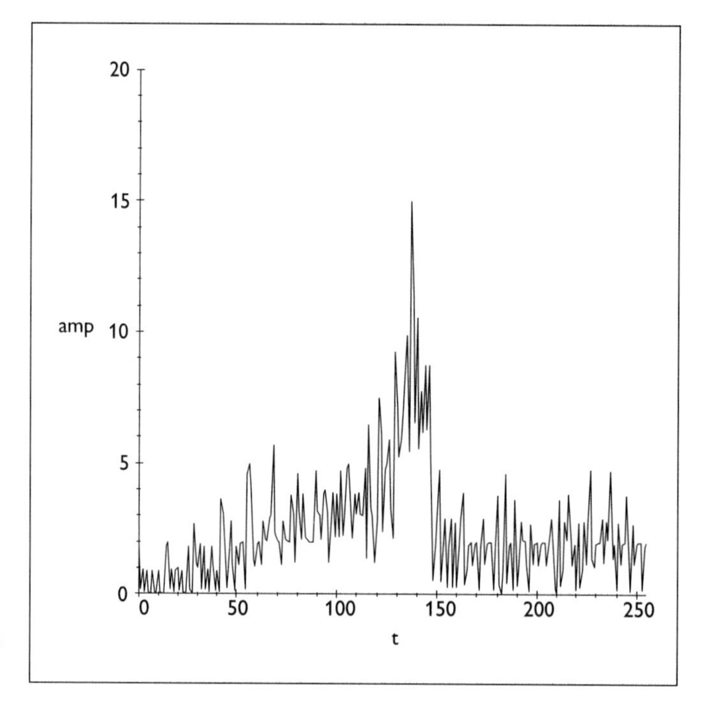

Figure 7.45
Filter weight
= –0.1.

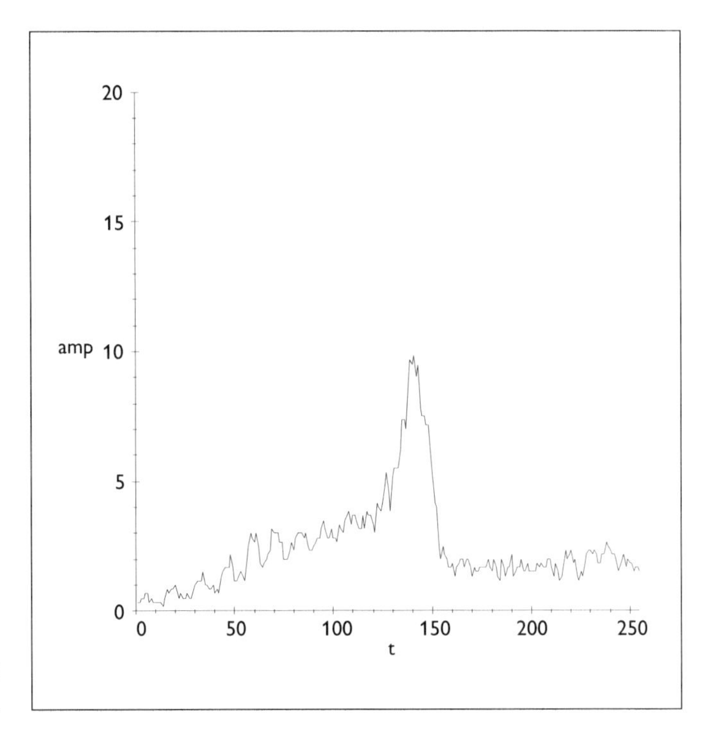

Figure 7.46
Window = 6.

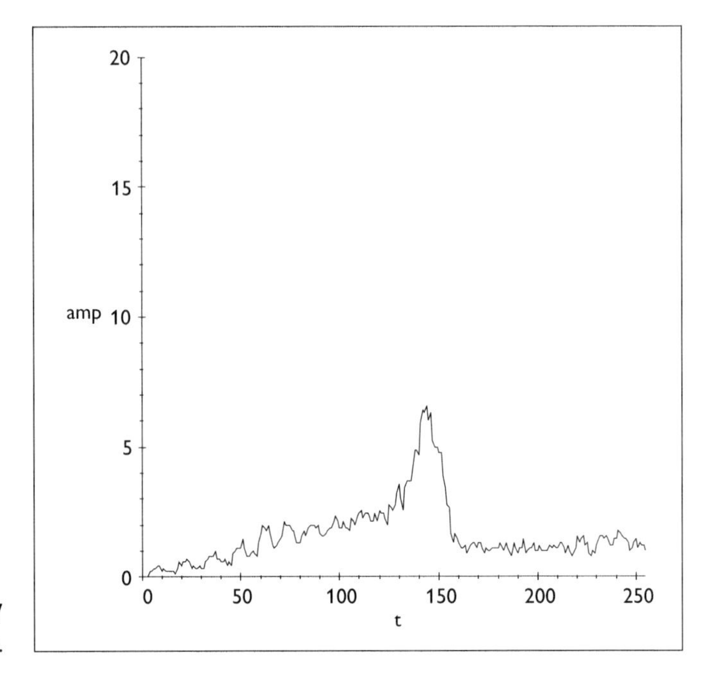

Figure 7.47
Window = 9.

The four filter configurations that have been discussed in this chapter are the differencing filter, the moving average filter, the moving mean filter, and the exponential filter. The differencing filter is good at helping to identify underlying trends and areas of rapid changes within data sets. The moving average and moving mean filters both exhibit good noise cancellation properties with the moving median filter possessing the best ability to track fast moving data. Finally, the exponential filter is a special implementation of a digital filter that has good noise reduction qualities and can track fast moving data adequately. All of the filters reduce the signal amplitude but this attenuation can be avoided through the application of normalization techniques.

Chapter 8

Switching topologies

GENERALLY, SWITCHING topologies are used for optimal power handling efficiency (conventional switching power supplies) and/or certain types of signal processing (e.g., precision analog multipliers/dividers, phase detectors). Two immediate and common applications involving switching circuit topologies are (1) pulse width modulator (PWM) signal acquisition and/or PWM drivers, and (2) switching power supplies. Both of these circuit topologies utilize switching to minimize the amount of varying conduction time, which results in excessive heating, experienced by the controlling device(s). Also, if it is a signal processing application, then varying conduction times cause mathematical errors via excessive noise (amplitude variance versus phase information variations).

In this chapter we explore two methods for analyzing switching output waveforms. The first method assumes the system is at steady state and each period exhibits the same response. The second method uses Fourier analysis to show the transient startup behavior of the switching topology.

For further reading on the solution for these and other switching networks, [1–4] are strongly recommended.

Steady-state method

Pulse width modulator driver

Figure 8.1 shows a typical PWM input waveform with a period, T. The second part of this design requires a filter that simply produces the average of the PWM waveform. Figure 8.2 shows a simple RC filter that achieves this result. Further, we know what the output voltage will approximate, so we can redraw Figure 8.1 with the filtered output superimposed as shown in Figure 8.3.

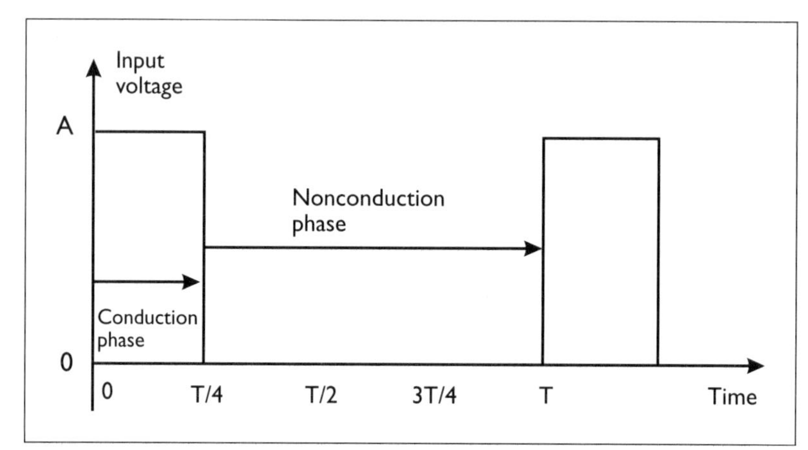

Figure 8.1 PWM input waveform.

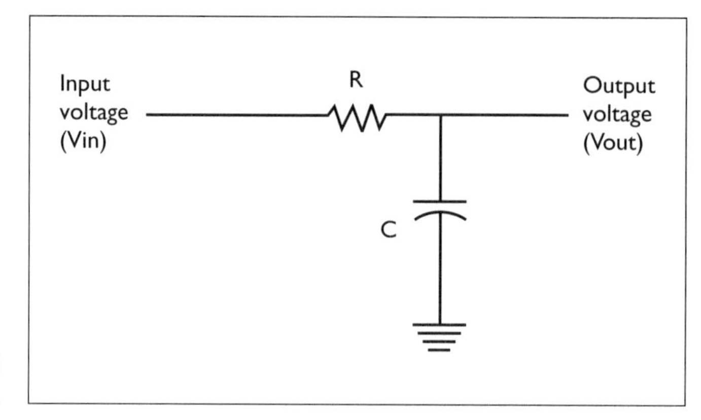

Figure 8.2 RC filter.

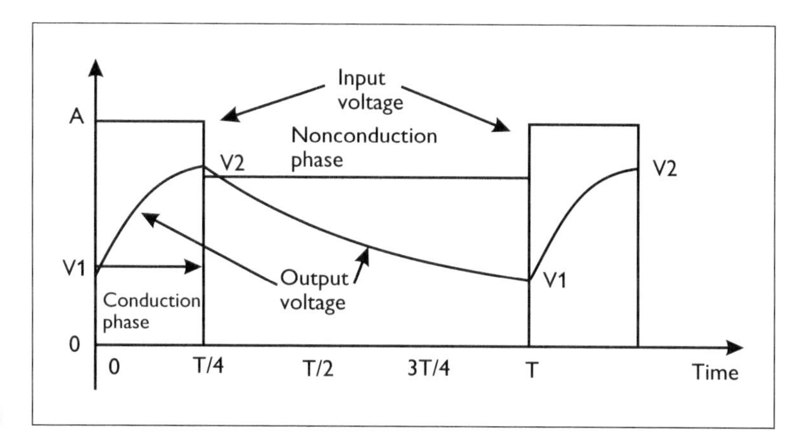

Figure 8.3

Note that we have assumed a steady-state solution for the network for a given set of conditions. In short, the output voltage waveform will exhibit a periodic form once transients have gone to zero. Therefore, we may derive the voltages $V1$ and $V2$ and from these values determine the entire steady-state response of the PWM circuit.

Perhaps the most important aspect of this driver is what kind of output can we tolerate in terms of ripple or switching noise associated with the switching operation? By deriving the $V1$ and $V2$ values in Figure 8.3, we not only solve the steady-state dynamics, but can also reverse engineer the process to determine the maximum allowable ripple (i.e., $|V1-V2|$) for any given set of circuit parameters.

To determine this and other important design aspects, let's start a Maple session that enters the waveforms and RC filter topology. Starting with the required Maple libraries,

```
with(inttrans)::
with(plots):
```

we enter the topology and switching input waveform values:

```
Freq := 10^4:
R := 1000:
C := 10^(-6):
alfa := .5:
Vin := 10/s:
T := 1/Freq:
```

where

Freq	$\rightarrow$	switching frequency (hertz)
R	$\rightarrow$	resistor value (ohms)
C	$\rightarrow$	capacitor value (farads)
alfa	$\rightarrow$	duty cycle (dimensionless)
Vin	$\rightarrow$	input voltage (volts)
T	$\rightarrow$	switching period (seconds)

Next, we take the inverse Laplace transforms associated with two continuous regimes, i.e.,

$$\text{Regime 1} \quad 0 \le t < \alpha T$$
$$\text{Regime 2} \quad \alpha T \le t < T$$

Let's define the output under the two regimes as

$$\text{Regime 1} \quad V_{\text{OUT}} \rightarrow V^1_{\text{OUT}} \rightarrow Vout\_1\_Time$$
$$\text{Regime 2} \quad V_{\text{OUT}} \rightarrow V^2_{\text{OUT}} \rightarrow Vout\_2\_Time$$

Therefore, in Maple syntax,

```
Vout_1_Time := invlaplace((Vin+s*R*C*V1/s)/(s*R*C+1),s,t);
Vout_2_Time := invlaplace((s*R*C*V2/s)/(s*R*C+1),s,t);
```

$$Vout\_1\_Time := 10 - 10\, e^{(-1000\, t)} + V1\, e^{(-1000\, t)}$$
$$Vout\_2\_Time := V2\, e^{(-1000\, t)}$$

Figure 8.3 indicates that at steady state the boundary conditions under the two regimes are

$$V^1_{\text{OUT}}\,(t = \alpha T) = V_2$$
$$V^2_{\text{OUT}}\,(t = T - \alpha T) = V_1$$

We cannot simply state the second boundary condition as

$$V_{\mathrm{OUT}}^{2}\ (t = T) = V_1$$

because as far as the output voltage under the second regime is concerned, the dynamics are functionally dependent on a time duration as opposed to any specific time on an arbitrary time axis. Consequently, we have two equations and two unknowns and can get the unique solution in Maple as follows: First, we solve the following boundary conditions just stated as:

```
Vout_1_Time_AlfaT := evalf(subs(t=alfa*T,Vout_1_Time));
Vout_2_Time_TminusAlfaT := evalf(subs(t=T-alfa*T,
Vout_2_Time));
```

$$Vout\_1\_Time\_AlfaT := .487705755 + .9512294245\ V1$$
$$Vout\_2\_Time\_TminusAlfaT := .9512294245\ V2$$

Then using Maple's `solve` command,

```
Solutions : =
solve({Vout_1_Time_AlfaT=V2,Vout_2_Time_TminusAlfaT=V1},
{V1,V2});
```

$$Solutions := \{V2 = 5.124973963\ ,\ V1 = 4.875026033\}$$

Converting the result into an order list for resubstitution, we perform the following:

```
XX := subs(Solutions,[V1,V2]):
V1 := XX[1];
V2 := XX[2];
```

$$V1 = 4.875026033$$
$$V2 = 5.124973963$$

Now, we need to convert the output voltage under the second regime ($\alpha T \leq t < T$) to reflect the time duration by sliding the time scale relative to the $V_{\mathrm{OUT}}^{2}(t)$ expression as mentioned earlier. Hence,

```
Vout_2_Time_Duration : = subs(t = t-alfa*T,Vout_2_Time):
```

In this way, we can substitute the graphic values of time into the second regime's output expression directly as we see them in the graph.

Plotting the periodic voltage output along with the average value requires that we first compute the average value of the periodic output as

$$V_{AVG} = \frac{1}{T}\left(\int_{0}^{\alpha T} V_{OUT}^{1}(t)dt + \int_{\alpha T}^{T} V_{OUT}^{2}(t)dt \right)$$

In Maple,

```
Output_Average := 1/T*(int(Vout_1_Time,t=0..alfa*T)+int
(Vout_2_Time_Duration,t=alfa*T..T));
```

$$\text{Output\_Average} := 5.000000000$$

This result is not surprising considering that we are using a 50% duty cycle with a 10-volt PWM square wave.

Now plotting the average value along with the entire two-regime time response over a per period of T,

```
Plot_1 := plot(Vout_1_Time,t=0..alfa*T,color = black,
style=point,symbol=cross):
Plot_2 := plot(Vout_2_Time_Duration,t=alfa*T..T,
color=black,style=point,symbol=circle):
Plot_3 := plot (Output_Average,t=0..T,color=black):
display({Plot_1,Plot_2,Plot_3},axes=boxed,color=black,
labels=[Time_seconds,Voltage]);
```

In Figure 8.4, we point plotted the regimes differently, so that the reader can see the individual solutions on one plot. Regime 1 or $0 \le t < \alpha T$ is plotted with crosses, whereas regime 2 or $\alpha T \le t < T$ is plotted with circles.

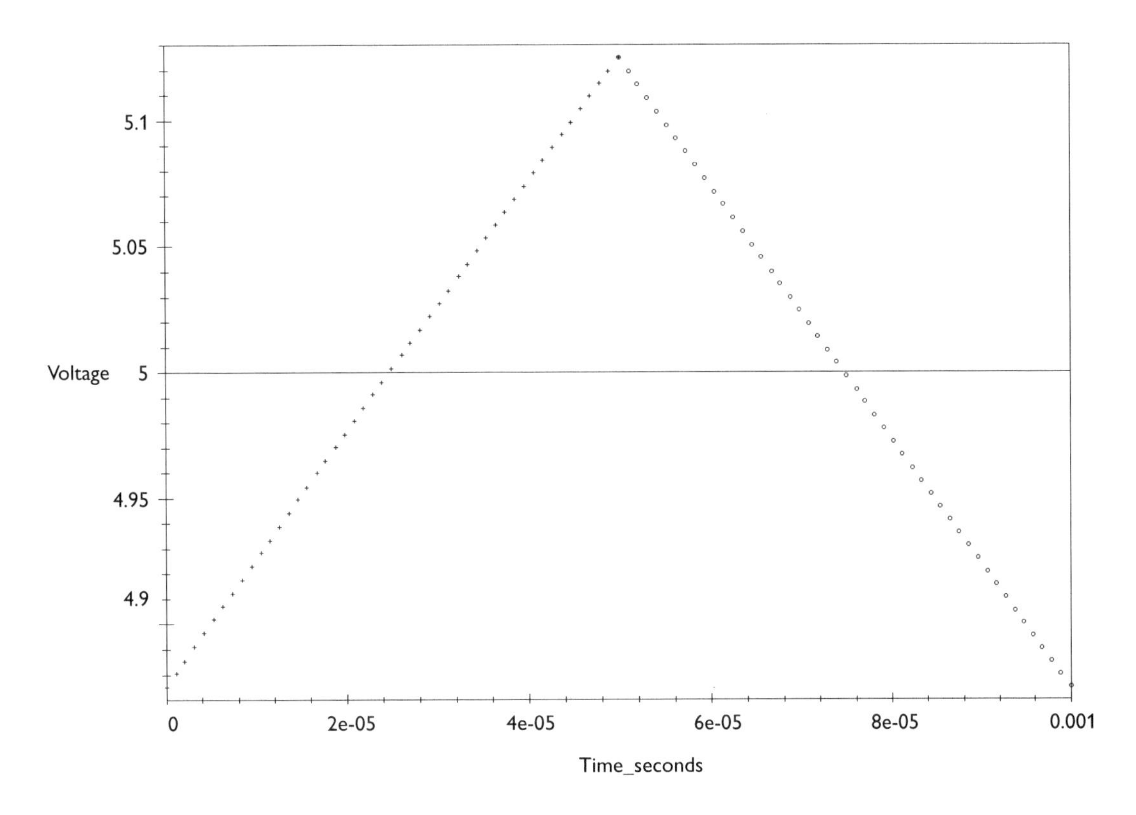

Figure 8.4 Output voltage plot of filtered PWM signal (*R* = 1000ohms, *C* = 1 μF, frequency = 10 kHz, duty cycle = 50%, *V*in = 10V).

Now we can play around with the variables to see the effects of parameter variation. For instance, what if we were to increase the RC filter's bandwidth by two orders of magnitude? We would expect to see more of the traditional RC roll-off characteristics on a per-period basis. Therefore, change the filter capacitor to 0.1 μF and resistance to 100 Ω.

```
with(plots):
with(inttrans):
Freq := 10^4:
R := 100:
C := 10^(-7):
alfa := .5:
Vin := 10/s:
T := 1/Freq:
Vout_1_Time := invlaplace((Vin+s*R*C*V1/s)/(s*R*C+1),s,t):
Vout_2_Time := invlaplace((s*R*C*V2/s)/(s*R*C+1),s,t):
```

```
Vout_1_Time_AlfaT := evalf(subs(t=alfa*T,Vout_1_Time)):
Vout_2_Time_TminusAlfaT := evalf(subs(t=T-alfa*T,
Vout_2_Time)):
Solutions := solve({Vout_1_Time_AlfaT=V2,
 Vout_2_Time_TminusAlfaT=V1},{V1,V2}):
XX := subs(Solutions,[V1,V2]):
V1 := XX[1]:
V2 := XX[2]:
Vout_2_Time_Duration := subs(t=t-alfa*T,Vout_2_Time):
Output_Average := 1/T*(int(Vout_1_Time,t=0..alfa*T)+
 int(Vout_2_Time_Duration,t=alfa*T..T)):
Plot_1 := plot(Vout_1_Time,t=0..alfa*T,color=black,
style=point,symbol=cross):
Plot_2 := plot(Vout_2_Time_Duration,t=alfa*T..T,
color=black,style=point,symbol=circle):
Plot_3 := plot (Output_Average,t=0..T,color=black):
display({Plot_1,Plot_2,Plot_3},axes=boxed,color=black,
labels=[Time_seconds,Voltage]);
```

Figure 8.5 shows that the low-pass RC filter has allowed the ripple to nearly be the full square-wave swing or 10V peak to peak.

Now let's set the duty cycle to 25% under the two previously performed filter values:

R = 1000 Ω and C = 1 μF

```
with(plots):
with(inttrans):
Freq := 10^4:
R := 1000:
C := 10^(-6):
alfa := .25:
Vin := 10/s:
T := 1/Freq:
Vout_1_Time := invlaplace((Vin+s*R*C*V1/s)/(s*R*C+1),s,t):
Vout_2_Time := invlaplace((s*R*C*V2/s)/(s*R*C+1),s,t):
Vout_1_Time_AlfaT := evalf(subs(t=alfa*T,Vout_1_Time)):
Vout_2_Time_TminusAlfaT := evalf(subs(t=T-alfa*T,
Vout_2_Time)):
Solutions := solve({Vout_1_Time_AlfaT=V2,
Vout_2_Time_ TminusAlfaT=V1},{V1,V2}):
```

```
XX := subs(Solutions,[V1,V2]):
V1 := XX[1]:
V2 := XX[2]:
Vout_2_Time_Duration := subs(t=t-alfa*T,Vout_2_Time):
Output_Average := 1/T*(int(Vout_1_Time,t=0..alfa*T)+
 int(Vout_2_Time_Duration,t=alfa*T..T)):
Plot_1 := plot(Vout_1_Time,t=0..alfa*T,color=black,
style=point,symbol=cross):
Plot _2 := plot(Vout_2_Time_Duration,t=alfa*T..T,
color=black,style=point,symbol=circle):
Plot_3  := plot (Output_Average,t=0..T,color=black):
display({Plot_1,Plot_2,Plot_3},axes=boxed,color=black,
labels=[Time_seconds,Voltage]);
```

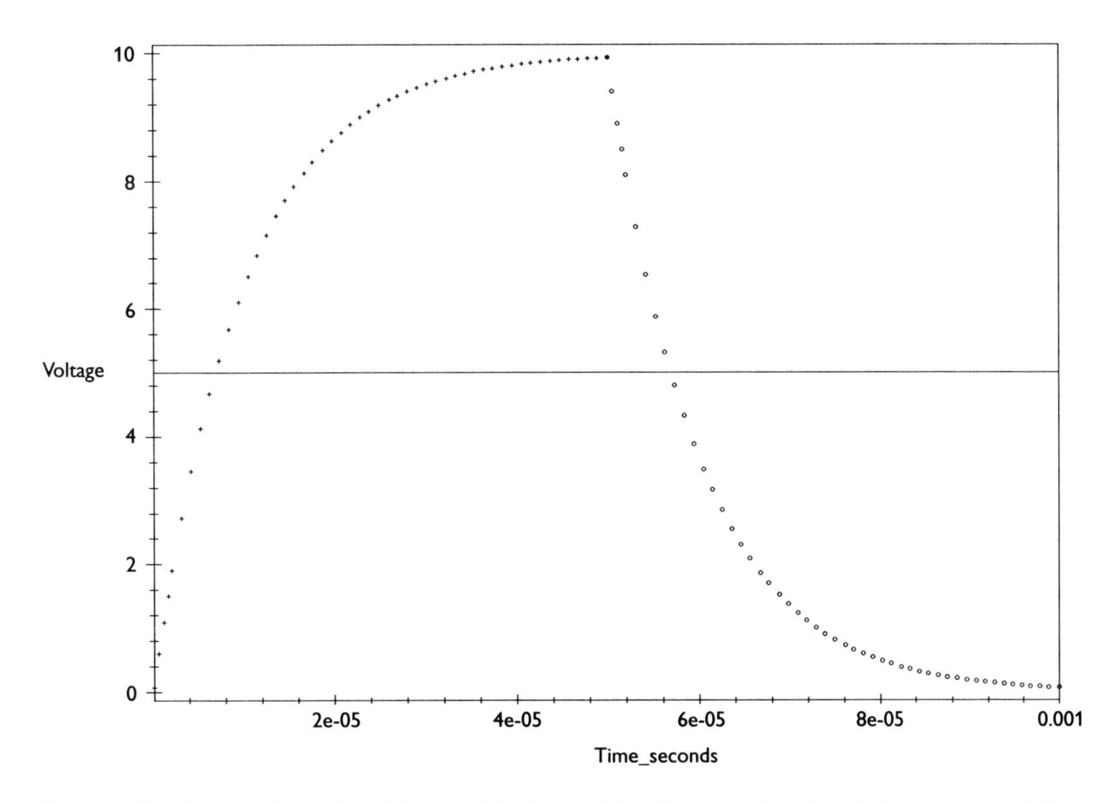

Figure 8.5 Output voltage plot of filtered PWM signal (*R* = 100 ohms, *C* = 0.1 μF, frequency = 10 kHz, duty cycle = 50%, *V*in = 10V).

R = 100 Ω and C = 0.1 μF

```
with(plots):
with(inttrans):
Freq := 10^4:
R := 100:
C := 10^(-7):
alfa := .25:
Vin := 10/s:
T := 1/Freq:
Vout_1_Time := invlaplace((Vin+s*R*C*V1/s)/(s*R*C+1),s,t):
Vout_2_Time := invlaplace((s*R*C*V2/s)/(s*R*C+1),s,t):
Vout_1_Time_AlfaT := evalf(subs(t=alfa*T,Vout_1_Time)):
Vout_2_Time_TminusAlfaT := evalf(subs(t=T-alfa*T,
 Vout_2_Time)):
Solutions := solve({Vout_1_Time_AlfaT=V2,
 Vout_2_Time_TminusAlfaT=V1},{V1,V2}):
XX := subs(Solutions,[V1,V2]):
V1 :=XX[1]:
V2 := XX[2]:
Vout_2_Time_Duration := subs(t=t-alfa*T,Vout_2_Time):
Output_Average := 1/T*(int(Vout_1_Time,t=0..alfa*T)+
 int(Vout_2_Time_Duration,t=alfa*T..T)):
Plot_1 := plot(Vout_1_Time,t=0..alfa*T,color=black,
 style =point,symbol=cross):
Plot_2 := plot(Vout_2_Time_Duration,t=alfa*T..T,
 color=black,style=point,symbol=circle):
Plot_3 := plot (Output_Average,t=0..T,color=black):
display({Plot_1,Plot_2,Plot_3},axes=boxed,color=black,
 labels=[Time_seconds,Voltage]);
```

From Figures 8.6 and 8.7 we immediately see that having a low-pass filter whose cutoff frequency

$$f_{\text{CUTOFF}} = \frac{1}{2\pi RC}$$

is well below the switching frequency provides a reasonably good averaging of the input waveform. The cutoff frequencies associated with the first and second cases were

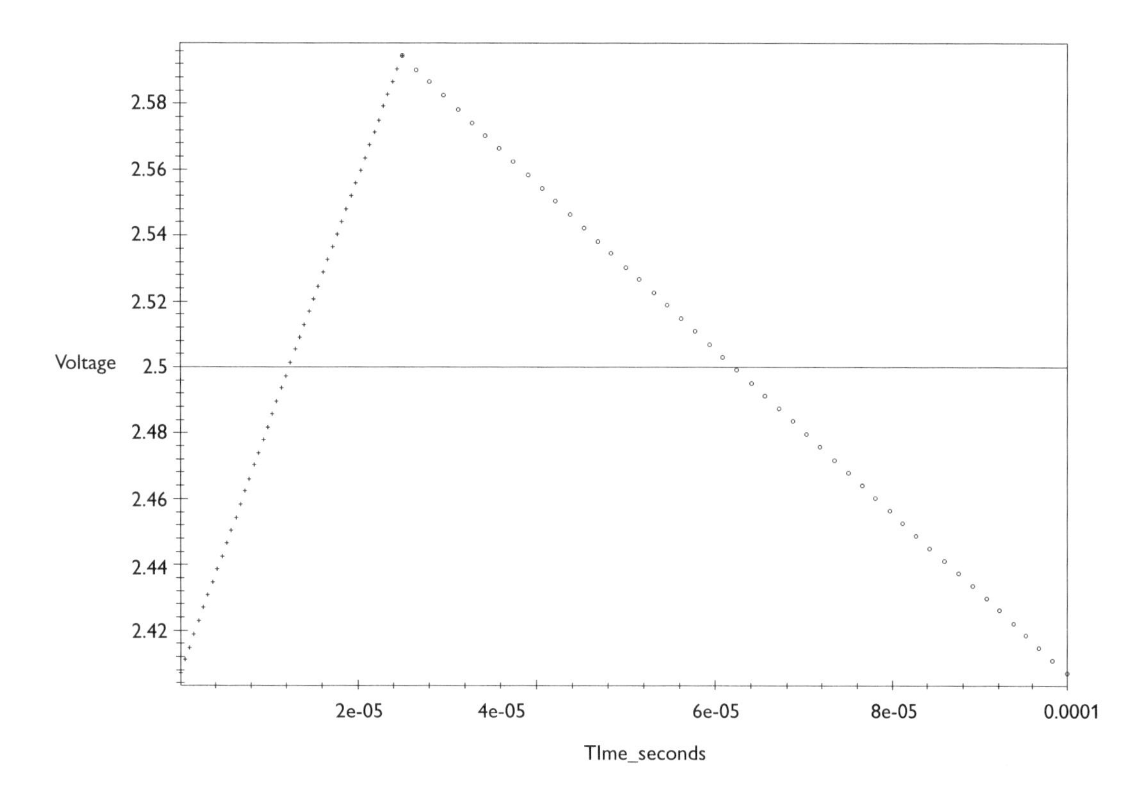

Figure 8.6 Output voltage plot of filtered PWM signal (*R* = 1000 ohms, *C* = 1 μF, frequency = 10 kHz, duty cycle = 25%, *V*in = 10V).

$$\text{Case 1} \quad f_{\text{CUTOFF}} = \frac{1}{2\pi RC} = \frac{1}{2\pi(1000)(1\mu F)} = 159.15 \text{ Hz}$$

$$\text{Case 2} \quad f_{\text{CUTOFF}} = \frac{1}{2\pi RC} = \frac{1}{2\pi(100)(.1\mu F)} = 15.915 \text{ kHz}$$

With a switching frequency of 10 kHz, most of the switching frequency was removed in the first case, and almost all of the switching information was passed in the second case as is evident from the plots. Now, let's reverse the process by specifying a certain ripple $|V_1 - V_2| \leq \zeta$ for a given load under fixed conditions of duty cycle and input switching voltage. In this way, whether the circuit is being used for signal processing or power transduction, we will know the switching ripple's worst case

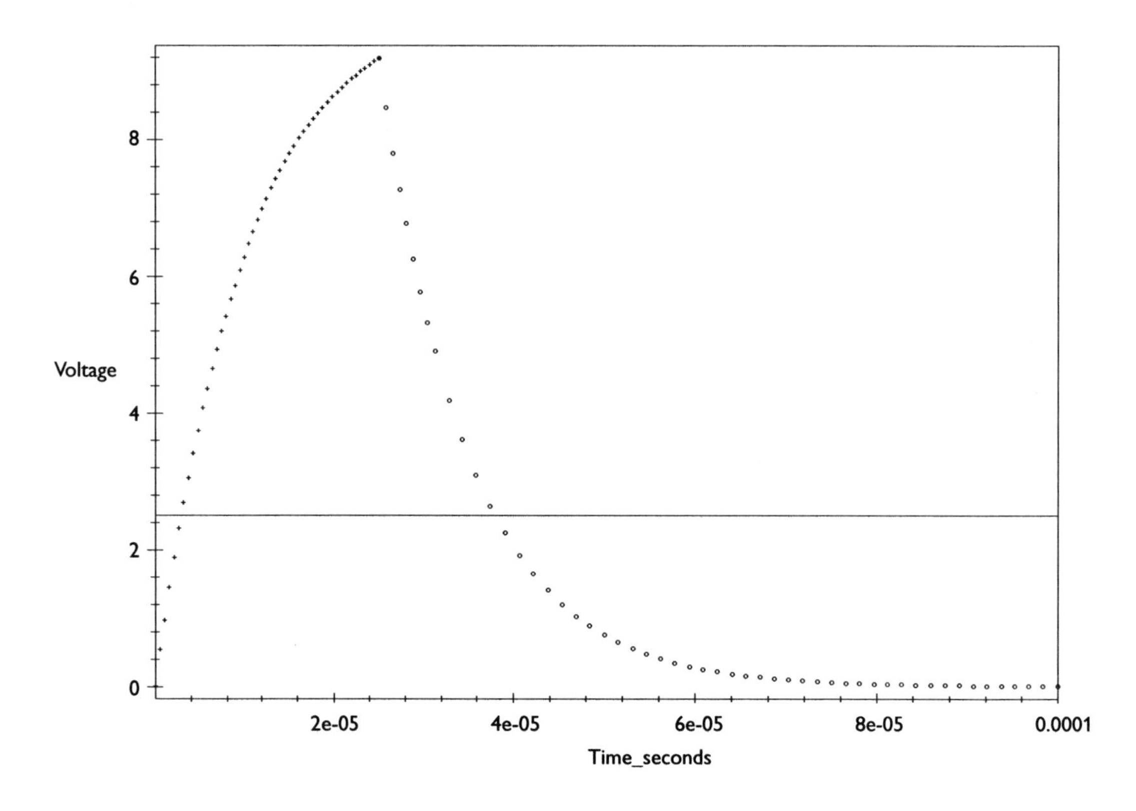

Figure 8.7 Output voltage plot of filtered PWM signal (R = 100 ohms, C = 0.1 μF, frequency = 10 kHz, duty cycle = 25%, Vin = 10V).

scenario. In signal processing applications, the ripple can detract from the averaging filter's dynamic range, whereas for a power application (say, heating coils), the ripple connotes the excessive voltages (especially V_1) beyond the desired average value stressing the elements.

If we constrain the ripple to a maximum of 1V peak to peak $V_1 - V_2 \leq 1$ under a duty cycle of 50% with an applied input switching voltage of 50V at 5 kHz, then we can determine the minimal RC product needed to ensure this specification. First we enter the knowns of the problem along with the appropriate Maple libraries:

```
with(inttrans):
Freq := 5*10^3:
alfa := .50:
Vin := 50/s:
T := 1/Freq:
Vout_1_Time := invlaplace((Vin+s*RC*V1/s)/(s*RC+1),s,t):
Vout_2_Time := invlaplace((s*RC*V2/s)/(s*RC+1),s,t):
Vout_1_Time_AlfaT := evalf(subs(t=alfa*T,Vout_1_Time)):
Vout_2_Time_TminusAlfaT := evalf(subs(t=T-alfa*T,
 Vout_2_Time)):
Solutions := solve({Vout_1_Time_AlfaT=V2,
 Vout_2_Time_TminusAlfaT=V1},{V1,V2}):
XX := subs(Solutions,[V1,V2]):
V1 := XX[1]:
V2 := XX[2]:
Peak_To_Peak_Ripple :=abs(V2-V1);
```

$$\text{Peak\_To\_Peak\_Ripple} :=$$

$$\left| 50.\ \frac{1}{e^{\left(-.0001000000000\ \frac{1}{\mathrm{RC}}\right)}+1.} - 50.\ \frac{e^{\left(-.0001000000000\ \frac{1}{\mathrm{RC}}\right)}}{e^{\left(-.0001000000000\ \frac{1}{\mathrm{RC}}\right)}+1} \right|$$

Now that we have the expression relating the RC product to the peak-to-peak ripple, we solve the expression using Maple's `solve` command:

```
Ripple_Result : = solve(Peak_To_Peak_Ripple = 1,RC);
```

$$\text{Ripple\_Result} := -.002499666603\ ,\ .002499666629$$

Since the RC product must be positive and nonzero, the correct value for the problem is

$$\mathrm{RC}_{\mathrm{MINIMUM}} = .002499666629$$

From this, we can simply play a balancing act between available resistor and capacitance values to obtain the minimal RC product to ensure the previous design requirements.

From an analysis point of view, the question now becomes what is the allowable range of duty cycle variance to ensure that the peak-to-peak ripple does not exceed 1V? This question is best answered by looking at the dependence of output ripple voltage as a function of duty cycle, hence, a plot of these variables in Maple becomes

```
with(plots):
with(inttrans):
Freq : =5*10^3:
RC : =.002499666629:
Vin : =50/s:
T :=1/Freq:
Vout_1_Time := invlaplace((Vin+s*RC*V1/s)/(s*RC+1),s,t):
Vout_2_Time : =invlaplace((s*RC*V2/s)/(s*RC+1),s,t):
Vout_1_Time_AlfaT := evalf(subs(t=alfa*T,Vout_1_Time)):
Vout_2_Time_TminusAlfaT := evalf(subs(t=T-alfa*T,
 Vout_2_Time)):
Solutions := solve({Vout_1_Time_AlfaT=V2,
 Vout_2_Time_TminusAlfaT=V1},{V1,V2}):
XX := subs(Solutions,[V1,V2]):
V1 :=XX[1]:
V2 := XX[2]:
Peak_To_Peak_Ripple := abs(V2-V1):
plot(Peak_To_Peak_Ripple,alfa=.3..+.7,axes=boxed,
 color=black,labels=[D uty_Cycle,Ripple]);
```

From Figure 8.8, it appears that the original computation of the minimal RC product was given at the maximal ripple duty cycle value (50%). Therefore, to ensure a range of acceptable duty cycle variations, we need only specify the minimum RC product sufficient to ensure that we do not exceed the 1V peak-to-peak ripple specification at 50%.

However, if we are in the signal processing business and we are concerned over the relative signal-to-noise ratio (SNR), then the picture is different. For instance, if we operate the PWM system at a low duty cycle, incurring a low average value, then the recovered information (average value) might not be sufficiently larger than that passed through switching frequency ripple.

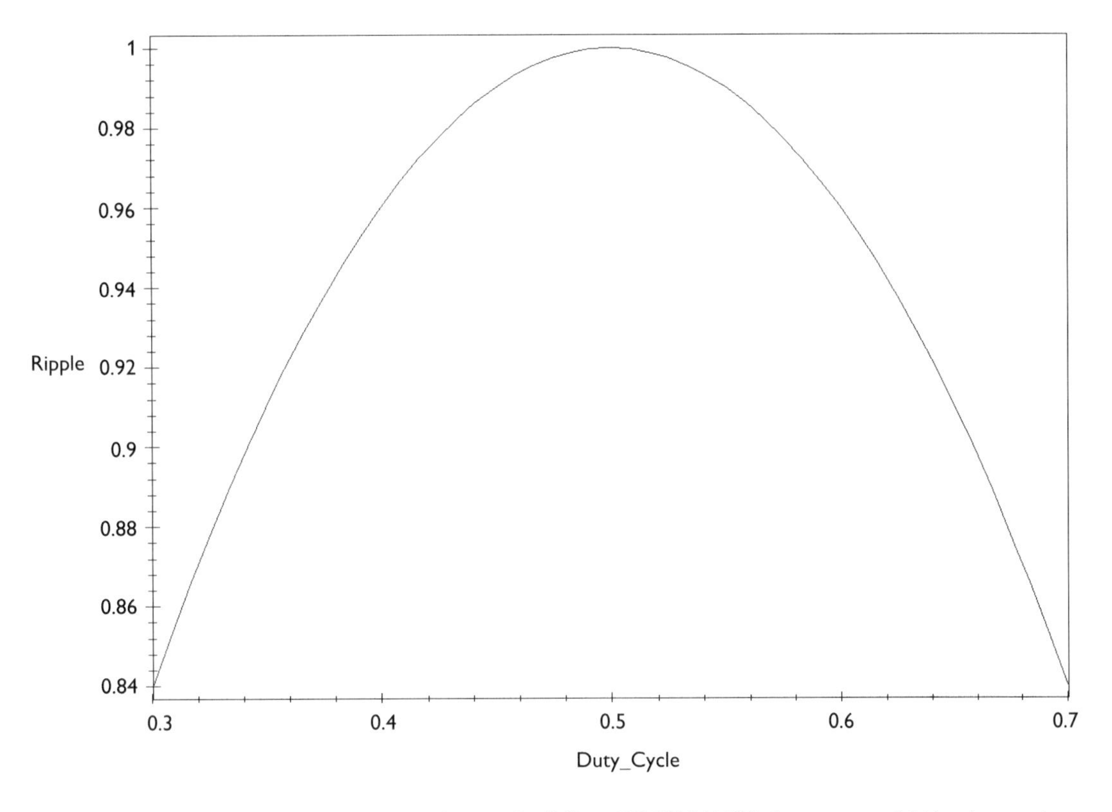

Figure 8.8 Peak-to-peak ripple versus duty cycle (RC = .002499666629, frequency = 5 kHz, duty cycle range = 30% to 70%, *Vin* = 50V).

The relative SNR merit function, SNR_{REL}, might be

$$SNR_{REL} = \frac{\text{Average output}}{\text{Peak–to–peak ripple output}}$$

Since we have computed these functions previously, we can plot this merit function as a function of duty cycle. Thus,

```
with(plots):
with(inttrans):
Freq := 5*10^3:
RC := .002499666629:
Vin := 50/s:
T := /Freq:
```

```
Vout_1_Time := invlaplace((Vin+s*RC*V1/s)/(s*RC+1),s,t):
Vout_2_Time := invlaplace((s*RC*V2/s)/(s*RC+1),s,t):
Vout_1_Time_AlfaT := evalf(subs(t=alfa*T,Vout_1_Time)):
Vout_2_Time_TminusAlfaT := evalf(subs(t=T-alfa*T,
 Vout_2_Time)):
Solutions := solve({Vout_1_Time_AlfaT=V2,Vout_2_Time_
 TminusAlfaT=V1},{V1,V2}):
XX := subs(Solutions,[V1,V2]):
V1 := XX[1]:
V2 := XX[2]:
Vout_2_Time_Duration := subs(t=t-alfa*T,Vout_2_Time):
Output_Average := 1/T*(int(Vout_1_Time,t=0..alfa*T)+
 int(Vout_2_Time_Duration,t=alfa*T..T)):
Peak_To_Peak_Ripple := abs(V2-V1):
SNR_Rel := Output_Average/ Peak_To_Peak_Ripple:
plot(SNR_Rel,alfa=+.1..+.9,axes=boxed,color=black,
 labels= [Duty_Cycle,SNR]);
```

As evident in Figure 8.9, the lower duty cycles have a much lower SNR, even though Figure 8.8 indicated that a 50% duty cycle gave the maximal ripple. Consequently, depending on the application of this type of signal handling, one needs to determine which aspect of the PWM recovered information is of importance, in other words, which quiescent operating point should be employed.

Switching power supply

In those applications where the regulated output dc voltage is less than the input unregulated dc voltage, a certain switching regulator is utilized. Since the output is less than the input voltage, it is called a buck-type converter. If the output voltage is greater than the input, then the switching regulator is called a boost-type converter. We will be dealing with the design and analysis of a buck-type dc-to-dc converter.

Figure 8.10 depicts the general topology associated with a buck-type converter, which can be functionally reduced to Figure 8.11 for analysis purposes. From Figure 8.11, one notices the second-order low-pass filter (LPF) with the inductor, capacitor, and resistive load circuitry. The topology is simply a series resonant circuit with resistive dissipation. Further simplification is obtained when the switching transistor model (Q1) and PWM controller blocks are modeled and removed by incorporation of their function into the input signal as shown in Figure 8.12.

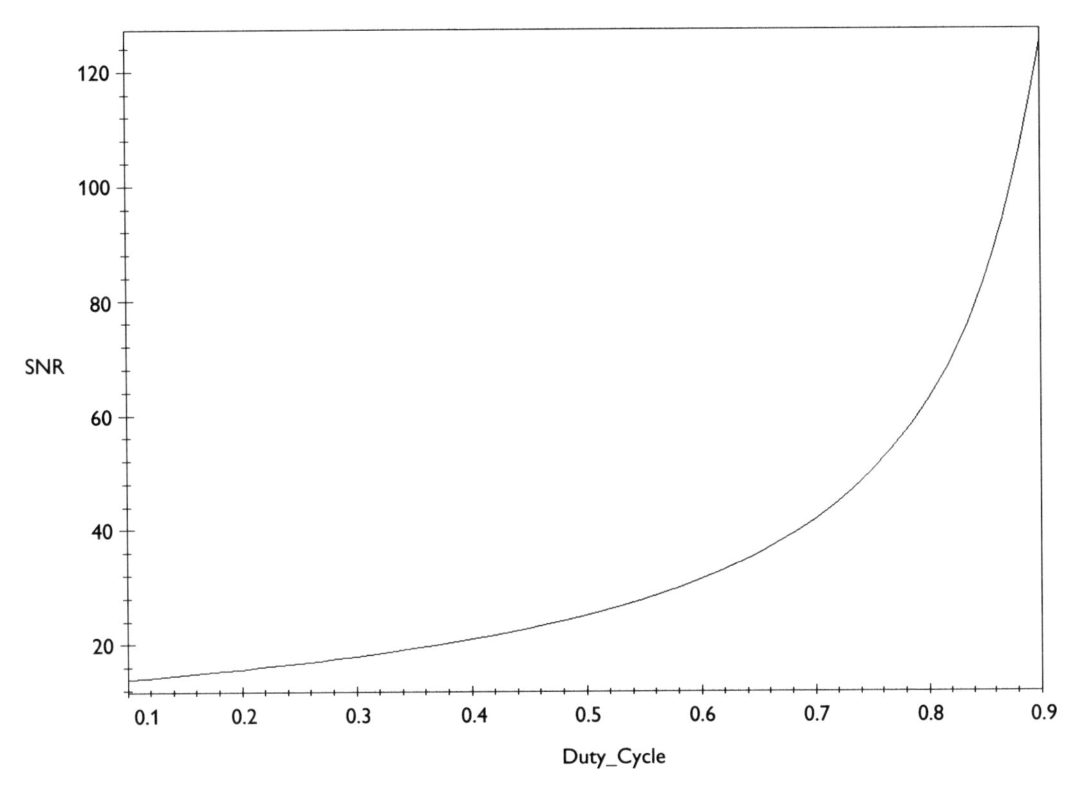

Figure 8.9 Relative SNR versus duty cycle (RC = .002499666629, frequency = 5kHz, *V*in = 50V).

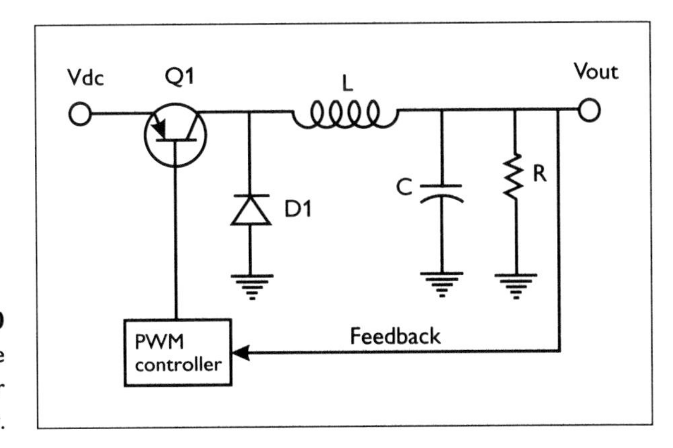

Figure 8.10 Simple buck-type switching power supply.

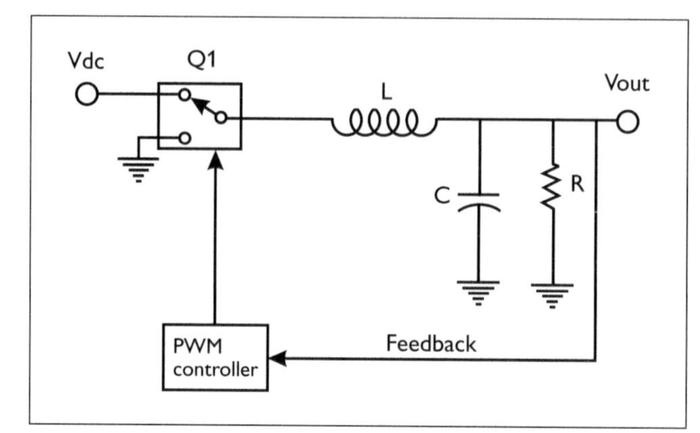

Figure 8.11 Base buck-type converter topology.

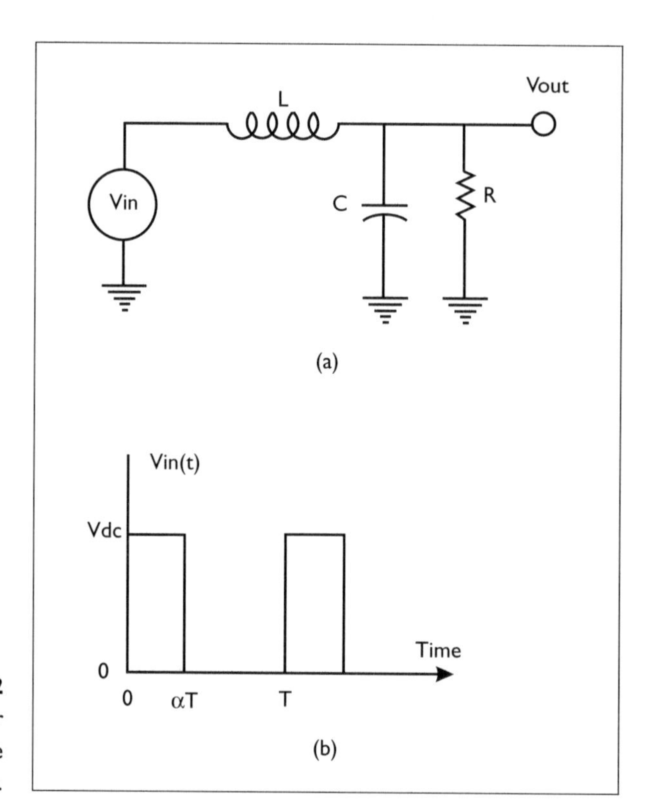

Figure 8.12 Basic model for the buck-type converter.

The average dc output voltage (*V*out, Figure 8.12(a)) dependence on duty cycle ($\alpha \rightarrow$ alfa, Figure 8.12(b)) transfer function is more complex and exhibits different dynamics than the previously described PWM design case. The same is also true for the output voltage ripple computations. Again, the recovered output, relative to the input waveform, will appear as shown in Figure 8.13.

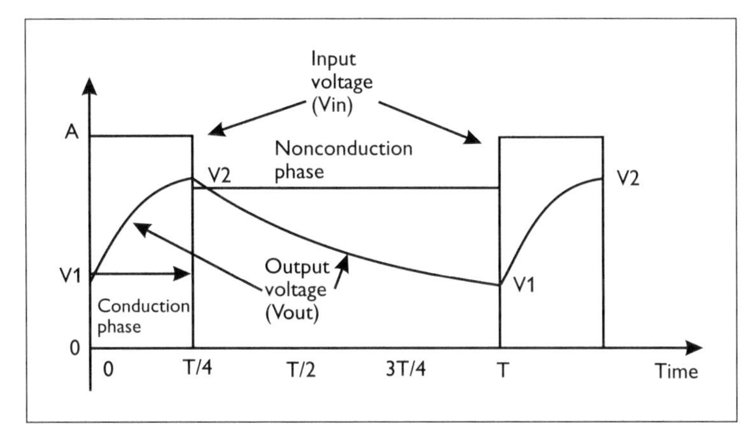

Figure 8.13
Output and input waveforms on a per-period basis.

However, there is another initial condition associated with this second-order circuit, namely, the inductor's (L) core current along with the output capacitor's initial voltage at each time interface. These initial boundary conditions are shown in Figure 8.14 and apply identically to each period once the steady-state dynamics have been reached.

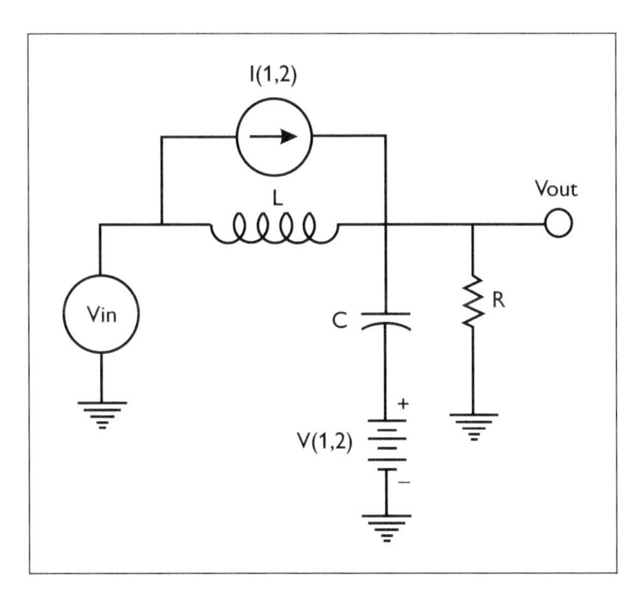

Figure 8.14
Boundary initial conditions elements.

The initial current source in parallel with the inductor has two dc states as does the initial voltage source in series with the output capacitor. The nomenclature is as follows:

Regime 1 $(0 \leq t < \alpha T)$	Regime 2 $(\alpha T \leq t < T)$
Inductor current = $I1$	Inductor current = $I2$
Capacitor voltage = $V1$	Capacitor voltage = $V2$

Therefore, there are two circuits that are being solved and matched at the time boundaries indicated earlier. Figures 8.15(a) and 8.15(b) show the two topologies within the two time regimes. From these circuits, one derives the appropriate equations for each regime.

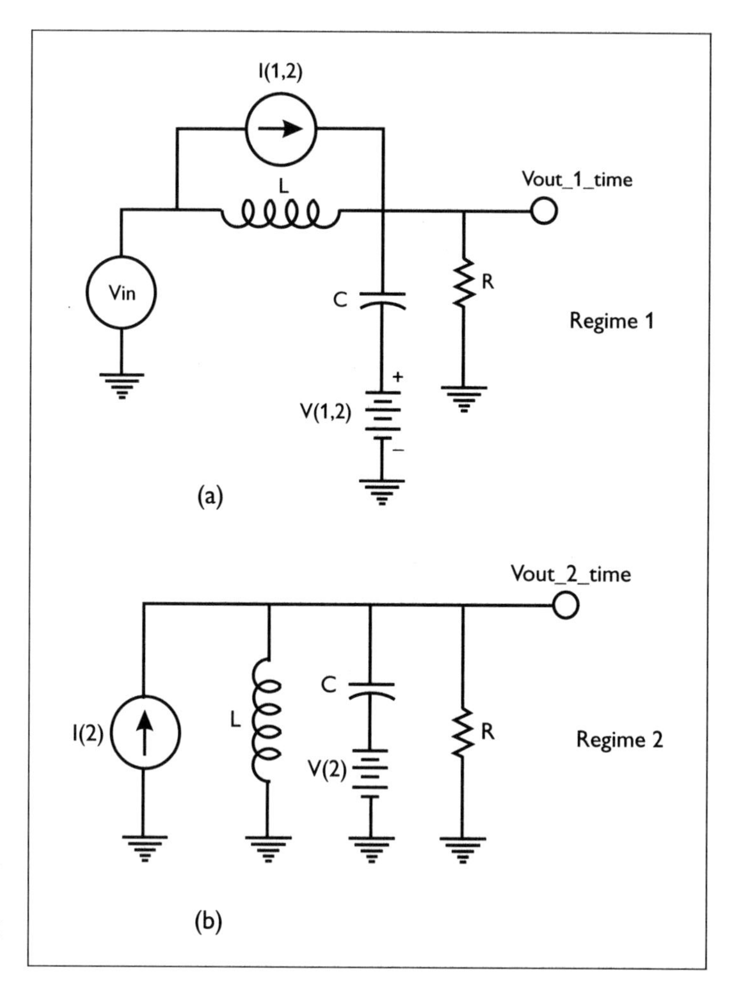

Figure 8.15 Two models associated with the two switching regimes.

Let's begin by initiating the appropriate Maple libraries and component values into a session:

```
with(inttrans):
Freq := 40*10^3:
R := 50:
L := 10^(-2):
C := 10^(-3):
Vin := 24/s:
alfa := .50:
T := 1/Freq:
```

Computing the time-domain forms over the two regimes per period as before,

```
Vout_1_Time :=invlaplace((Vin+s*L*C*V1+L*I1)/
(L*C*s^2+s*L/R+1),s,t);
Vout_2_Time :=invlaplace((s*L*C*V2+L*I2)/
(L*C*s^2+s*L/R+1),s,t);
```

$$
\begin{aligned}
Vout\_1\_Time := {} & 24 - \frac{8}{111} e^{(-10\ t)} \sin(30\ \sqrt{3}\ \sqrt{37}\ t)\ \sqrt{3}\ \sqrt{37} \\
& + \frac{100}{333} I1\ e^{(-10\ t)} \sin(30\ \sqrt{3}\ \sqrt{37}\ t)\ \sqrt{3}\ \sqrt{37} \\
& - 24\ e^{(-10\ t)} \cos(30\ \sqrt{3}\ \sqrt{37}\ t) \\
& - \frac{1}{333} V1\ e^{(-10\ t)} \sin(30\ \sqrt{3}\ \sqrt{37}\ t)\ \sqrt{3}\ \sqrt{37} \\
& + V1\ e^{(-10\ t)} \cos(30\ \sqrt{3}\ \sqrt{37}\ t)
\end{aligned}
$$

$$
\begin{aligned}
Vout\_2\_Time := {} & -\frac{1}{333} V2\ e^{(-10\ t)} \sin(30\ \sqrt{3}\ \sqrt{37}\ t)\ \sqrt{3}\ \sqrt{37} \\
& + V2\ e^{(-10\ t)} \cos(30\ \sqrt{3}\ \sqrt{37}\ t) \\
& + \frac{100}{333} I2\ e^{(-10\ t)} \sin(30\ \sqrt{3}\ \sqrt{37}\ t)\ \sqrt{3}\ \sqrt{37}
\end{aligned}
$$

From these results, we compute a few boundary conditions:

$V_{OUT}\_1\_Time(t = 0) = V_{OUT}\_1\_Time\_Zero = V_1$
(Second-to-first regime boundary check)

$V_{OUT}\_2\_Time(t = 0) = V_{OUT}\_2\_Time\_Zero = V_2$
(First-to-second regime boundary check)

$$V_{OUT}\_1\_Time(t = \alpha T) = V_{OUT}\_1\_Time\_AlfaT$$
$$V_{OUT}\_2\_Time(t = T - \alpha T) = V_{OUT}\_2\_Time\_AlfaT$$

Continuing,

```
Vout_1_Time_Zero := evalf(subs(t=0,Vout_1_Time));
Vout_2_Time_Zero := evalf(subs(t=0,Vout_2_Time));
```

$$V_{OUT}\_1\_Time\_Zero := 1.\ V_1$$
$$V_{OUT}\_2\_Time\_Zero := 1.\ V_2$$

They check out; now continuing to generate the two basis equations,

```
Vout_1_Time_AlfaT := evalf(subs(t = alfa*T,Vout_1_Time));
Vout_2_Time_TminusAlfaT := evalf(subs(t = T-alfa*T,
Vout_2_Time));
```

$$V_{OUT}\_1\_Time\_AlfaT := .00018748 + .01249840508\ I1 + .9997422200\ V1$$
$$V_{OUT}\_2\_Time\_TminusAlfaT := .9997422200\ V2 + .01249840508\ I2$$

Maple nearly has all the basis equations, but we still need two more conditions to solve uniquely. Note in Figures 8.15(a) and 8.15(b) that for any boundary (again, assuming steady state) that the following statements must hold:

$$I1 = \frac{V_1}{R}$$
$$I2 = \frac{V_2}{R}$$

this is true because the boundary exhibits zero time duration, hence all inductors and capacitors have infinite and zero impedance, respectively.

Now Maple has a sufficient number of equations to solve for unique solutions at the boundaries.

```
Solutions := solve({Vout_2_Time_TminusAlfaT=V1,
 Vout_1_Time_AlfaT=V2,I1=V1/R,I2=V2/R},{V1,V2,I1,I2});
```

$$\text{Solutions} := \{V2 = 11.99971838\ ,\ V1 = 11.99962464\ ,\\ I2 = .2399943676\ ,\ I1 = .2399924928\ \}$$

Reassigning the variables explicitly since Maple's output is in a set form,

```
XX :=subs(Solutions,[V1,V2,I1,I2]);
```

$$XX := [11.99962464\ ,\ 11.99971838\ ,\ .2399924928\ ,\ .2399943676\ ]$$

then associating and abstracting the numerical solutions to the variables of interest,

```
V1 := XX[1];
V2 := XX[2];
I1 := XX[3];
I2 := X[4];
```

$$\begin{aligned} V1 &:= 11.99962464 \\ V2 &:= 11.99971838 \\ I1 &:= .2399924928 \\ I2 &:= .2399943676 \end{aligned}$$

Note that the values are pretty close to each other since we are filtering at

$$f_{\text{CUTOFF}} = \frac{1}{2\pi\sqrt{LC}} = \frac{1}{2\pi\sqrt{(10\ mH)(1000\ uF)}} = 50\text{Hz}$$

which is 2.90 orders below the switching frequency (40 kHz) with a second-order LPF. This will account for approximately a 640,000:1 attenuation of the 24V dc square-wave input. A quick reality check indicates that the V_1, V_2 differential of peak-to-peak ripple is about 94 μV. Dividing the 24V dc drive (or 24V peak to peak) by 640,000 yields an expected peak to peak value of 37.5 μV. The ratiometric difference between the Maple computation and the theoretical estimate is due to the effective Q gain associated with the series resonant circuit. This would indicate a Q of around 2.5.

Now we can state the final time-domain expression for the two time regimes on a per-period basis:

```
Vout_1_Time_Final := evalf(subs(V1=V1,Vout_1_Time));
Vout_2_Time_Duration := evalf(subs(V2=V2,Vout_2_Time)):
Vout_2_Time_Final := evalf(subs(t=t-alfa*T,Vout_2_Time_
 Duration));
```

$$Vout\_1\_Time\_Final := 24. - .3796750740\ e^{(-10.\ t)}\ \sin(316.0696125\ t) - 12.00037536\ e^{(-10.\ t)}\ \cos(316.0696125\ t)$$

$$Vout\_2\_Time\_Final := .3796542881\ e^{(-10.\ t\ +.0001250000000\ )}\ \sin(316.0696125\ t - .003950870156\ ) + 11.99971838\ e^{(-10.\ t\ +.0001250000000\ )}\ \cos(316.0696125\ t - .003950870156\ )$$

Now to create plots of both regimes (i.e., output for $0 \le t < T$) with the corresponding average dc value, we enter the following to display pertinent output voltage information of the converter under the given constraints:

```
Output_Average := 1/T*(int(Vout_1_Time_Final,t=0..alfa*T)+
 int(Vout_2_Time_Final,t=alfa*T..T));
Output_Peak_to_Peak_Ripple := abs(V1-V2);
```

$$Output\_Average := 11.99967152$$
$$Output\_Peak\_to\_Peak\_Ripple := .00009374$$

and to generate the compound plot (Figure 8.16):

```
with(plots):
Plot_1 := plot(Vout_1_Time_Final,t=0..alfa*T,color = black,
 style=point,symbol=cross):
Plot_2 := plot(Vout_2_Time_Final,t=alfa*T..T,color = black,
 style=point,symbol=circle):
Plot_3 := plot(Output_Average,t=0..T,color=black):
display({Plot_1,Plot_2,Plot_3},axes=boxed,labels=
 [Time_seconds,Voltage]);
```

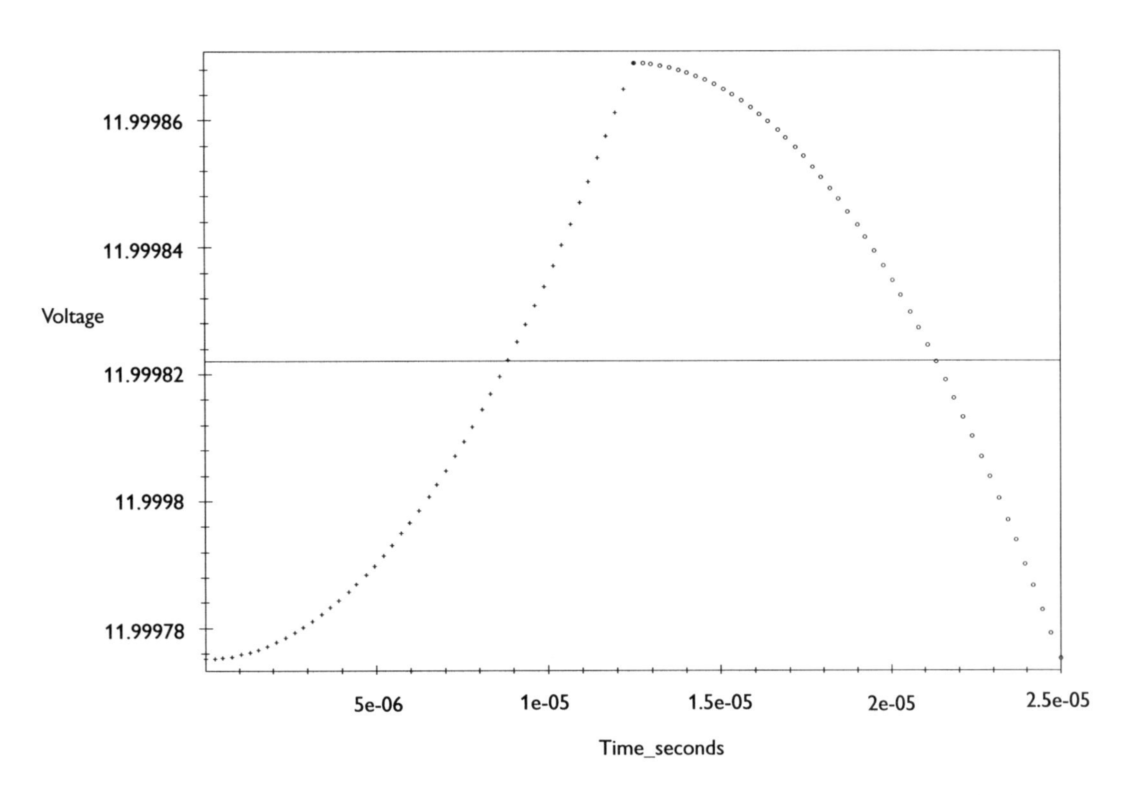

Figure 8.16 Buck converter output waveform (R = 50 ohms, L = 10 mH, C = 1000 μF, Vin = 24V, duty cycle = 50%).

Now let's try some different component values and another duty cycle:

```
with(plots):
with(inttrans):
Freq := 40*10^3:
R := 50:
L := 10^(-3):
C := 10^(-5):
Vin := 24/s:
alfa := .25:
T := 1/Freq:
Vout_1_Time := invlaplace((Vin+s*L*C*V1+L*I1)/
 (L*C*s^2+s*L/R+1),s,t):
Vout_2_Time := invlaplace((s*L*C*V2+L*I2)/
 (L*C*s^2+s*L/R+1),s,t):
Vout_1_Time_Zero := evalf(subs(t=0,Vout_1_Time)):
Vout_2_Time_Zero := evalf(subs(t=0,Vout_2_Time)):
```

```
Vout_1_Time_AlfaT := evalf(subs(t=alfa*T,Vout_1_Time)):
Vout_2_Time_TminusAlfaT :=evalf(subs(t=T-alfa*T,
 Vout_2_Time)):
Solutions := solve({Vout_2_Time_TminusAlfaT=V1,
 Vout_1_Time_AlfaT=V2,I1=V1/R,I2=V2/R},{V1,V2,I1,I2}):
XX := subs(Solutions,[V1,V2,I1,I2]):
V1 := XX[1]:
V2 := XX[2]:
I1 := XX[3]:
I2 := XX[4]:
Vout_1_Time_Final := evalf(subs(V1=V1,Vout_1_Time)):
Vout_2_Time_Duration := evalf(subs(V2=V2,Vout_2_Time)):
Vout_2_Time_Final := evalf(subs(t=t-alfa*T,Vout_2_Time_
 Duration)):
Output_Average := 1/T*(int(Vout_1_Time_Final,t=0..alfa*T)+
 int(Vout_2_Time_Final,t=alfa*T..T));
Output_Peak_to_Peak_Ripple :=abs(V1-V2);
Plot_1 := plot(Vout_1_Time_Final,t=0..alfa*T,color=black,
 style=point,symbol=cross):
Plot_2 := plot(Vout_2_Time_Final,t=alfa*T..T,color=black,
 style=point,symbol=cir cle):
Plot_3 := plot(Output_Average,t=0..T,color=black):
display({Plot_1,Plot_2,Plot_3},axes=boxed,labels =
 [Time_seconds,Voltage]);
```

Output_Average := 2.410331814
Output_Peak_to_Peak_Ripple := .042026107

Figure 8.17 exhibits a much larger peak-to-peak ripple than previously. This is the result of a much smaller amount of filtering (note the reduction in capacitance and inductance). Also note that the dc or average recovered voltage is not linearly related to the duty cycle as was the PWM case. If this system were linear, we would have expected to see an average dc value of 6.0V (25% of 24V) instead of the computed 2.41V.

To see the dependence of the average output dc voltage on duty cycle, we need to set up the equation relating these two items and then plot them; thus:

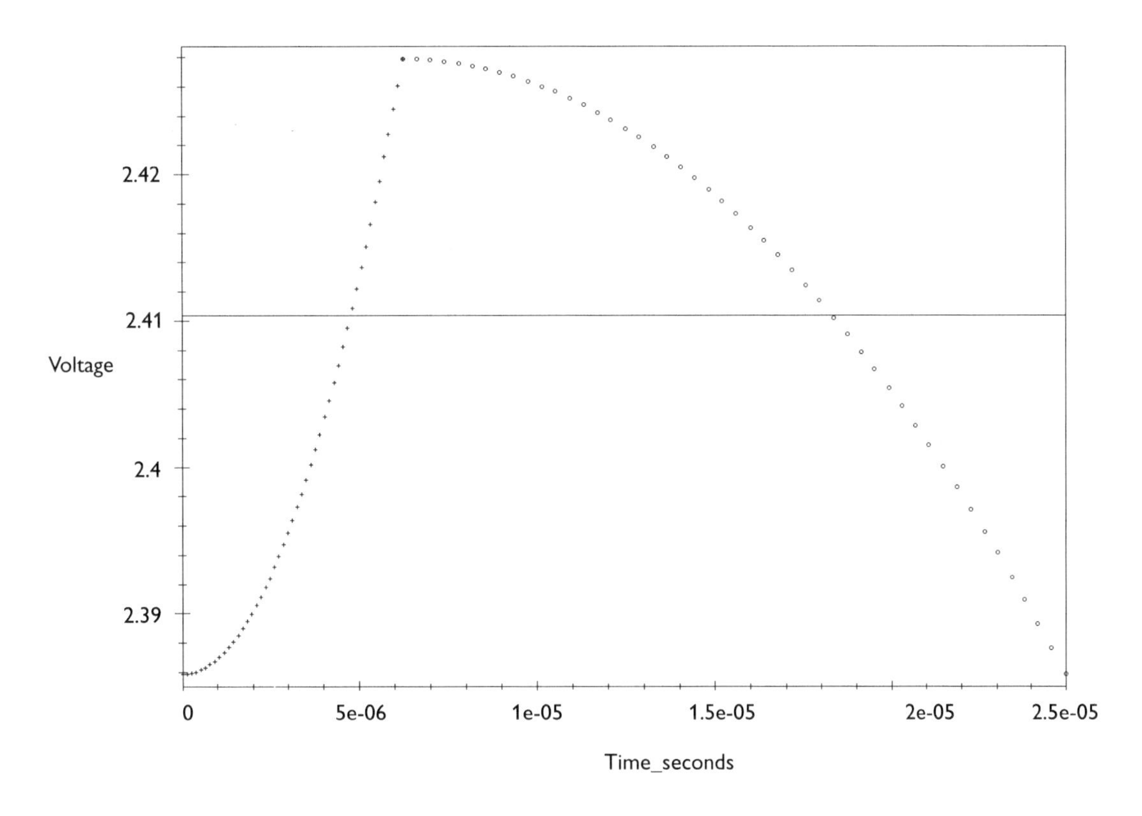

Figure 8.17 Buck converter output waveform (R = 50 ohms, L = 1 mH, C = 10 μF, Vin = 24V, duty cycle = 25%, frequency = 40 kHz).

```
with(plots):
with(inttrans):
Freq := 40*10^3:
R := 50:
L := 10^(-3):
C := 10^(-5):
Vin := 24/s:
T := 1/Freq:
Vout_1_Time := invlaplace((Vin+s*L*C*V1+L*I1)/
 (L*C*s^2+s*L/R+1),s,t):
Vout_2_Time := invlaplace((s*L*C*V2+L*I2)/
 (L*C*s^2+s*L/R+1),s,t):
Vout_1_Time_Zero := evalf(subs(t=0,Vout_1_Time)):
Vout_2_Time_Zero := evalf(subs(t=0,Vout_2_Time)):
Vout_1_Time_AlfaT := evalf(subs(t=alfa*T,Vout_1_Time)):
Vout_2_Time_TminusAlfaT := evalf(subs(t=T-alfa*T,
```

```
 Vout_2_Time)):
Solutions := solve({Vout_2_Time_TminusAlfaT=V1,
 Vout_1_Time_AlfaT=V2,I1=V1/R,I2=V2/R},{V1,V2,I1,I2}):
XX := subs(Solutions,[V1,V2,I1,I2]):
V1 := XX[1]:
V2 := XX[2]:
I1 := XX[3]:
I2 := XX[4]:
Vout_1_Time_Final := evalf(subs(V1=V1,Vout_1_Time)):
Vout_2_Time_Duration := evalf(subs(V2=V2,Vout_2_Time)):
Vout_2_Time_Final := evalf(subs(t=t-alfa*T,Vout_2_Time_
 Duration)):
Output_Average := 1/T*(int(Vout_1_Time_Final,t=0..alfa*T)+
 int(Vout_2_Time_Final,t=alfa*T..T)):
plot(Output_Average,alfa=0..1,color=black,axes=boxed,
 labels=[Duty_Cycle,Voltage]);
```

Clearly, Figure 8.18 does not represent a linear relationship between duty cycle and dc output voltage. Consequently, this type of filtering would not be an advisable approach to signal processing (i.e., decoding phase information), especially if linearity were important. However, for power supply regulation it is desirable, because the efficiency of these switching converters can be as high as 95% (depending on the power levels involved). Further, considering the compressing of the encoded information (the dc value), one would want to operate the buck converter somewhere around the middle of the duty cycle curve. Otherwise, larger loop gains will be necessary to hold output voltage levels at values far removed from the 12V center (duty cycle = 50% at Vin = 24V) under varying parameter and environmental conditions. This could lead to instability and probable loss of the controlled output voltage variance specification.

To show Figure 8.18's S-shaped dependence on the value of inductance, let's produce a 3-D plot of average output voltage versus duty cycle and inductance value:

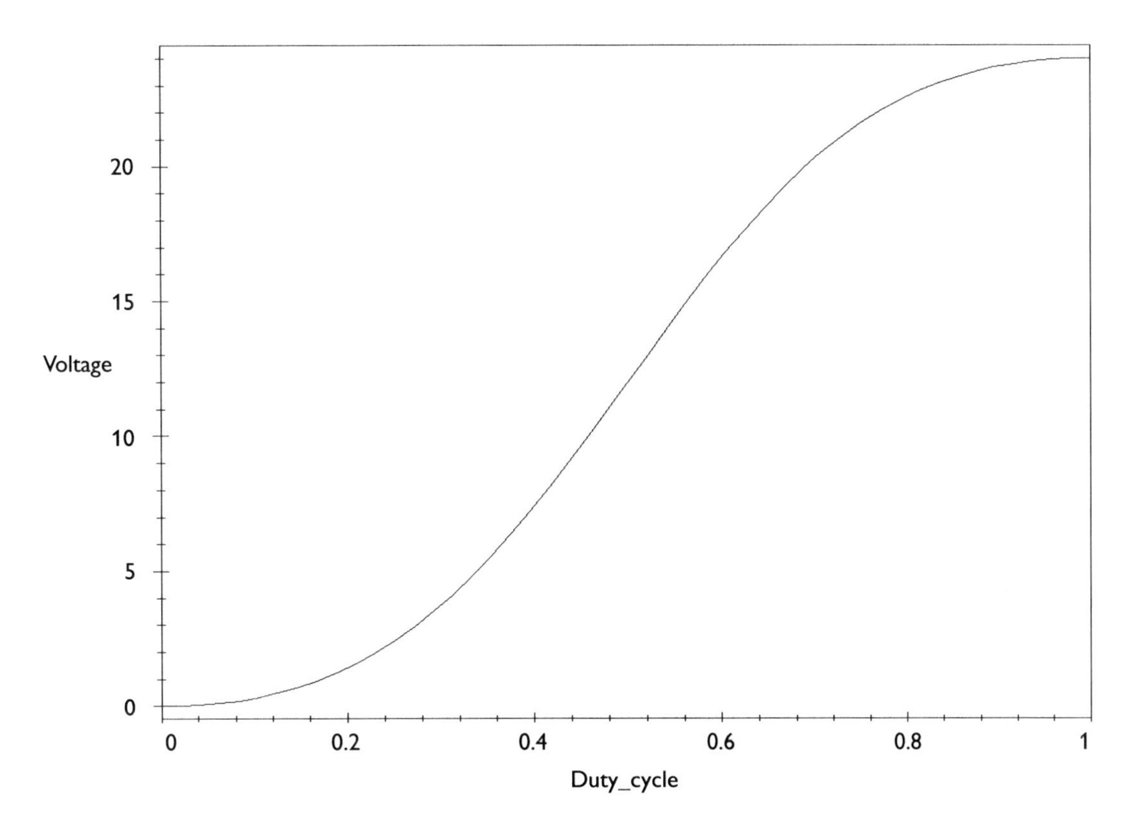

Figure 8.18 Output voltage versus duty cycle of buck converter (R = 50 ohms, L = 1 mH, C = 10 μF, Vin = 24V, frequency = 40 kHz).

```
with(plots):
with(inttrans):
Freq := 40*10^3:
R := 50:
C := 10^(-5):
Vin := 24/s:
T := 1/Freq:
Vout_1_Time := invlaplace((Vin+s*L*C*V1+L*I1)/
 (L*C*s^2+s*L/R+1),s,t):
Vout_2_Time := invlaplace((s*L*C*V2+L*I2)/
 (L*C*s^2+s*L/R+1),s,t):
Vout_1_Time_Zero := evalf(subs(t=0,Vout_1_Time)):
Vout_2_Time_Zero := evalf(subs(t=0,Vout_2_Time)):
Vout_1_Time_AlfaT := evalf(subs(t=alfa*T,Vout 1 Time)):
Vout_2_Time_TminusAlfaT := evalf(subs(t=T-alfa*T,
```

```
 Vout_2_Time)):
Solutions := solve({Vout_2_Time_TminusAlfaT=V1,
 Vout_1_Time_AlfaT=V2,I1=V1/R,I2=V2 /R},{V1,V2,I1,I2}):
XX := subs(Solutions,[V1,V2,I1,I2]):
V1 := XX[1]:
V2 := XX[2]:
I1 := XX[3]:
I2 := XX[4]:
Vout_1_Time_Final := evalf(subs(V1=V1,Vout_1_Time)):
Vout_2_Time_Duration := evalf(subs(V2=V2,Vout_2_Time)):
Vout_2_Time_Final := evalf(subs(t=t-alfa*T,Vout_2_Time_
 Duration)):
Output_Average := 1/T*(int(Vout_1_Time_Final,t=0..alfa*T)+
 int(Vout_2_Time_Final,t=alfa*T..T)):
```

Now plotting the three variable relationship,

```
plot 3d(Output_Average,alfa=0..1,L=10^(-9)..10^(-5),
 grid=[50,50], color=black,axes=boxed,
 style=hidden,orientation= [-45,60],labels=[Duty_Cycle,
 Inductance,Volts]);
```

Figure 8.19 shows that as the inductance value (scale on the right-hand side baseline) decreases, the output voltage transfer becomes linear with the duty cycle (dc). The peaking taking place during the lower inductance values reflects the fact that some resonant behavior is getting through to the output along with a dramatic increase in switching ripple (look closely at the jagged nature of the straight line at low inductance on the left-hand wall of the 3-D plot). Clearly, the inductor affords us a great deal of filtering, but does so at the cost of linearity to the output voltage's dependence upon duty cycle.

Fourier method

Now that we have seen the exact solution at steady state on a per-period basis, let's look at another method for determining the output voltage characteristics. However, this method approximates the input PWM signal with its approximate Fourier series. Also, this method does not use any time sliding to implement two separate time-domain solutions to reconstruct the complete periodic solution; instead, this approach works on the premise of a running time average.

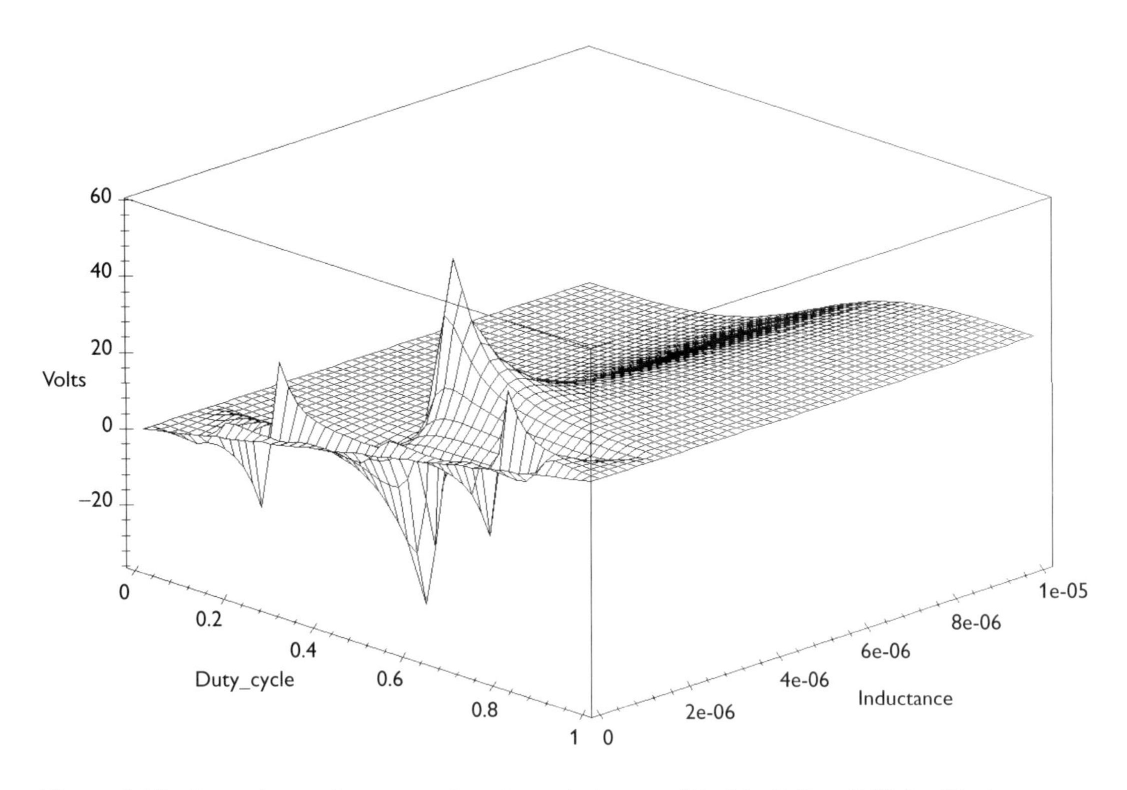

Figure 8.19 Dependence of converter linearity on Inductance (C = 10 μF, Vin = 24V, R = 50 ohms, frequency = 40 kHz).

Before we start generating Maple code, let's review the Fourier series that will approximate the PWM equivalent input as shown in Figure 8.1. Simply stated, the Fourier series can approximate any periodic function with a finite number of discontinuities that converge from both the left- and right-hand side of any given discontinuity. Further, the Fourier series approximation quality depends on how many expansion terms one wants to handle during the analysis. Consequently, creating a PWM waveform (a general square wave) is generally considered reasonably approximated with the first 10 harmonics. More harmonics are better, but this can become computationally prohibitive, as we shall see during our forthcoming revisit of the buck-type switching power supply.

The Fourier series is defined by the following set of constituent relations:

$$f(t) = a_0 + \sum_{n=1}^{\infty} \left(a_n \cos(n\omega_0 t) + b_n \sin(n\omega_0 t) \right)$$

where the Fourier coefficients are

$$a_0 = \frac{1}{T} \int_{-T/2}^{T/2} f(t)dt$$

$$a_0 = \frac{2}{T} \int_{-T/2}^{T/2} f(t) \cos(n\omega_0 t)dt$$

$$b_n = \frac{2}{T} \int_{-T/2}^{T/2} f(t) \sin(n\omega_0 t)dt$$

Revisiting Figure 8.1 again, by inspection, we can state the limits of integration:

$$a_0 = \frac{1}{T} \int_{0}^{\alpha T} f(t)dt$$

$$a_0 = \frac{2}{T} \int_{0}^{\alpha T} f(t) \cos(n\omega_0 t)dt$$

$$b_n = \frac{2}{T} \int_{0}^{\alpha T} f(t) \sin(n\omega_0 t)dt$$

where

$$f(t) = \begin{cases} A & \text{for } 0 \leq \pm < \alpha T \\ 0 & \text{for } \alpha T \leq \pm < T \end{cases}$$

and where

$$A \rightarrow \text{Peak amplitude of input square wave}$$
$$T \rightarrow \text{Switching period}$$
$$\alpha \rightarrow \text{Duty cycle}$$
$$\omega_0 = 2\pi f_0 \rightarrow \text{Switching frequency}$$

Substituting this dc voltage value, $f(t)$, into the Fourier relations yields the following Fourier coefficients:

$$a_0 = A\alpha$$

$$a_n = \frac{2A}{T}\frac{\sin(n\omega_0\alpha T)}{n\omega_0}$$

$$b_n = \frac{2A}{T}\frac{\left(1 - \cos(n\omega_0\alpha T)\right)}{n\omega_0}$$

Resubstituting this into the overall Fourier series form gives us

$$f(t) = A\alpha + \frac{2A}{\omega_0 T}\sum_{n=1}^{\infty}\left(\frac{\sin(n_0\alpha T)\cos(n\omega_0 t) + (1 - \cos(n\omega_0\alpha T))\sin(n\omega_0 t)}{n}\right)$$

Since we know that $\omega_0 = 2\pi f = 2\pi/T$, then the series can be restated as

$$f(t) = A\alpha + \frac{A}{\pi}\sum_{n=1}^{\infty}\left(\frac{\sin(2\pi n\alpha)\cos\left(\frac{2\pi nt}{T}\right) + (1 - \cos(2\pi n\alpha))\sin\left(\frac{2\pi nt}{T}\right)}{n}\right)$$

This is the input voltage function form we will use in our Maple session to compute the output voltage waveform after passing through a second-order low-pass passive RLC circuit. Put graphically, Figure 8.20 shows the process we will now analyze.

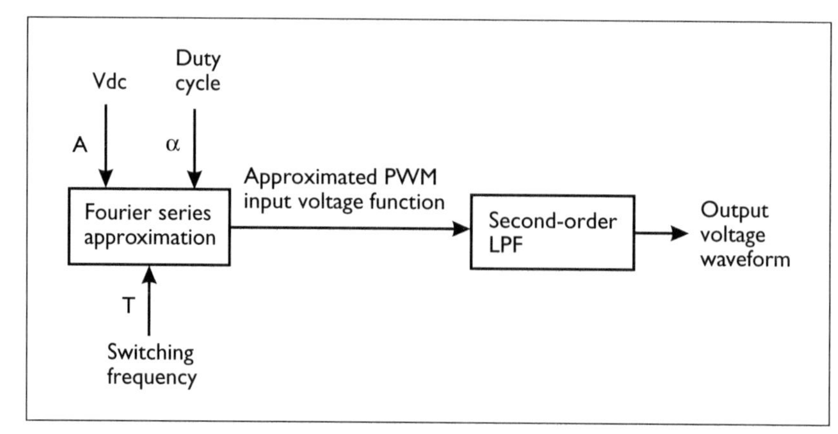

Figure 8.20
Overall Fourier analysis process.

Starting the Maple session with the appropriate libraries,

```
with(inttrans):
with (plots):
```

we continue by entering the circuit components and input switching waveform shown in Figures 8.12(a) and 8.12(b), respectively:

```
Switching_Freq := 40*10^3:
Vin := 24:
L := 10^(-2):
C := 10^(-3):
R := 50:
alfa := .50:
T := 1/Switching_Freq:
```

Computing the Fourier coefficients,

```
Ao := 1/T*int(Vin,t=0..alfa*T);
An := 2/T*int(Vin*cos(2*n*Pi*t/T),t=0..alfa*T);
Bn := 2/T*int(Vin*sin(2*n*Pi*t/T),t=0..alfa*T);
```

$$Ao := 12.00000000$$

$$An := 7.639437266\,\frac{\sin(3.141592654\;n)}{n}$$

$$Bn := -7.639437266\,\frac{\cos(3.141592654\;n)}{n} + 24\,\frac{1}{n\;\pi}$$

The *Ao* term represents the dc component of the Fourier series and simply evaluates from the integral to the following:

$$Ao = Vin \times \text{alfa} = 24\text{V} \times 50\% = 12\text{V}$$

Now substituting the coefficients into the approximating series,

```
Vin_Fourier : =
  Ao+sum(An*cos(2*n*Pi*t/T)+Bn*sin(2*n*Pi*t/T),
  n = 1..10);
```

$$\begin{aligned}
Vin_Fourier := {} & 12.00000000 - .3133748821\ 10^{-8}\cos(251327.4123\ t) \\
& + 15.27887453\ \sin(251327.4123\ t) \\
& + .3133748821\ 10^{-8}\ \cos(502654.8246\ t) \\
& - .3133748822\ 10^{-8}\ \cos(753982.2370\ t) \\
& + 5.092958178\ \sin(753982.2370\ t) \\
& + .1077318609\ 10^{-7}\ \cos\left(.1005309649\ 10^{7}\ t\right) \\
& - .3133748821\ 10^{-8}\cos\left(.1256637062\ 10^{7}\ t\right) \\
& + 3.055774906\ \sin\left(.1256637062\ 10^{7}\ t\right) \\
& - .1959209356\ 10^{-8}\ \cos\left(.1507964474\ 10^{7}\ t\right) \\
& - .5316445184\ 10^{-8}\ \cos\left(.1759291886\ 10^{7}\ t\right) \\
& + 2.182696362\ \sin\left(.1759291886\ 10^{7}\ t\right) \\
& + .1223889504\ 10^{-8}\ \cos\left(.2010619299\ 10^{7}\ t\right) \\
& - .6529054273\ 10^{-8}\ \cos\left(.2261946711\ 10^{7}\ t\right) \\
& + 1.697652726\ \sin\left(.2261946711\ 10^{7}\ t\right) \\
& + .3133748821\ 10^{-8}\ \cos\left(.2513274123\ 10^{7}\ t\right)
\end{aligned}$$

If we plot this Fourier approximation, we can compare the series approximation (at least subjectively) against the exact input waveform shown in Figure 8.12(b). The following Maple structure will generate two complete switching waveform periods:

```
Fourier_Plot :=plot(Vin_Fourier,t=0..2*T,color=black):
piece_1 :=piecewise(t=0,Vin,0):
piece_2 :=piecewise(t=alfa*T,-Vin,0):
piece_3 :=piecewise(t=T,Vin,0):
piece_4 :=piecewise(t=T*(1+alfa),-Vin,0):
Exact_Plot :=plot(piece_1+piece_2+piece_3+piece_4,
 t=0..2*T,color=black):
display({Fourier_Plot,Exact_Plot},axes=boxed,labels=
 [Time_seconds,Voltage]);
```

Figure 8.21 indicates that the first 10 harmonics (plus dc term) seem to give a fairly reasonable approximation to the 50% duty cycle modulated input waveform.

If we were to sacrifice some computer resources and wait a bit (this computation can get really long and cause dangerously low system resources, so save your work before you execute the Maple command), we can increase the approximation to the first 30 harmonics:

```
Vin_Fourier_30 :=Ao+sum(An*cos(2*n*Pi*t/T)+Bn*
 sin(2*n*Pi*t/T),n=1..30):
Fo urier_Plot_30
:=plot(Vin_Fourier_30,t=0..2*T,color=black):
display({Fourier_Plot_30,Exact_Plot},axes=boxed,
 labels=[Time_seconds,Voltage]);
```

Clearly, the more harmonics we incorporate into the input PWM synthesis, the better the approximation (Figure 8.22). The limit here, of course, is that of time and system resources for the extra harmonic terms associated with the Fourier series expansion. It is for this reason that we will use the approximation depicted in Figure 8.21, which only uses the first 10 harmonics.

Later, we take the Laplace transform of each of these harmonics. One can readily see that this computational approach can get extremely intensive, especially if the use of a large number of harmonics is required for a more accurate result.

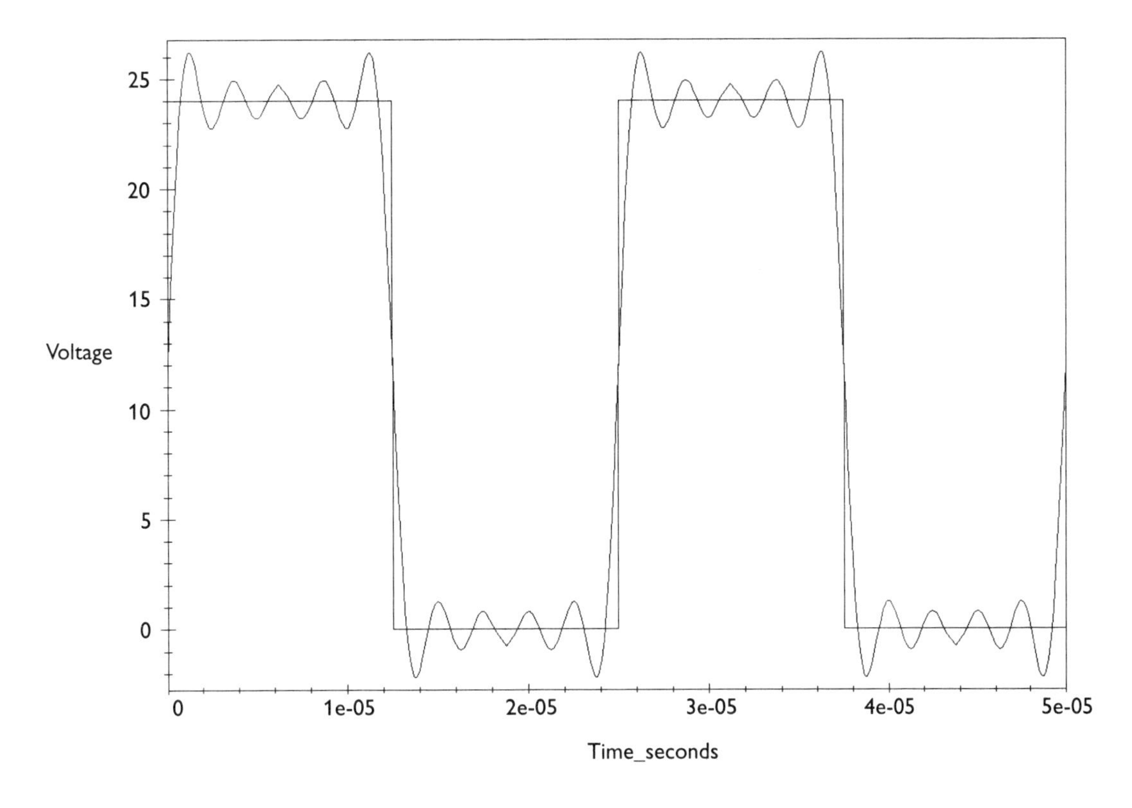

Figure 8.21 Comparison of Fourier approximation to exact input waveform (10 harmonics plus dc with a 50% duty cycle).

Now continuing by taking the Laplace transform of the Fourier approximated input voltage (Figure 8.20, `Vin_Fourier`) and stating the second-order LCR filter's (Figure 8.12(a)) transform,

```
Vin_Laplace :=laplace(Vin_Fourier,t,s):
LPF_Transfer :=1/(L*C*s^2+L/R*s+1):
Output_Voltage_Laplace :=LPF_Transfer*Vin_Laplace:
Output_Voltage_Time :=invlaplace(Output_Voltage_Laplace,
 s,t):
```

and then plotting the output as a function of time over the first 10,000 switching periods. The result we obtain is as follows (the superimposed diamond plot of Figure 8.23 represents the final average dc output value):

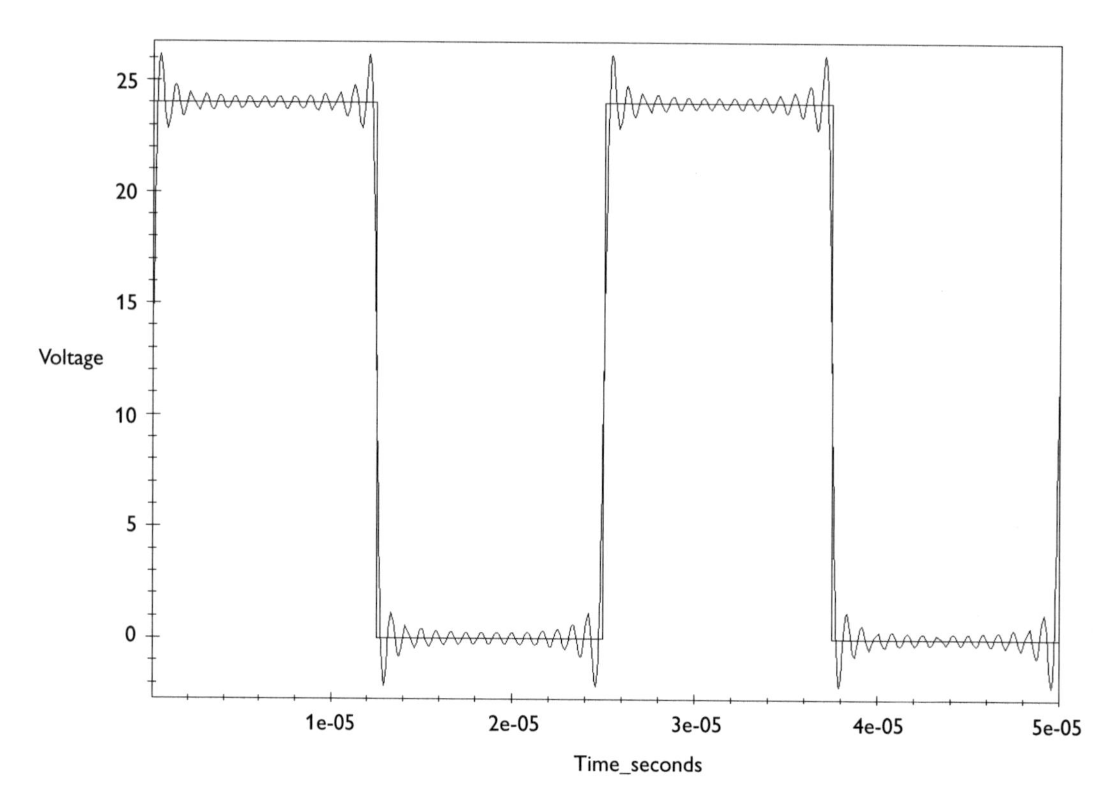

Figure 8.22 Comparison of the Fourier approximation with the exact input waveform (30 harmonics plus dc).

```
Output_Plot := plot(Output_Voltage_Time,t=0..10000*T,
 color=black):
Output_DC_Plot := plot(Ao,t=0..10000*T,color=black,
 style=point,symbol=diamond):
display({Output_Plot, Output_DC_Plot},axes=boxed,
 labels=[Time_seconds,Voltage]);
```

Figure 8.23 shows a typical underdamped second-order effect to a 12V step input. As stated earlier, the 12V average is the dc term associated with the Fourier series approximation.

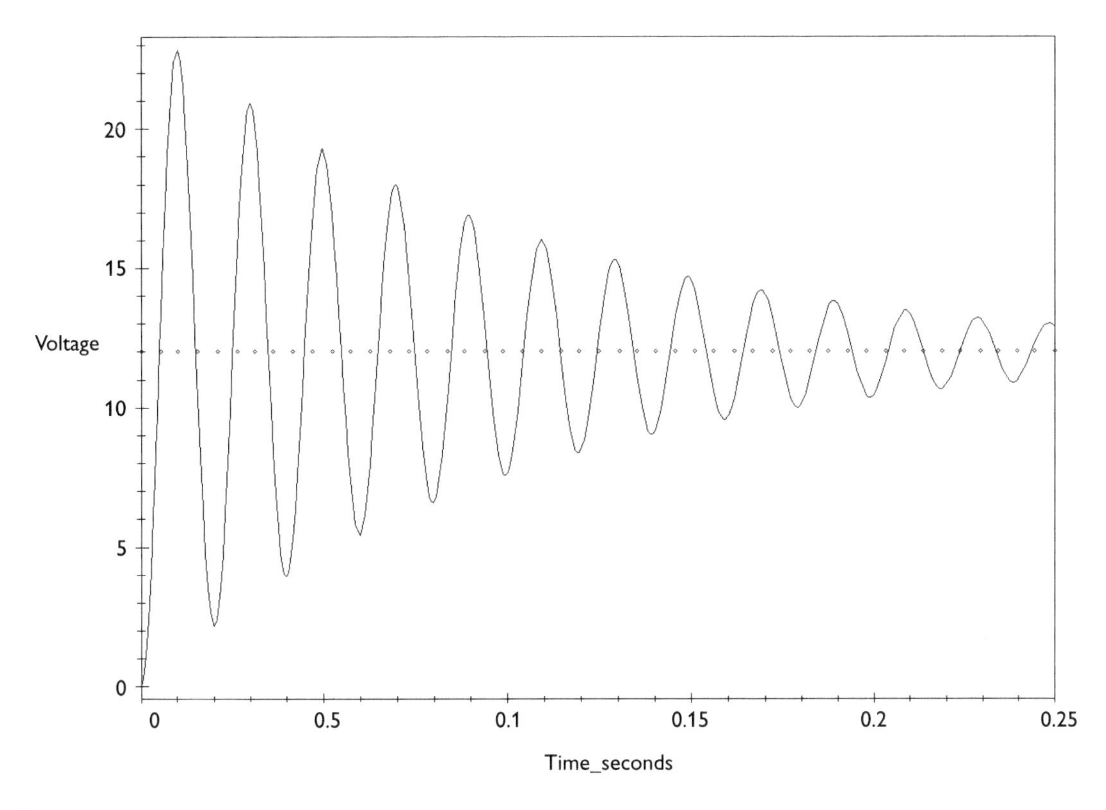

Figure 8.23 Output voltage derived from Fourier series analysis (Vin = 24V, L = 10 mH, C = 1000 μF, R = 50 ohms, duty cycle = 50%, frequency = 40 kHz).

In the exact method, we derived the boundary voltages to solve the output waveform for any period. This method also gave us the peak-to- peak ripple by simply subtracting the two boundary voltages. The output ripple value can be determined from the Fourier method, but only after we perform an RMS (`Root-Mean-Square`) derivation of the approximated output function (`Output_Voltage_Time`).

We start by going well out in time, where the output has settled down, and integrating over one period. In this case, the choice of translating out to 100,000 switching periods was used, which, at 40 kHz, equates to about 2.5 sec after startup. Further increases in this period measure decreased the ripple computation but did not change the ripple value appreciably.

```
Output_RMS_Ripple := evalf(sqrt(1/T*(int((Output_
 Voltage_Time-Ao)^2,t=100000*T..100001*T))));
Output_PP_Ripple := evalf(2*Output_RMS_Ripple*sqrt(2));
```

$$Output\_RMS\_Ripple := .00001711662531$$
$$Output\_PP\_Ripple := .00004841312730$$

Consequently, the peak-to-peak ripple with the Fourier method was about half the value computed using the exact method (peak-to-peak ripple = 94 μV). The reason for the lower ripple value in the Fourier method is due to the fact that we only used the first 10 harmonics, whereas in the exact method, all harmonics were present, hence leading to a higher residual ripple value.

At this point, we can play around with some component values and watch the effects at the output. For instance, let's change the duty cycle, inductor, and capacitance values to the following:

$$\text{alfa} \rightarrow .25$$
$$C \rightarrow 10\,\mu\text{F}$$
$$L \rightarrow 1\,\text{mH}$$

Starting the Maple session:

```
with(inttrans):
with (plots):
```

The component values are entered:

```
Switching_Freq := 40*10^3:
Vin := 24:
L := 10^(-3):
C := 10^(-5):
R := 50:
alfa := .25:
T := 1/Switching_Freq:
```

The Fourier coefficients are computed:

```
Ao := 1/T*int(Vin,t=0..alfa*T):
An := 2/T*int(Vin*cos(2*n*Pi*t/T),t=0..alfa*T):
Bn := 2/T*int(Vin*sin(2*n*Pi*t/T),t=0..alfa*T):
```

The Fourier approximation using the first 10 harmonics is computed:

```
Vin_Fourier := Ao+sum(An*cos(2*n*Pi*t/T)+Bn*
 sin(2*n*Pi*t/T),n=1..10):
```

Compare the Fourier approximation with the exact PWM input waveform (Figure 8.24):

```
Fourier_Plot := plot(Vin_Fourier,t=0..2*T,color=black):
piece_1 := piecewise(t=0,Vin,0):
piece_2 := piecewise(t=alfa*T,-Vin,0):
piece_3 := piecewise(t=T,Vin,0):
piece_4 := piecewise(t=T*(1+alfa),-Vin,0):
Exact_Plot :=plot(piece_1+piece_2+piece_3+piece_4,
 t=0..2*T,color=black):
display({Fourier_Plot,Exact_Plot},axes=boxed,labels=
 [Time_seconds,Voltage]);
```

The the appropriate Laplace transforms are taken and the time-domain output is computed:

```
Vin_Laplace := laplace(Vin_Fourier,t,s):
LPF_Transfer := 1/(L*C*s^2+L/R*s+1):
Output_Voltage_Laplace := LPF_Transfer*Vin_Laplace:
Output_Voltage_Time : = invlaplace (Output_Voltage_Laplace,
 s,t):
```

We then plot (Figure 8.25) the output's time-domain response (line) along with the average value (diamond):

```
Output_Plot :=plot(Output_Voltage_Time,t=0..100*T,
 color=black):
Output_DC_Plot :=plot(Ao,t=0..100*T,color=black,
 style=point,symbol=diamond):
display({Output_Plot, Output_DC_Plot},axes=boxed,
 labels=[Time_seconds,Voltage]);
```

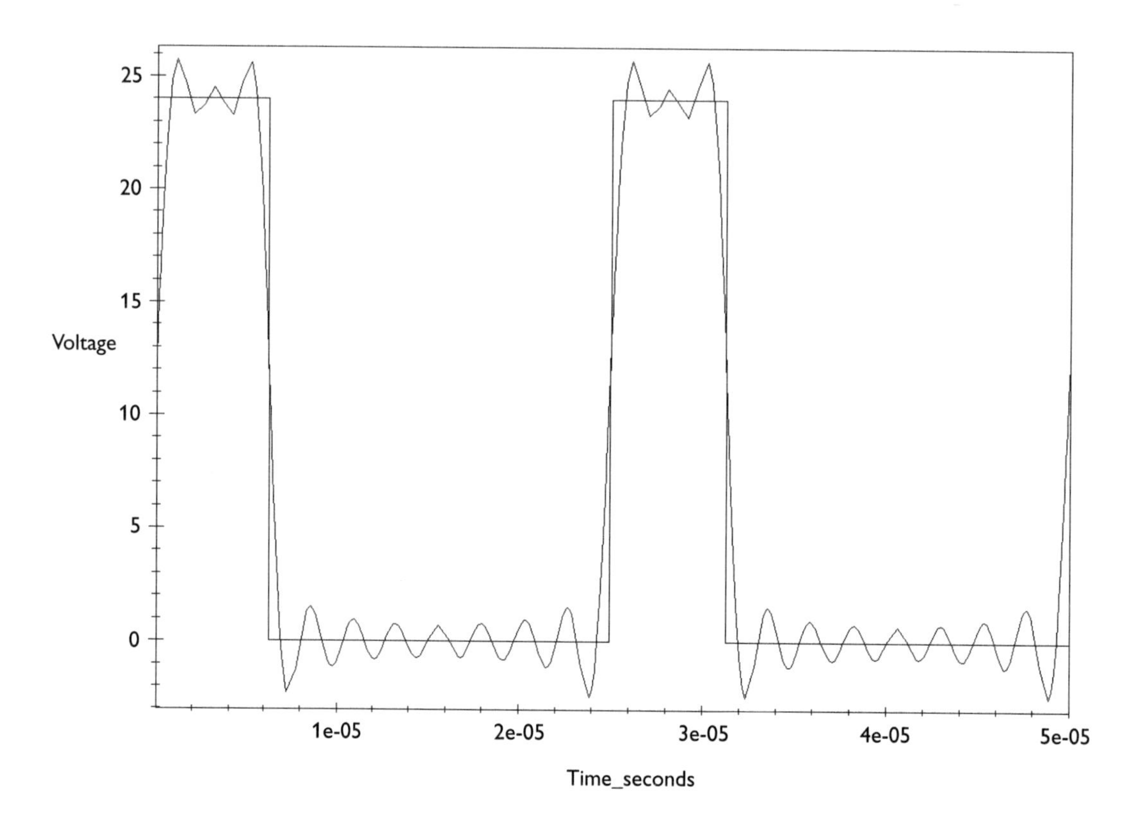

Figure 8.24 Comparison of Fourier approximation to exact input waveform (10 harmonics plus dc with a 25% duty cycle).

We now compute the output ripple's RMS and peak-to-peak value:

```
Output_RMS_Ripple := evalf(sqrt(1/T*(int
  ((Output_Voltage_Time-Ao)^2,t=10000*T..10001*T))));
Output_PP_Ripple := evalf(2*Output_RMS_Ripple*sqrt(2));
```

$$\text{Output\_RMS\_Ripple} := .01230936550$$
$$\text{Output\_PP\_Ripple} := .03481614326$$

The peak-to-peak ripple is comparable to the exact method computation (peak to peak = 42 mV), though, as before, it is lower with the Fourier method due to the presence of only the first 10 harmonics. However, what is interesting here is that the average value appears to be about

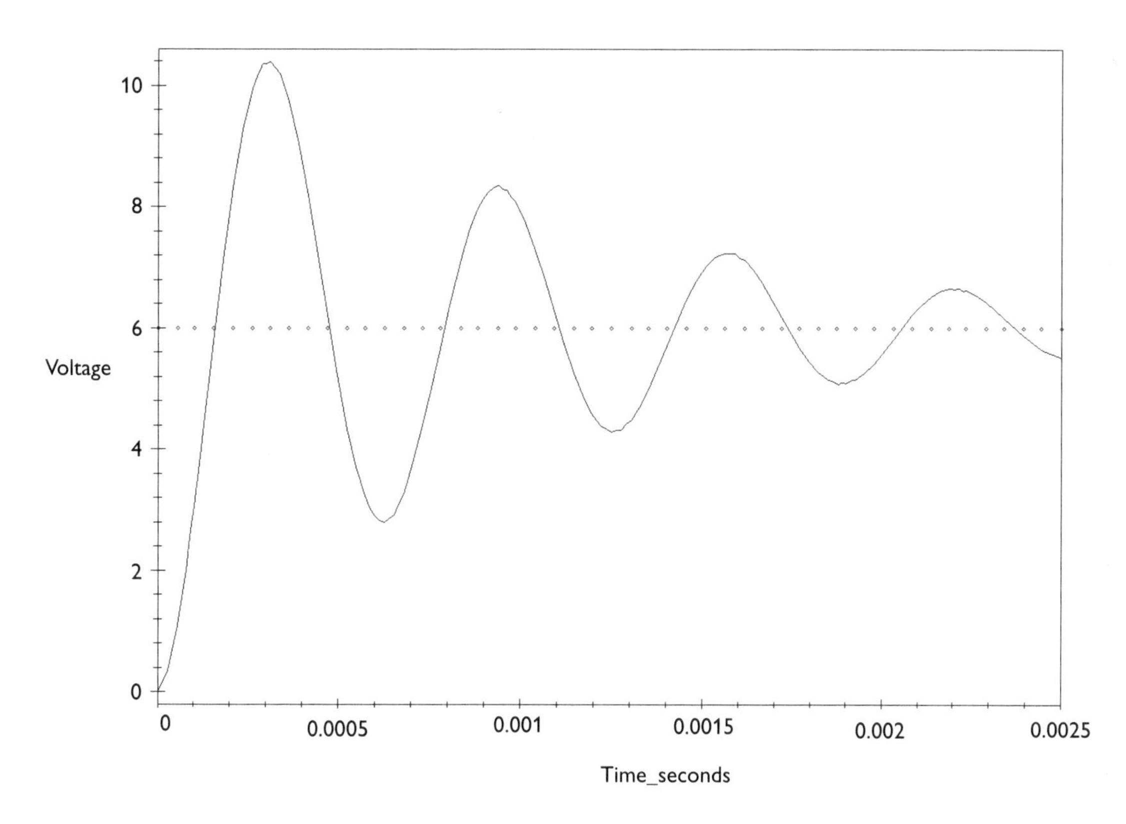

Figure 8.25 Output voltage derived from Fourier series analysis (*Vin* = 24V, *L* = 1 mH, *C* = 10 μF, *R* = 50 ohms, duty cycle = 25%, frequency = 40 kHz).

6.0V (25% of 24V), whereas the exact method gave us around 2.41V. Why is this? From Figure 8.18 we saw that some sort of output voltage compression was taking place, and Figure 8.19 indicated the compression was dependent on the existence of the inductor. Consequently, the phenomenon is not strictly a function of the duty cycle, but the interactive presence between the inductor and output capacitor (a second-order effect). Therefore, is the Fourier method somewhat incorrect? No, but understand that the approximation eliminates an infinite number of harmonics that are really present in the input to the second-order filter. Remember, the lower significant harmonic contents of a PWM waveform are maximal at a 50% duty cycle and, by symmetrical reasoning, minimal at duty cycles above and below the 50% point. Consequently, this approximation will contribute further to the output's compressed transfer and, in this case, to the nonlinear output voltage versus duty cycle transfer function.

REFERENCES

[1] Pressman, A. I., *Switching and Linear Power Supply, Power Converter Design*, Hayden Book Co., 1977.

[2] Chryssis, George, *High-Frequency Switching Power Supplies*, New York: McGraw-Hill, 1984.

[3] Close, Charles M., *The Analysis of Linear Circuits*, New York: Harcourt, Brace & World, 1966.

[4] Chua, Leon O., *Introduction to Nonlinear Network Theory*, New York: McGraw-Hill, 1969.

Appendix A

Define the state equation

$$\dot{\mathbf{x}} = \mathbf{A}(t)\mathbf{x}$$

where $\mathbf{A}(t)$ contains one or more time-dependent elements. Then looking at a reduced scalar formulation (i.e., dealing with only one state variable),

$$\dot{x} = a(t)x$$

A solution to this scalar form is

$$x(t) = e^{b(t)}x(\tau)$$

where τ is defined as the initial time condition. Letting $\tau = 0$ will not invalidate this analysis, therefore we rewrite the scalar solution form as

$$x(t) = e^{b(t)}x(0)$$

and

$$b(t) = \int_{\tau=0}^{t} a(\lambda)d\lambda$$

Substituting this result into the vector/matrix form, we obtain:

$$\mathbf{x(t)} = \left[e^{\int_0^t \mathbf{A}(\lambda)d\lambda}\right]\mathbf{x(0)}$$

However, if we substitute this expression into $\dot{\mathbf{x}} = \mathbf{A}(t)\mathbf{x}$, we see that this expression holds true if and only if

$$\frac{d}{dt}e^{\mathbf{B}(t)} = \frac{dB(t)}{dt}e^{\mathbf{B}(t)}$$

where

$$\mathbf{B}(t) = \int_0^t \mathbf{A}(\lambda)d\lambda$$

Unfortunately, these last two mathematical statements are rarely true *except* when the **A** matrix is either a constant or diagonal matrix. In fact, it can be shown that the **A** matrix must have the following commutative property [1] if we are to use the conventional scalar approach for the state variable solutions:

$$\mathbf{A}(t_1)\mathbf{A}(t_2) = \mathbf{A}(t_2)\mathbf{A}(t_1)$$

Needless to say, when dealing with matrices, this equation can only be true for a constant coefficient or for diagonal matrices.

Hence, if we desire to solve this type of state variable system, then let's perform a diagonalization of the **A** matrix using what is known as a *similarity* transformation. The similarity transformation requires the user to obtain the eigenvectors and redefine the states as follows:

$$\mathbf{x} = \mathbf{Pq}$$

Then assuming a constant coefficient transformation matrix, **P**,

$$\dot{\mathbf{x}} = \mathbf{P}\dot{\mathbf{q}}$$

substituting this expression into the state space formulation,

$$\dot{\mathbf{x}} = \mathbf{Ax} + \mathbf{Bu} \xleftrightarrow{\text{Transforms}} \mathbf{P}\dot{\mathbf{q}} = \mathbf{A}(\mathbf{Pq}) + \mathbf{Bu}$$

and premultiplying both sides by $\mathbf{P}^{-1}$ yields

$$\dot{\mathbf{q}} = \mathbf{P}^{-1}\mathbf{APq} + \mathbf{P}^{-1}\mathbf{Bu}$$

The matrix expression $\mathbf{P}^{-1}\mathbf{AP}$ is exactly of the form

$$\mathbf{M}^{-1}\mathbf{AM} \rightarrow \mathbf{D}$$

Hence,

$$\mathbf{P} = \mathbf{M}$$

where **M** is defined as the *modal* matrix and is composed of the **A**'s eigenvectors. The **D** matrix product is special and is denoted as the *spectral* or *canonic* matrix associated with the **A** matrix. By way of definition, the matrix **A** is said to be *similar* to a matrix **D** if there exists a nonsingular matrix **M** such that $\mathbf{M}^{-1}\mathbf{AM} \rightarrow \mathbf{D}$ holds true.

The **D** matrix has the property

$$\mathbf{D} = \begin{bmatrix} \lambda_1 & 0 & 0 & \cdot & 0 \\ 0 & \lambda_2 & 0 & \cdot & 0 \\ 0 & 0 & \lambda_3 & \cdot & \cdot \\ \cdot & \cdot & \cdot & \cdot & 0 \\ 0 & 0 & \cdot & 0 & \lambda_N \end{bmatrix}$$

where it only has diagonal elements, which are the **A** matrix's distinct eigenvalues (nonrepeated characteristics roots). Consequently, substituting this expression into our transformed state space formulation yields

$$\dot{\mathbf{q}} = \mathbf{Dq} + \mathbf{M}^{-1}\mathbf{Bu}$$

Then let's redefine this state equation in more familiar terms of **A** and **B**:

$$\dot{\mathbf{q}} = \mathbf{A_N q} + \mathbf{B_N u}$$

where $\mathbf{A_N} = \mathbf{M}^{-1}\mathbf{AM}$ and $\mathbf{B_N} = \mathbf{M}^{-1}\mathbf{B}$. At this point, the user has to remember what the new state variables, **q**, represent in terms of the original state variables, **x**, i.e.,

$$\mathbf{x} = \mathbf{Pq} = \mathbf{Mq}$$

Hence, on obtaining the solution set in **q**, one resubstitutes and premultiplies by the modal matrix results in the solutions for the original states. Also, the output matrix formulation was given as

$$\mathbf{y} = \mathbf{Cx} + \mathbf{Du}$$

Hence,

$$\mathbf{y} = \mathbf{C(Mq)} + \mathbf{Du} \rightarrow \mathbf{C_N q} + \mathbf{Du}$$

where $\mathbf{C_N} = \mathbf{CM}$.

The process of obtaining the transition matrix and continuing onto the final complete solution to the states is identical to the process described in Chapter 5.

Reference

[1] Hirsch, M., and S. Smale, *Differential Equations, Dynamical Systems, and Linear Algebra,* New York: Academic Press, 1974.

Appendix B

Data Structure Keyword*	User Plot Option	Description
STYLE(POINT | LINE | **PATCH** | PATCHNOGRID | ***HIDDEN*** | CONTOUR | *PATCHCONTOUR*).	style= POINT | LINE | PATCH | PATCHNOGRID| *HIDDEN* | CONTOUR | *PATCHCONTOUR.*	Sets the plot style of the nontext objects in the image. The options given in italics are for 3-D plots.
THICKNESS(**0** | n), where n is a positive integer.	thickness=0 | n.	Controls the thickness of any line segments, other than the axes, resulting from a graphics primitives in the image. The higher the value of n the thicker the line. The special value 0 corresponds to the default thickness for the selected output device.
LINESTYLE(**0** | n), where n is a positive integer.	linestyle=**0** | n.	Sets the dash pattern to be used when drawing line segments in the image. The special value 0 selects the device default whereas the value 1 selects a solid line.
COLOUR† (RGB, r, g, b) | COLOUR(HUE, c) | COLOUR(HSV, c, s, h) | *COLOUR(type)*, where the variables r, g, b, c, s and h are floating-point numbers in the range 0..1 and type can be ***XYZSHADING*** | *XYSHADING* | *ZSHADING* | *ZHUE* | *ZGREYSCALE*.	color=expr, where expr is either a color name‡ (red, blue, green, etc.), a value (0 .. 1), a procedure returning a valid color desciptor, or a valid color expression.	Specifies the color of the graphics object. The options in italics are applicable to the rendering of three-dimensional surfaces and apply color information according to a particular attribute of the plot in question.
AXESSTYLE(BOX | FRAME | NORMAL | NONE | **DEFAULT**).	axesstyle= BOX | FRAME | NORMAL | NONE.	Selects the axes style for the image.

* All of the valid arguments to each structure and option are listed with the default being emboldened. Where a single argument must be selected from a list the elements of the list are seperated using |, for example, op_1 | op_2 | op_3 is shorthand for select op_1 or op_2 or op_3.

† Due to Maple's Canadian origins, many spellings are UK-English instead of American-English.

‡ The color name must be known to Maple. For more help see ?color.

Data Structure Keyword*	User Plot Option	Description
AXESTICKS(xvals, yvals, *zvals*), where xvals, yvals and *zvals*, in the three-dimensional case can be **N**, | [n_1, n_2, .., n_i] | [eqn_1, eqn_2, .., eqn_i] | **DEFAULT**, where N is an integer, n_i are numbers and eqn_i have the form n_i=$label_i$. The value $label_i$ is a string.	xtickmarks=xvals and ytickmarks=yvals, or axisticks=[xvals, yvals, zvals] in the three-dimensional case.	Allows the number, location and labeling of the tick marks on an axis to be specified. An integer value causes the axis ticks to be chosen such that there is at least the number requested, a list of numbers causes the ticks and labels to be those specified in the list and no more while a list of equations produces tick marks at the position specified by the left-hand side of the equation with the label specified by the right-hand side.
SCALING(DEFAULT | CONSTRAINED | **UNCONSTRAINED**).	scaling=CONSTRAINED | UNCONSTRAINED.	Scaling is used to determine whether the x and y axes scaling is the same: CONSTRAINED, independent: UNCONSTRAINED. Selecting DEFAULT normally is the same as selecting UNCONSTRAINED.
SYMBOL(BOX | CROSS | CIRCLE | POINT | DIAMOND | **DEFAULT**).	symbol=BOX|CROSS|CIRCLE|POINT|DIAMOND.	Allows the point plot symbol accordingly.
TITLE(**Null string** | title string).	title=title string.	Enables a title to be set for the plot.
AXESLABELS(**Null string** | x-label string, **Null string** | y-label string).	axeslabels=**Null string** | x-label string, **Null string** | y-label string.	Enables axes labels to be set for the plot. If used both label specifications must be present.

* All of the valid arguments to each structure and option are listed with the default being emboldened. Where a single argument must be selected from a list the elements of the list are seperated using \|, for example, op_1 \| op_2 \| op_3 is shorthand for select op_1 or op_2 or op_3.

Data Structure Keyword[*]	User Plot Option	Description
FONT(family, typeface, size) where family can be TIMES | **COURIER** | HELVETICA, typeface can be ROMAN | **DEFAULT** | BOLD | ITALIC | BOLDITALIC | OBLIQUE | BOLDOBLIQUE and size is the point size.[§]	font=[family, typeface, size], axesfont=[family, typeface, size], labelfont=[family, typeface, size], titlefont=[family, typeface, size].	Allows the font, used in rendering TEXT objects, to be specified. The font specification of a text object is set either directly using FONT or via the plot options shown.[‖]
VIEW($x_1..x_2$ | **DEFAULT**, $y_1..y_2$ | **DEFAULT**), where x_i and y_i are numbers.	view=[x1..x2, y1..y2].	Allows regions of a plot to be viewed. The ranges $x_1..x_2$ and $y_1..y_2$ specify the subrange of the x-y plane that is to be displayed. The special value DEFAULT forces a view to be selected such that all of the elements in the plot object are displayed.
GRIDSTYLE(TRIANGULAR | **RECTANGULAR**).	gridstyle=TRIANGULAR | **RECTANGULAR**.	Allows either a rectangular or triangular grid to be used when a three-dimensional surface is rendered.
ORIENTATION(**45** | θ, **45** | ϕ), where . θ and θ are angles of rotation and inclination respectively in degrees.	orientation=[theta,phi].	Set the view angle of a three-dimensional plot.
AMBIENTLIGHT(r, g, b), where the entries r, g and b have numeric values between **0** and 1.	ambientlight=[r, g, b].	Sets the ambient light of a three-dimensional plot in terms of the intensity of the red, green, and blue components of the light.

* All of the valid arguments to each structure and option are listed with the default being emboldened. Where a single argument must be selected from a list the elements of the list are seperated using |, for example, op_1 | op_2 | op_3 is shorthand for select op_1 or op_2 or op_3.

§ For more information see ?plot[options].

‖ It should be noted that not all combinations are possible.

Data Structure Keyword*	User Plot Option	Description
LIGHT(**45** | θ, **45** | ϕ, r, g, b), where θ and ϕ specify the direction to the light in polar coordinates (angles specified in degrees) and r, g and b have numeric values as in AMBIENTLIGHT defined above.	light=[θ, ϕ, r, g, b].	Allows the direction and intensity of a directed light shining on a three-dimensional surface to be specified.
LIGHTMODEL(USER | **LIGHT_1** | LIGHT_2 | LIGHT_3 | LIGHT_4).	lightmodel= USER | LIGHT_1 | LIGHT_2 | LIGHT_3 | LIGHT_4.	Allows a lighting scheme, from those available, to be selected. If USER is specified the light definitions given in the LIGHT and AMBIENTLIGHT options are used.

* All of the valid arguments to each structure and option are listed with the default being emboldened. Where a single argument must be selected from a list the elements of the list are seperated using |, for example, op_1 | op_2 | op_3 is shorthand for select op_1 or op_2 or op_3.

Glossary

Term	Description	Syntax
!	The factorial of a expression.	expr! 5! = 120
"	Shorthand for the previously evaluated expression. The evaluation stack is three deep (i.e., ", "", and """ can be used).	a:=2: " 2
$	The infix form of seq.	expr$range a$4 a^n$n=1..4

Term	Description	Syntax
&^	Part of the liesymm package, this infix operator computes the wedge product of its arguments.	expr1&^expr2
-	The arrow operator used in defining functions.	f:=(args)-body f:=x-x^2 f:=(a,b)-a*b
::	Infix operator used to assign a type definition to a procedure argument.	proc(argn::typen) proc(x::numeric)
?	Return the help page for the specified topic or function.	?topic ?topic[sub-topic] ?help ?plots[animate]
??	Return the calling sequence for the specified function.	??function ??package[function] ??sin ??plots[animate]
???	Return the examples section for the specified function.	???function ???package[function] ???sin ???plots[animate]
@	The repeated composition operator that applies the expression f to the expression g *n* times.	(f@@n)(g) (sin@@2)(x) a@@y
A matrix	Used in state-space analysis, the elements of the A matrix are determined by the system's dynamics.	
abs	Calculate the absolute value of an expression.	abs(expr) abs(-2) abs(n)
alias	Create an alias to an expression.	alias(aliasi=expri) alias(sin=s) alias(a_matrix=linalg[matrix])

Term	Description	Syntax
aliasing	Aliasing is caused when a signal is sampled at a rate less than the Nyquist rate. This causes the higher frequency components of the signal to be folded down or aliased to lower frequences. This effect accounts for the wagon wheels "running" backwards in old films.	
allvalues	Evaluate all of the possible values of an expression containing RootOfs.	allvalues(expr) allvalues(expr, dependent)
analytic solution	A solution that has been obtained by resorting to numerical methods.	
animate	Create a two-dimensional animation of the functions over the range r with the frame parameter t. This function is found in the plots package.	animate(funci, r, t, opts) animate(sin(x+p), x=0..10, p=0..5)
animate3d	Create a three-dimensional animation of the functions over the ranges r1, r2 with the frame parameter t. This function is found in the plots package.	animate3d(funci, r1, r2, t, opts) animate(sin(x*y+p), x=0..10, y=0..3, p=0..5)
arbitrary precision arithmetic	The ability to perform floating point calculations with a selected number of digits.	
args	A global variable containing the list of arguments passed to the current procedure.	
array	A Maple array is defined by setting the row and column dimensions and the elements.	array(dim) array(dim, elems) array(dim1, dim2) array(dim1, dim2,, elems) array(0..2) array(0..1, 1..2) array(1..2, [1,2]) array(1..2, 1..2, [[a, b], [1, 2]])
assigned	A Boolean test to determine whether a name has a value other than its name.	assigned(name) assigned(f)

Term	Description	Syntax
attenuator	A device that reduces a signals amplitude by a preset ratio.	
axis jw	Also called the frequency axis, it is the vertical axis on the s-plane.	
B matrix	Used in state-space analysis, the elements of the B matrix represent the input gains of the system.	
Bilinear transform	The bilinear transform is conformal mapping which translates the *j* axis of the s-plane onto the unit circle of the z-plane.	$s = \frac{1 - \frac{1}{z}}{T}$ $s = 2\frac{1 - z}{T(1 + z)}$
bit depth	The number of bits available to represent all of the intensity levels in an image.	
Boolean valued expression	An expression that can only take true or false.	
C matrix	Used in state-space analysis the elements of the C matrix represent the output gains of the system.	
controllable	State matrices in a form that are well suited to the design of state variable feedback controllers.	
Jordan	The system is decoupled, as A matrix is diagonal.	
observable	In the observable canonical form, the system coefficients appear in the first column of the A matrix.	
canonical form	An expression is said to be in a canonical form when it is the most concise form.	
CAS	Computer algerba system.	

Term	Description	Syntax
cat	The concatonation function.	cat(expri) cat('a', ' b', ' 1')
characteristic equation	The characteristic equation in l of a matrix is $\mid A-I\lambda\mid=0$.	
coeff	Return the coefficient of the term in xn in the polynomial p.	coeff(p,x) coeff(p,x,n) coeff(p,x^n) coeff(x+x^2, x, 2) coeff(x+x^2, x^2) coeff(x+x^2, x)
coeffs	Return all of the coefficients of the polynomial in x.	coeffs(p, x)
color function	A function that determines the coloring of an image.	
color, COLOUR	The wrapper for an image's color information.	COLOUR(format, data) COLOUR(RGB,0,1,0)
color, default map	Maple's default color map for coloring a surface. Color is applied at a point on the surface as a function of that point's x-y-z coordinates.	
color, false	The application of unnatural colors to an image for the purpose of image enhancement.	
color, HVS	The Maple keyword determing that the color information is in the hue-intensity-saturation format.	
color, mapping functions	See color function.	
color, plane, 15	A single layer of color information used in a composite color image.	

Term	Description	Syntax
color, primary	A set of pure colors from which all other colors can be made.	RGB CYM
color, RGB	The Maple keyword determing that the color information is in the red-green-blue format.	
conditional statement	See if.	
conformal mapping	A function that defines the mathematical relationship between two systems.	
conjugate	Return the conjugat of a complex number.	conjugate(expr) conjugate(a+I*b) conjugate(3/4+I*0.56)
constants	A global variable containing the constants known to Maple.	
convert, +	Add the elements of a list or set.	convert(expr, '+') convert([a, b, c], '+') convert({1, 2, 3}, '+')
convert, confrac	Transform an expression into its continued fraction form.	convert(expr, confrac, var) convert(1/(s^2*(s+1)), confrac, s) convert(exp(y), confrac, y)
convert, exp	Convert a trigonometric expression into one comprising exponentials.	convert(expr, exp) convert(sin(y)+cos(y), exp)convert(sin(y)+cos(y), exp) convert(tan(t), exp)
convert, helper functions	Conversion helper functions are implemented by the user to extend the convert function. Helper functions are either functions or procedures but all conform to the same naming convention: "convert\convert_tag."	'convert\convert_tag':= ...

Term	Description	Syntax
convert, parfrac	Decompose a rational function f in the variable var into partial fractions.	convert(expr, parfrac, var) convert(1/(s^2*(s+1)), parfrac, s) convert(exp(y)/(y+1), parfrac, y)
convert, polynom	Convert a series data structure into a polynomial by removing the order term.	convert(expr, polynom) convert(series(exp(x), x), polynom)
convert, rational	Convert a floating-point number to an approximate rational number.	convert(num, rational, opts) convert(3.142, rational) convert(3.142, rational, 5)
convert, trig	Convert all exponentials in expr to trigonometric and hyperbolic trigonometric functions.	convert(expr, trig) convert(exp(I*t)+I*exp(t))
convert	The mechanism by which Maple objects and data structures can be converted to different formats.	
CURVES	A wrapper inside a plot structure containing data points that are to be plotted as a curve.	CURVE(data) CURVE([[0, 0], [10, 100]])
D matrix	Used in state-space analysis, the D matrix is the disturbance matrix.	
D	The Maple differential operator.	
dc gain	The gain of a system at zero hertz.	
DC	A signal with a frequency of zero hertz.	
defform	Define the basic variables used in a computation or define the exterior derivative of an expression using the equations eqni.	defform(eqni) defform(a=const,b=scalar)
degree	Return the degree (highest power of the free variable) of a polynomial.	degree(poly, var) degree(x^7-y^5*x-1=0, x)

Term	Description	Syntax
denom	Return the denominator of a quotient.	denom(expr) denom((1+s)/(1-s)
DEplot	See ODE, odeplot.	
derivative, finite difference	Approximate the derivative of a function at a point (x, y) using the horizontal and vertical differences between it and a previous point; $\frac{dy}{dx} \approx \frac{\text{inc in y}}{\text{inc in x}}$	
DESol	A representation of a solution of a differential equation.	
DEtools	The Maple package containing the differeintial equation manipulation utilities.	
dfieldplot	Found in the Detools package, this plots the direction field of one- or two-dimensional systems of differential equations over the range t. This function uses numerical techniques.	dfieldplot(eqn$_i$, var$_i$, t, opts dfieldplot(y^2*sin(x),[x,y],-5..5);)
diff	Return the derivative, with respect to the variable var, of an expression.	diff(expr, var) diff(sin(x), x)
difference equation	See recurrence relation.	
differential operator	See D.	
difforms	The Maple package containing differential equation manipulation utilities.	
digital control	The method of manipulating the behavior of dynamical systems using digital techniques.	
digital signal processing	The method of manipulating continuous and discrete signals using digital techniques.	

Term	Description	Syntax
discont	The optional argument to plot allowing any discontinuities encountered to be displayed.	plot(func, range, discont=true, opts) plot(tan(x), x=-10..10, discont=true)
discrete transfer function	The expression linking the input and output of a discrete system.	
display	Found in the plots package display is used to rerender plot structures.	display(plot, opts) display({ploti}, opts) display(pp) display({plot(sin), plot(cos)})
ditto	See ".	
dverk78	A setting for dsolve's optional argument method which sets the method to be used to be the gear method.	dsolve({diff(y(t),t)=y(t), y(0)=3}, y(t), method=dverk78)
dsolve, gear	A setting for dsolve's optional argument method, which sets the method to be used to be the gear method.	dsolve({diff(y(t),t)=y(t), y(0)=3}, y(t), method=gear)
dsolve, lsode	A setting for dsolve's optional argument method, which sets the solution method to lsode.	dsolve({diff(y(t),t)=y(t), y(0)=3}, y(t), method=lsode)
dsolve, mgear	A setting for dsolve's optional argument method, which sets the solution method to mgear.	dsolve({diff(y(t),t)=y(t), y(0)=3}, y(t), method=mgear)
dsolve, numeric	A setting for dsolve's optional argument method, which sets the solution method to numerical.	dsolve({diff(y(t),t)=y(t), y(0)=3}, y(t), numeric)
dsolve, Runga-Kutta	The default numeric method used by dsolve.	dsolve({diff(y(t),t)=y(t), y(0)=3}, y(t), numeric)
dsolve, taylorseries	A setting for dsolve's optional argument method, which sets the solution method to Taylor series.	dsolve({diff(y(t),t)=y(t), y(0)=3}, y(t), method=taylorseries)

Term	Description	Syntax
dsolve	Maple's ordinary differential equation solver.	dsolve({eqns}, {vars}, opts) dsolve(diff(y(t),t)=y(t), y(t)) dsolve({diff(y(t),t)=y(t), y(0)=3}, y(t))
DSP	Digital signal processing.	
dynamical system	A system that contains both energy storage and enegry disipation elements.	
eigenvalues	Values indicating the behavior of a dynamical system.	
erf	Return the error function of expr.	erf(expr) erf(Pi/2)
ERROR	Force an error condition resulting in control being returned to the top-most level.	ERROR() ERROR(expr) ERROR('this is wrong') ERROR(0)
eval	Force the evaluation of an expression.	eval(expr) eval(sin)
evalf	Evaluate expr as a floating point quantity.	evalf(expr) evalf(expr, digits) evalf(Pi/2) evalf(Pi/2, 500)
evalm	Evaluate expr as a matrix.	evalm(expr) evalm(A &* B)
example	See ???.	
exp	Return the exponential of expr.	exp(expr) expr(1)
expand	Expand the subterms in expr.	expand(expr) expand((x+1)*(a+b))
exterior derivative	Creates an explicit differential form of a multivariate expression.	

Term	Description	Syntax
factor	Factorize an expression.	factor(expr) factor((x^2+1))
Fehlberg four-five order Runga-Kutta	The default method used by Maple to solve ODEs numerically.	
filter, difference	A filter whose output is the difference between the previous two input data values.	
filter, digital	A filter, predominantly microprocessor based, which operates on numeric representations of continuous signals.	
filter, exponential	A digital filter whose next output is dependent upon the next data and previous output values.	
filter, forcing function	The input signal to the filter.	
filter, hystersis	See filter memory.	
filter, linear	A filter whose output is a linear weighted combination of its input and output values.	
filter, low-pass	A filter that attenuates all frequency components above a certian cut-off frequency.	
filter, memory	The filter's ability to store previous output values.	
filter, moving average	The filter output is the average value of the windowed data.	
filter, moving median	The filter output is the median value of the windowed data.	

Term	Description	Syntax
filter, weight	Used with the exponential filter, it determines the level of emphasis applied to previous filter output values.	
filter, window size	This number of data points manipulated at once to compute the filter's next output.	
filter, window	A method by which a fixed number of data points are isolated, prior to the filter's output being computed, from a stream of data.	
floor	Return the nearest integer that is less than or equal to expr.	floor(expr) floor(3.4)
FONT	A wrapper containing font information in a plot structure.	FONT(family, face, size)
for	One of Maple's looping constructs.	for var to val do expr od for var to val by inc do expr od for var from init to val by inc do expr od for var in expr1 do expr2 od for x to 10 do x+1 od for y to 4 by 0.1 do y^2 od for z from -2 to 2 by .5 do z od for a in [1,2,3] do a^2 od
force balance equation	For a body in equilibium, the sum of the forces acting upon it must equal zero.	
forcing function, discontinuous	A forcing function containing discontinuities, such as a step or a pulse.	
forcing function, pulse train	A forcing function comprising a sequence of pulses.	
forcing function, pulse	A forcing function comprising a single pulse.	

Term	Description	Syntax
forcing function, step	A forcing function comprising step change.	
forcing function	The function that is used to excite a dynamical system via its input terminal.	
friendly files	Text files that only contain valid Maple expressions.	
fsolve	Maple's numeric equation solver.	fsolve(eqn,) fsolve(eqn, var) fsolve(x+1=0) fsolve(sin(y)=cos(y), y)
function, body	The expressions comprising the function.	
function, pure	A function without a name.	
function	A Maple expression capable of manipulating expressions.	f:=x-x^2 f:=(a,b)-a*b
fundamential frequency	The lowest frequency component present in a complex waveform.	
GAMMA	Return either the complete or the incomplete G function.	GAMMA(expr) GAMMA(expr, expr) GAMMA(2.3) GAMMA(2.3+I, 3/4)
greyscale, default	The default map used by Maple to represent a color image in monochrome, invoked by setting shading=GREYSCALE.	
greyscale	A map often used to represent a color image in monochrome.	
GRID	Plot structure representing a surface defined by a uniform sampling over an aligned rectangular region.	GRID(a..b,c..d,[[z11,...z1n],[z21,...z2n],...[zm1...zmn]]) GRID(1..2, 1..2, [[1, 2], [2, 2], [1, 4], [4, 8]])

Term	Description	Syntax
has, exp	Determine whether an expression contains any exponential components.	has(expr, exp) has(sin(f)+5, exp) has(exp(sin(y))*y, exp)
Heaviside	A step function with a value of unity for t < 0 and zero for t.	
help	The procedural form of ?.	help(topic) help(topic, sub-topic)
histogram	A graphical representation of quantized data showing the data values and their frequencies.	
hyperlinks	Live links embeded in a document that allow easy navigation through it.	
Hz	The unit of frequency, hertz.	
I	The Maple constant corresponding to the imaginary constant Ã-1.	
if ... fi	The conditional construct, where test is a Boolean valued expression.	if test then expr fi if test then expr1 else expr2 fi if test1 then expr1 elif test2 then expr2 else expr3 fi
ifactor	Return the integer factors of expr.	ifactor(expr) ifactor((2^64+4)/5)
image processing	An operation whereby an image is altered to produce a new one.	
image	The data being processed.	
impulse, response	The time response of a dynamical system that has been excited using an impulse.	
impulse, weighted	An impulse with a finite amplitude.	

Term	Description	Syntax
impulse	A pulse of infinite height and infinitely narrow width with an area equal to unity.	
indets	Return the unassigned variables in an expression.	indets(expr) indets(sin(x)+a/b+7)
infinity	The Maple constant corresponding to ∞.	
infix form	The form that a procedure or function takes when it is invoked by placing it between its arguments.	arg1 func arg2 arg1 proc arg2
insequence	An optional argument to display that when set equal to true causes a sequence ploti to be displayed as an animation.	display({ploti}, insequence=true, opts)
int	Return the intergral of an expression.	int(expr, var) int(expr, range) int(sin(y), y) int(sin(y), y=-1..alpha)
integer	The Maple type integer.	type(3, integer)
integration trapezoid rule	A method of approximating the area under a function.	
intensity	The value of brightness of an image at the point (x,y).	
intersect	Return the intersetion of two sets.	intersect(set1, set2) intersect({1,2,3,a}, {a})
inttrans	The Maple package containing the integral transformations.	
inverse Z transform, direct method	Obtaining the time response of a discrete system by obtaining the weighted sum of the system's previous outputs and the system's previous (and current) inputs.	

Term	Description	Syntax
inverse Z transform, long division method	Obtaining the time response of a discrete system described by a polynomial in 1/z through the process of long polynomial division.	
inverse Z transform	The process of transforming an expression from the discrete domain to the continuous domain.	
invlaplace	Compute the inverse Laplace transform of expr.	invlaplace(expr, var1, var2)
invlaplace(1/(1-s), s, t) invztrans	Compute the inverse Z transform of expr.	invztrans(expr, var1, var2) invztrans(1/(1-z), z, t)
IRIS	The Maple display engine.	
isolate	Isolate an expression from an equation. This function is readlib defined.	isolate(eqn, expr) isolate(x*y-4=0, y)
ithprime	Return the ith prime number.	ithprime(i) ithprime(501)
j	The complex constant $\sqrt{-1}$, used by engineers.	
jordan	Compute the Jordan form of a matrix. The transition matrix is stored in the optional name. This function is found in the linalg package.	jordan(mat) jordan(mat, name) jordan(matrix(2,2,[1,2,3,4])) jordan(matrix(2,2,[1,2,3,4]), 'trans')
kernel	The part of the Maple system that is compiled for reasons of efficiency.	

Term	Description	Syntax
labels	The optional argument to plot and plot3d allowing axes labels to be set.	plot(func, range, labels=[namex, namey], opts) plot3d(func, range1, range2, labels=[namex, namey, namez], opts) plot(sin, labels=['x', 'amp']) plot(sin, labels=['x', 'amp']) plot3d(sin(x/y), x=0..1, y=1..2. labels=['x', 'y' ,' z'])
Laplace transform	The process of transforming a function of time into a function of the Laplace operators.	
laplace	Return the Laplace transform in var2 of the function func in var1. The Laplace transform pair is found in the inttrans package.	laplace(func, var1, var2) laplace(sin(x)^2, x, g)
last name evaluation	The process whereby the last name only is evaluated. This applies particularly to matrices, tables, and arays. It is done to save screen real estate.	
lcoeff	Return the leading coefficient of a polynomial.	lcoeff(poly) lcoeff(poly, var) lcoeff(2*x^2+x-1) lcoeff(a*x^2+x-1, a)
lhs	Obtain the left-hand side of an equation.	lhs(eqn) lhs(a) lhs(x^2+1=a+b)
library	The part of the Maple system that contains approximately 95% of Maple's functionality.	
Liesymm	The Maple package containing functions for determining equations leading to the similarity solutions of a system of partial differential equations.	
Limit	The inert form of limit.	Limit(expr, lim, opts)

Term	Description	Syntax
limit	Return the limit of an expression in the limit lim.	limit(expr, lim, opts) limit(sin(x)/x, x=0) limit(cos(x)/x, x=0, left)
linalg	The Maple package containing the linear and matrix algebra functions.	
linalg[band]	Define a band matrix.	linalg[band]([elem], size); linalg[band]([A], 3);
linalg[crossprod]	Return the vector product of two lists or vectors.	linalg[crossprod](list1, list2) linalg[crossprod]([1,2,3], [a,b,c])
linalg[det]	Return the determinant of a matrix.	linalg[det](mat) linalg[det](matrix(2,2, [1,2,3,4])
linalg[inverse]	Return the inverse of a matrix.	linalg[inverse](mat) linalg[inverse](matrix(2,2, [1,2,3,4])
linalg[iszero]	Determine whether a matrix is zero.	linalg[iszero](mat) linalg[iszero](matrix(2,2, [0,0,0,0,])
linalg[jordan]	See jordan.	
linalg[matrix]	Define an n-by-m matrix in the Maple system.	linalg[matrix](n, m, [elemi,j]) linalg[matrix](2,2, [1,2,3,4])
linalg[submatrix]	Select a submatrix from an already existing one.	linalg[submatrix](mat, rows, cols) linalg[submatrix](A, 1..2, 3..5)
linalg[transpose]	Return the transpose of a matrix.	linalg[transpose](mat) linalg[transpose](matrix(2,2, [1,2,3,4])
linalg[vector]	See vector.	

Term	Description	Syntax
list	A Maple type and data structure.	type([1,2,3], list) [1,2,3,4]
listlist	A Maple type, a list of lists.	type([[a],[b,c]], listlist)
listplot	Found in the plots package this will plot a curve defined using a list of points, [[x1, y1],[x2, y2], ... [xn, yn]].	listplot(list, opts)
log	Return the general logarithm of expr.	log(expr) log(a) log(123.456)
long name	A Maple function or procedure name that includes the package name. These can be used without the respective function or package being loaded using with.	linalg[matrix] liesymm[&^] plots[listplot]
map	Apply an operation or function to the elements comprising a compound expression.	map(f, expr, ops) map(x-x^2, [1,2,a]) map((x, y)-x+y, [1, r], 7)
map2	Apply a function, with the first parameter specified, to the operands of a compound expression.	map2(func, op1, expr) map2((x,y)-x^y, 10, [a, b])
matrixplot	Found in the plots package, matrixplot enables numerical matrices to be displayed as a three-dimensional surface.	matrixplot(mat, opts) matrixplot(matrix(2,2, [1,2,3,4])
mellin	Found in the inttrans package, mellin returns the Mellin transform in var2 of an expression in var1.	mellin(expr, var1, var2) mellin(sin(t), t, p)
member	Test for membership of a set or list. If the optional argument is used and member returns true, the position of the first occurance of elm is stored in it.	member(expr, elem) member(expr, elem, opts) member([1,2,3], 3) member({1,2,3,4,5,6}, 4, 'where')

Term	Description	Syntax
MESH	A wrapper inside a plot structure containing data points ([x, y, z]) that define a surface.	MESH(data) MESH([[[0, 0, 0], [10, 100, 1000]], [[0, 0, 0], [10, 100, 1000]]])
minus	Return the difference of two sets.	set1 minus set2 {1,2,3} minus {3,4}
mod	Return the expression evaluated over the integers modulo m.	expr mod m 12 mod 4
name	A name is a Maple string that can have data assigned to it. A name can be less than or equal to 500k characters in length.	a 'A_string'
nops	Return the number of operands in a compound expression.	nops(expr) nops(a) nops([1,2,3,4,5])
normal	Return a normalize or simplified rational expression. If the optional expanded is present, then the numerator and denominator will be a product of expanded polynomials.	normal(expr) normal(expr, expanded) normal(1/a+1/b) normal(1/(a*(a+1))+1/b, expanded)
normalize	The process whereby a set of data values are mapped onto the range 0..1.	
not	The Boolean negation operation.	not a
NULL	The Maple Null operator.	
numer	Return the numertor of a quotient.	numer(expr) numer((1+s)/(1-s))
numeric	A Maple type.	type(34, numeric) type(2/3, numeric) type(-67.9, numeric)

Term	Description	Syntax
numpoints	The optional argument to plot and plot3d allowing the number of plot points to be set.	plot(func, range, numpoints=value opts) plot3d(func, range1, range2, numpoints=value, opts) plot(sin, numpoints=100) plot3d(sin(x/y), x=0..1, y=1..2. numpoints=40^2)
ODE, analytical solution	Normally an exact solution obtained without resulting to numerical methods.	
ODE, coupled	An nth-order ODE described with a set of simultaneous lower order ODEs.	
ODE, initial conditions	Values from which the constants of integration can be obtained.	
ODE, numerical solution	An approximate solution to the ODE obtained by using one of the numerical methods known to Maple.	
ODE, odeplot	A plotter capable of plotting an ODE using the procedures supplied by dsolve(…, numeric).	odeplot(proc, vars, r1, r2, opts);
ODE	Ordinary differential equation.	
op	Return a single or list of operands from a compound expression.	op(expr) op(num, expr) op(range, expr) op([num1, num2], expr) op(2^a) op(2, [1, 2, 3]) op(2..3, [1, 2, 3]) op([2, 2], [1, [2^a], 3])
operator, delay	See z.	
operator, differential	$\frac{d}{dt}$	

Term	Description	Syntax
operator Laplace	See s.	
operator, z	A delay of one sample period is introduced when an expression is divided by the delay operator z. Using this technique, continuous signals can be discretized and the temperal information maintined.	
orientation	The optional argument to plot3d allowing the viewing orientation (q, j) to be set.	plot3d(func, range1, range2, orientation=[q, j], opts) plot3d(sin(x/y), x=0..1, y=1..2. orientation=[10, -150])
overshoot	The amount by which the output of a dynamical system initially overshoots the steady-state value following the application of a step input.	
partial differential equations	A differential equation in more than a single variable.	
partial fraction expansion	A process whereby a rational function f in the variable x is decomposed into partial fractions.	
phase-plane	A graphical method of approximating solutions to first- and second-order differential equations.	
phaseportrait	Plot the phase portrait or approximate solutions to one- or two-dimensional systems of differential equations. This uses numerical methods.	phaseportrait(eqni, vars, range, ics, opts)
Pi	The Maple representation of [1].	
piecewise	Construct a function using segments.	piecewise(eqni) piecewise(x, sin(x), x3, cos(x), 3)

Term	Description	Syntax
PLOT	An unevaluated function that forms the data structure of a two-dimensional plot	PLOT(data) PLOT(CURVES([[1,1],[2,3]]), TITLE('A plot'))
plot	Generate a two-dimensional plot of the functions over the range r.	plot(func) plot(func, r, opts) plot({func}, r, opts) plot([pts], opts) plot(sin) plot(sin(t), t=0..Pi) plot({sun(y), cos(y)}, y=-5..5) plot([[1, 1],[2, 4]])
PLOT3D	An unevaluated function that forms the data structure of a three-dimensional plot.	PLOT3D(data) PLOT(MESH([[[1,1,1],[2,3,4]],[[1,2,3],[4,3,6]]]), TITLE('A 3d plot'))
plot3d	Generate a two-dimensional plot of the functions over the range r.	plot(func) plot(func, r, opts) plot({func}, r, opts) plot([pts], opts) plot(sin) plot(sin(t), t=0..Pi) plot({sun(y), cos(y)}, y=-5..5) plot([[1, 1],[2, 4]])
plots, 10Z	The Maple package containing the graphing functions and utilities.	
plots[display]	See display.	
plots[listplot]	See listplot.	
plots[matrixplot]	See matrixplot.	
plots[odeplot]	See ODE, odeplot.	
plots[surfdata]	See surfdata.	
plots[textplot]	See textplot.	

Term	Description	Syntax
plottools	The Maple package containing the plotting, utilities, and graphical objects.	
plottools[disk]	Plots a disk.	plottools[disk]([x, y], rad, opts) plottools[disk]([0, 1], 5)
POINTS	A setting for STYLE determining that all nontext objects will be rendered as points.	
pole	The roots of the denominator of the system transfer function when expressed in either the s- or z-plane.	
POLYGONS	A wrapper inside a plot structure containing data that are to be plotted as polygons.	POLYGONS(data) POLYGONS([[1,1], [2,2], [3,3]])
polynomial degree of	See degree.	
posint	A Maple type.	type(3, posint)
print	Print an expression to the current output device.	print(expr) print('This is printed') print(sin)
proc	The Maple keyword used in a procedure definition.	f:=proc(args) body end f:=proc(x) x^2 end f:=proc(a,b) a*b end
procedure, body	The expressions comprising the procedure.	
procedure, pure	A procedure without a name.	
procedure	A Maple expression for manipulating expressions.	f:=proc(args) body end f:=proc(x) x^2 end f:=proc(a,b) a*b end

Term	Description	Syntax
product	Compute the product of expr.	product(expr) product(expr, r) product(1/n) product(n^s, s=0..3)
pulse transfer function	The transfer function of a discretized system.	
quantization	The process whereby a signal with an infinite number of levels is represented by a fixed number, for example, 256.	
read	Read the contents of file into the current Maple session.	read(file) read('data.ms') read(data1)
readlib	Read a readlib-defined function or procedure into the current Maple session. Readlib-defined functions and procedures are not stored using the Maple package structure.	readlib(func) readlib(isolate)
recurrence relationship	A relationship equating the current output to a weighted sum of previous outputs.	y(n)=y(n-1)+y(n-2)
related	Return the "see also" section of specified help page.	related(topic) related(sin)
repetition frequency	The frequency at which an object or operation is repeated.	
RETURN	Return control to the previous level.	RETURN() RETURN(expr) RETURN('data invalid') RETURN(0)
rhs	Obtain the right-hand side of an equation.	rhs(eqn) rhs(a) rhs(x^2+1=a+b)
RootOf	The placeholder for the roots of a polynomial.	RootOf(x^2-1)

Term	Description	Syntax
round	Return the nearest integer to expr.	round(expr) round(3.4)
rsolve	The Maple recurrence relationship solver, which solves the equations eqni for the functions funci.	rsolve(eqns, fcns) rsolve(y(n-1)=y(n), y(n))
s	The Laplace operator where $s = \frac{d}{dt}$ and $\frac{1}{s} = \int dt$.	
s-plane	The plane on which the trajectories of a continuous system's poles are plotted.	
sample instance	The point at which a sample is valid.	
sample period	The period between successive samples of a continuous system.	
sampling period	See sample period.	
sampling rate	See sample period.	
sampling	The process by which a continuous system is converted into a discrete one.	
sec	Return the secant of expr.	sec(expr) sec(5.46)
select	Select all elements of a specified type from an object.	select(oper, expr) select(oper, expr, opts) select(isprime, [$1..20]) select(type, [$10..20], even)
seq	The Maple sequence operator. A sequence is formed using expr over the range r=x..y such that r=x, x+1, x+2, ... for r≤y.	seq(expr, r) seq(a^n, n=1..3) seq(a*n, n=[1,2,3,s])

Term	Description	Syntax
series, Fourier	The Fourier series has the following form: $y = a_0 + \sum_{n=1}^{\infty} b_n \sin(n\omega t) + \sum_{n=1}^{\infty} c_n \cos(n\omega t)$	
series, Frobenius	The Frobenius series has the following form: y = xc (a0 + a1x + a2x2 + a3x3 + ... + arxr + ...) where a0 is the first nonzero coefficient.	
series, Taylor	The Taylor series has the following form; $y = f(0) + f'(0)t + \frac{f''(0)t^2}{2} + \frac{f'''(0)t^2}{6} + \ldots + \frac{f^n(0)t^n}{n!}$	38
set	A Maple type and data structure.	type({1,2,3}, set) {a,b,c}
settling time	The time taken for a dynamical system to reach its steady-state ± 5% (sometimes ±2% is used).	
short name	A Maple function or procedure name without the package name included. These can only be used once the respective function or package has been loaded using with.	matrix &^ listplot
simplify	Return a simplified form of expr.	simplify(expr) simplify(expr, {siderels}) simplify(1+a+2*a) simplify(x+y+z, {y+z=A})
sin	Return a sine form of expr.	sin(expr) sin(a) sin(3.67)
single-input-single-output	A dynamical system with a single input terminal and a single output terminal.	

Term	Description	Syntax
single-shot	An event that only happens once after it has been triggered.	
solve	Maple's symbolic equation solver.	
sort	Return a sorted set of objects.	sort(obj) sort(q,w,e,r,t,y) sort(1,3,4,7,3,9,3,2,8,5,4)
spectrum, continuous	The frequency spectrum of a continuous system.	
spectrum, discrete	The frequency spectrum of a discrete system.	
spectrum	A graphical representation of a signal's frequency components.	
sqrt	Return the square root of expr.	sqrt(expr) sqrt(2.34)
stable system	A dynamical system whose output does not grow in an unbounded fashion with time in the absence of any stimuli.	
staircase plot	A two-dimensional plot that resembles a staircase. Often used to plot signals within a digital system that only change at the sampling instant.	
stared variables	Variables representing samples versions of a corresponding continuous signal.	
state variable feedback controller	A controller using all of the system states.	
state variables	The variables that define the state of a dynamical system at any time.	
state-space matrices	The matrices that define a system in state-space: $\dot{\underline{x}} = \underline{Ax} + \underline{Bu}$ $\underline{y} = \underline{Cx} + \underline{D}$	

Term	Description	Syntax
state-space	The *n*-dimensional space that contains the trajectories of the system's states.	
stats[describe, median]	Return the median of a sorted list of numbers.	stats[describe, median]([data]) stats[describe, median]([1,4,3,7,6,2,1])
steady state	The final state of a dynamical system reached when there is no further change in its input.	
step function	A function that abruptly changes from one amplitude to another.	
string	A Maple type.	'This is a string'
student[trapezoid]	A numerical method of approximating an intergral of func over the range r. The optional argument specifies the number of rectangles to use in the computation.	trapezoid(func, r) trapezoid(func, r, n) trapezoid(sin(t), t=0..5) trapezoid(sin(t), t=0..5, 10)
style	The optional argument to plot allowing the interpolation style to be set.	plot(func, range, style=value, opts) plot(tan(x), x=-10..10, style=line)
sub-matrix	A matrix produced by removing elements from another matrix.	
subs	Substitute the equations into the expression.	subs(eqni, expr) subs([eqni], expr) subs(a=3, b=t, a*sin(a*t)) subs([a=y, y=x], a*y)
subsop	Substitute for the specified operands in the expression.	subsop(opi, expr) subsop(0=g, f(t)) subsop(2=3, [a, b, c])
Sum	The inert form of sum.	Sum(expr, eqn,)

Term	Description	Syntax
sum	Return the definite or indefinite, (determined by eqn), sum of the expression.	sum(expr, eqn) sum(1/n^2, n) sum(1/n^2, n=1..5)
surfdata	Produce a three-dimensional surface from a list of amplitude points. This is found in the plots package.	surfdata(data, opts) surfdata([[1,2,3],[4,5,6],[1,2,3]])
system, continuous	A dynamical system whose output is valid for all time t.	
system, digital	A dynamical system whose output is valid at the sample instance.	
system, discrete	See system, digital.	
system, dynamics	The part of a system that determines how that system will react when it is excited.	
system, state	The condition of a dynamical system at a given moment in time.	
system, unstable	See unstable system.	
table	Define a Maple table using the equations eqn. The left-hand side of eqn is the index into the table, while the right-hand side is the table entry.	table[eqni]) table([a=1, b=2])
tanh	Return the hypobolic tangent of an expression.	tanh(expr) tanh(a) tanh(5/6)
taylor	Return the Taylor series approximation of an expression about the point given by eqn. By default six terms are returned.	taylor(expr, eqn, opts) taylor(sin(x), x) taylor(sin(x), x, 19) taylor(sin(x), x=h, 7)

Term	Description	Syntax
TEXT	The wrapper for an image's text information.	TEXT([x,y],'string',horiz,vert) TEXT([1,1],'This is text', ALIGNLEFT, ALIGNBELOW)
textplot	Render text on a graphic.	textplot(data, opts) textplot([1, 2, 'This is text'])
time constant	A measure of how quickly a dynamical system will respond to external stimuli.	
time response	The output of a dynamical system as a function of time.	
time series data	Data that has been gathered periodically over time.	
transfer function	A description relating a dynamical system's input and output.	
type, anything	Test for any valid Maple expression.	type(x, anything)
type, function	Test for a Maple function.	type(x-x^2, function)
type, numeric	Test for a numeric quantity.	type(56, numeric)
type, specfunc	Test for a specific function with a given type of argument.	type(sin(t), specfunc(name, sin))
type	Maple's type checker, which can be used in procedural programming because it is a Boolean valued function.	type(expr, type) type(sin, function) type(a, posint)
u matrix	Used in state-space analysis, the u matrix contains the system's inputs.	
unapply	Convert an expression to functional notation in var.	unapply(expr, var) unapply(x^2+x+1, x)

Term	Description	Syntax
unassigned	Determine whether a Maple expression is unassigned.	unassign(name)
unconditionally stable	See stable system.	
union	Return the union of two sets.	union(set1, set2) union({1,2,3,a}, {a})
unit circle	The region of unconditional stability on the z-plane.	
unit step	A step change in a signal of unit amplitude.	
unstable system	A dynamical system whose output grows in an unbounded fashion with time in the absence of any stimuli.	
Vandermonde matrix	A square matrix with its (i,j)th entry equal to L(j-1) where L is the matrices' second column.	$\begin{bmatrix} 1 & a & a^2 \\ 1 & b & b^2 \\ 1 & c & c^2 \end{bmatrix}$
vector	Define a vector in Maple.	vector(len) vector([elmi]) vector(3) vector([1,2,3,4])
view	The optional argument to plot and plot3d allowing a specific view to be set.	plot(func, range, view=[rangex, rangey], opts) plot3d(func, range1, range2, view=[rangex, rangey, rangez], opts) plot(sin, view=[1..4, DEFAULT plot3d(sin(x/y), x=0..1, y=1..2. view=[0.5..1, 1..3/2, -1..1])
volatility	A measure of how rapidly time series data is changing.	
wedge product	see &^.	

Term	Description	Syntax
whattype	Maple's interactive type checker.	whattype(expr) whattype(sin) whattype(6/8)
while	A form of the repetition construct supported by Maple. The body of the loop is evaluated while the loop test is true.	while test do ... od while true do x:=x+1 od
with	Load the package function's short names into the session's name space.	with(pack) with(pack, [funci]) with(linalg) witn(linalg, [matrix, vector])
Worksheet	Maple's graphical user interface.	
x matrix	Used in state-space analysis, this matrix contains the states of the system under observation.	
y matrix	Used in state-space analysis, this matrix contains the output signals of the system.	
Z transform, direct	A method of discretizing a system by finding a closed form solution to an infinite sum.	
Z transform, impulse-invariant	A method of discretizing a system in such a way that the discrete and continuous impulse responses are identical, at least at the sample instances.	
Z transform, step invatiant	A method of discretizing a system in such a way that the discrete and continuous step responses are identical, at least at the sample instances.	
Z transform, substitution method	A method of discretizing a system by substituting for s in the continuous transfer function.	
z-plane	The plane on which the trajectories of a continuous system's poles are plotted.	

Term	Description	Syntax
Z-transform	The process of discretizing a function of time.	
zero	The roots of the numerator of the system transfer function when expressed in either the s- or z-plane.	
zip	Combine the elemets of two lists or vectors according to some operation.	zip(f, expr1, expr2, opts) zip((x,y)-x+y, [1,2,3], [a,b,c]) zip((x,y)-x^y, [a,b,c], [1], r)
ztrans	Return the Z transform in var2 of the function func in var1.	ztrans(func, var1, var2) ztrans(sin(x)^2, x, z)

About the authors

Christopher S. Tocci is currently a senior projects engineer at Allen-Bradley Automation Company, Inc., in Chelmsford, Massachusetts. Dr. Tocci was also one of the cofounders of Applied Research Consortium, Inc., in Charlton, Massachusetts. His past technical experience has been in atomic spectroscopy, medical engineering, optical communications, optical device technology, and biophysics. He has been involved with industrial and military hardware designs and analysis of optics and optoelectronics as it applied to infrared sensors, two-dimensional signal processing, communication, and interconnection for massively parallel computing architectures. He has had senior positions at Baird Electronic, Raytheon, MIT Lincoln Labs, Augat Fiberoptics, and he has consulted for Ciba-Geigy Diagnostics on optical metrology for blood analysis. Dr. Tocci has had several patents in electro-optical device technology and over 35 publications in both trade and professional journals. He received his Ph.D. from Clarkson University in engineering science in 1985. Dr. Tocci is a member of OSA, Who's Who in the East in Science & Technology, AMS, AAAS, and the NRA.

Steven Adams is currently technical marketing director for TCI Software Research, Inc., in Las Cruces, New Mexico. Previously, Dr. Adams was the director of Technical Marketing for Harmonix Corporation, Woburn, Massachusetts. Dr. Adams has had a strong working relationship with Waterloo Maple Software in Waterloo, Canada, since 1993, when he was manager of their U.S. Technical Services group. Previous to his involvement with Maple Software, he was technical member of Wolfram Research's Applications Group, the developers of *Mathematica*. Dr. Adams has had several consulting and lecturing positions within industry and academia—most notably, senior lecturer at South Bank Polytechnic, London, U.K., in the Department of Electrical and Electronic Engineering. At South Bank Polytechnic, Steve lectured on robotics, modern control systems, and sensor design and analysis. Dr. Adams has also lectured in Bombay and Puna, India, as a sponsored representative of the British Council, and IPM in Moscow CIS, Russia, as sponsored by the British Royal Society. Steve received his Ph.D. in Electrical Engineering from King's College, University of London, School of Electronic Engineering, London, U.K., in 1984.

Index